Page left deliberately blank

Stay up to date with the latest research

Sign up for FREE Journal Email Alerts
http://online.sagepub.com/cgi/alerts

Register online at **SAGE Journals** and start receiving…

Content Alerts
Receive table of contents alerts when a new issue is published

OnlineFirst Alerts
Receive notification when forthcoming articles are published online before they are scheduled to appear in print

Announcements
Receive need-to-know information about a journal such as calls for papers, special issue notices, and events

CiteTrack Alerts
Receive notification for citations and corrections to selected articles

Search Alerts
Create custom fielded boolean keyword search alerts across one or more journals

VOLUME 660

JULY 2015

THE ANNALS

of The American Academy of Political
and Social Science

Residential Inequality in American
Neighborhoods and Communities

Special Editors:
BARRETT A. LEE
The Pennsylvania State University
GLENN FIREBAUGH
The Pennsylvania State University
JOHN ICELAND
The Pennsylvania State University
STEPHEN A. MATTHEWS
The Pennsylvania State University

⑤SAGE

Los Angeles | London | New Delhi
Singapore | Washington DC | Boston

Origin and Purpose. The Academy was organized December 14, 1889, to promote the progress of political and social science, especially through publications and meetings. The Academy does not take sides in controverted questions, but seeks to gather and present reliable information to assist the public in forming an intelligent and accurate judgment.

Meetings. The Academy occasionally holds a meeting in the spring extending over two days.

Publications. THE ANNALS of The American Academy of Political and Social Science is the bimonthly publication of the Academy. Each issue contains articles on some prominent social or political problem, written at the invitation of the editors. These volumes constitute important reference works on the topics with which they deal, and they are extensively cited by authorities throughout the United States and abroad.

Subscriptions. THE ANNALS of The American Academy of Political and Social Science (ISSN 0002-7162) (J295) is published bimonthly—in January, March, May, July, September, and November—by SAGE Publications, 2455 Teller Road, Thousand Oaks, CA 91320. Periodicals postage paid at Thousand Oaks, California, and at additional mailing offices. POSTMASTER: Send address changes to The Annals of The American Academy of Political and Social Science, c/o SAGE Publications, 2455 Teller Road, Thousand Oaks, CA 91320. Institutions may subscribe to THE ANNALS at the annual rate: $957 (clothbound, $1081). Individuals may subscribe to the ANNALS at the annual rate: $115 (clothbound, $170). Single issues of THE ANNALS may be obtained by individuals for $36 each (clothbound, $49). Single issues of THE ANNALS have proven to be excellent supplementary texts for classroom use. Direct inquiries regarding adoptions to THE ANNALS c/o SAGE Publications (address below).

All correspondence concerning membership in the Academy, dues renewals, inquiries about membership status, and/or purchase of single issues of THE ANNALS should be sent to THE ANNALS c/o SAGE Publications, 2455 Teller Road, Thousand Oaks, CA 91320. Telephone: (800) 818-SAGE (7243) and (805) 499-0721; Fax/Order line: (805) 375-1700; e-mail: journals@sagepub.com. *Please note that orders under $30 must be prepaid.* For all customers outside the Americas, please visit http://www.sagepub.co.uk/customerCare.nav for information.

Printed on acid-free paper

THE ANNALS

Editorial Office: 202 S. 36th Street, Philadelphia, PA 19104-3806
For information about individual and institutional subscriptions address:
SAGE Publications
2455 Teller Road
Thousand Oaks, CA 91320

For SAGE Publications: Peter Geraghty (Production) and Mimi Nguyen (Marketing)

From India and South Asia,
write to:
SAGE PUBLICATIONS INDIA Pvt Ltd
B-42 Panchsheel Enclave, P.O. Box 4109
New Delhi 110 017
INDIA

From Europe, the Middle East,
and Africa, write to:
SAGE PUBLICATIONS LTD
1 Oliver's Yard, 55 City Road
London EC1Y 1SP
UNITED KINGDOM

International Standard Serial Number ISSN 0002-7162
ISBN 978-1-5063-2455-5 (Vol. 660, 2015) paper
ISBN 978-1-5063-2456-2 (Vol. 660, 2015) cloth
Manufactured in the United States of America. First printing, July 2015

Please visit http://ann.sagepub.com and under the "More about this journal" menu on the right-hand side, click on the Abstracting/Indexing link to view a full list of databases in which this journal is indexed.

Information about membership rates, institutional subscriptions, and back issue prices may be found on the facing page.

Advertising. Current rates and specifications may be obtained by writing to The Annals Advertising and Promotion Manager at the Thousand Oaks office (address above). Acceptance of advertising in this journal in no way implies endorsement of the advertised product or service by SAGE or the journal's affiliated society(ies) or the journal editor(s). No endorsement is intended or implied. SAGE reserves the right to reject any advertising it deems as inappropriate for this journal.

Claims. Claims for undelivered copies must be made no later than six months following month of publication. The publisher will supply replacement issues when losses have been sustained in transit and when the reserve stock will permit.

Change of Address. Six weeks' advance notice must be given when notifying of change of address. Please send the old address label along with the new address to the SAGE office address above to ensure proper identification. Please specify the name of the journal.

THE ANNALS

OF THE AMERICAN ACADEMY OF POLITICAL AND SOCIAL SCIENCE

Volume 660   July 2015

IN THIS ISSUE:

Residential Inequality in American Neighborhoods and Communities

Special Editors: BARRETT A. LEE, GLENN FIREBAUGH, JOHN ICELAND, and STEPHEN A. MATTHEWS

Introduction

Racial and Ethnic Segregation

The Income Divide

Locational Attainment and Housing Insecurity

Understanding Residential Moves

Neighborhood Change

Conclusion

FORTHCOMING

Race, Racial Inequality, and Biological Determinism in the Genetic and Genomic Era
Special Editors: W. CARSON BYRD and MATTHEW W. HUGHEY

*Intermarriage and Integration Revisited: International Experiences and Cross-Disciplinary
Approaches*
Special Editor: DAN RODRÍGUEZ-GARCÍA

Introduction

Keywords: residential inequality; spatial sorting; life chances; race/ethnicity; income

Residential Inequality: Orientation and Overview

By
BARRETT A. LEE,
STEPHEN A. MATTHEWS,
JOHN ICELAND,
and
GLENN FIREBAUGH

W**here people live reflects and affects their** position in society. This tenet is implicit in the American Dream, which promises access to desirable homes, neighborhoods, and communities for those willing to work hard enough. As recent events remind us, however, effort alone does not guarantee fulfillment of the dream. Natural disaster, recession, mortgage foreclosure, and escalating (and plummeting) housing prices are among the forces that have already thwarted residential aspirations in the United States during the new century. The difficulty that many immigrants to the nation face in achieving their housing and neighborhood goals is another current concern. Over a longer period, the discriminatory practices of real estate agents and lenders, preferences for neighbors similar to oneself, marked income inequality, and government policies and programs have perpetuated spatial divides by race and class.

This volume is motivated by the post-2000 interplay of these forces, which suggests that the multifaceted relationship between social

Barrett A. Lee is a professor of sociology and demography at The Pennsylvania State University. He studies community diversity, racial segregation, neighborhood change, residential mobility and displacement, and urban homelessness. A general interest in spatial manifestations of social inequality runs throughout his research.

Stephen A. Matthews is a professor of sociology, anthropology, and demography and director of the Graduate Program in Demography at The Pennsylvania State University. He has published in the leading journals in demography, sociology, public health, and geography on topics such as residential segregation, neighborhood change, access to health care, and food environments.

DOI: 10.1177/0002716215579832

inequality and the residential environment—long a staple of urban scholars—needs to be revisited. As an initial step toward that end, the editors of this volume convened a two-day research conference at The Pennsylvania State University in September 2014. The articles compiled here, all delivered in draft form at the conference, represent the best work from a large pool of submissions generated through invitation and a widely circulated call for papers. While the contributions to the volume are primarily substantive in nature, they shed light on issues of both theoretical and policy importance. By design, their authors span a range of disciplinary perspectives and career stages.

What unifies this volume is a return to basics, albeit with fresh eyes. Since the heyday of the Chicago School, social scientists have documented the ethnoracial and socioeconomic differentiation of community residential landscapes. Robust lines of inquiry attempt to understand why distinct racial and income groups so often occupy separate geographic areas and housing types and how such separation impacts the well-being of group members across a variety of domains. The articles in the volume consider the same issues through a focus on four fundamental types of spatial sorting: *segregation, locational attainment* (the extent to which people are able to secure favorable housing and community conditions for themselves), *residential mobility*, and *neighborhood change*. Of particular interest is whether these sorting phenomena still operate as they once did, with similar antecedents and outcomes. In the pages that follow, our conference participants utilize innovative data and rigorous methods to examine the sorting-inequality nexus during an era of unprecedented demographic diversity, rising economic uncertainty, and shifting public policy.

Life Chances

Scholarship on the various forms of spatial sorting has been driven by the belief that differences in residential circumstances influence people's life chances: that is, the ability to take advantage of the opportunities offered by a society and to

John Iceland is head of the Department of Sociology and a professor of sociology and demography at The Pennsylvania State University. His research focuses on social demography, poverty, residential segregation, and immigration. He has written three books on these issues and published numerous articles in top journals in sociology and demography.

Glenn Firebaugh is Roy Buck Professor of Sociology and Demography at The Pennsylvania State University and author of Seven Rules for Social Research *(Princeton 2008). Currently, he is using U.S. census data to investigate neighborhood inequality and data on cause of death to investigate sources of change in life expectancy.*

NOTE: The articles in this volume were first presented at a conference on "Residential Inequality in American Neighborhoods and Communities" held September 12–13, 2014, at Penn State's University Park campus. We thank the authors and attendees for their contributions to a stimulating event. Generous support for the conference was provided by the Penn State Department of Sociology, College of the Liberal Arts, Population Research Institute, and Social Science Research Institute; and by the American Academy of Political and Social Science.

avoid its pitfalls. Neighborhood effects research examines the positive and negative sides of the life chances "coin," with more attention devoted to the latter than
the former (Newberger, Birch, and Wachter 2011; Sampson 2012). When sorting
produces neighborhoods marked by high poverty rates, large minority proportions, and other indicators of concentrated disadvantage linked to social and
economic stratification, residents tend to suffer from deficits in health, safety,
academic performance, employment, environmental quality, and public services.
They are also at greater risk of experiencing shortfalls in social capital, collective
efficacy, and political clout. Simply put, residents of disadvantaged locations have
heightened exposure to problems and reduced access to valuable resources.

The locations that carry such costs—and those that confer benefits—are not
limited to neighborhoods. From a place stratification perspective, larger geographic areas may carry fateful implications for life chances as well (Logan and
Molotch 1987). Consider, for example, the potential contrasts in economic well-
being for workers in metropolitan regions with growing or contracting labor
markets, or the education provided by school districts in affluent or cash-strapped
municipalities. Whatever the scale, the impact of a community setting depends
on the duration and timing of one's residence in that context. Indeed, growing up
in a disadvantaged community can register effects that endure throughout adulthood and beyond. Sharkey (2013) makes this point in dramatic fashion. His
research on the residential trajectories of African American families over the past
four decades documents the low likelihood of residents escaping harmful neighborhood conditions not only within their lifetimes but also across generations.

This longer-term view of residential inequality warns against causal oversimplification, reminding us that life chances may be antecedents to as well as consequences of spatial sorting. The socioeconomic status of parents (and of parental
neighborhoods) significantly determines how offspring will fare in terms of
health, cognitive skills, and educational and occupational pursuits. The extent to
which these life chances are realized should shape the probability that adult offspring wind up in better or worse housing units and communities. Of course,
knowledge of and proximity to alternative destinations—or the lack thereof—
structure the process of residential selection (Krysan and Bader 2009) as do race
and other ascriptive characteristics of homeseekers. Yet our larger point stands:
life chances are among the factors that operate via selection to influence the
character of places (specifically, the local mix of resources and problems) and
where they rank in the place hierarchy. Hence, the relationship between inequality and spatial sorting is most appropriately conceived as bidirectional, a principle
recognized throughout this volume.

Spatial Sorting

We use the term *spatial sorting* to refer to the differential distribution of groups
across meaningful residential sites such as metropolitan areas, cities, suburbs,
small towns, neighborhoods, and housing units. The groups of primary concern
here are defined by income and ethnoracial identity. Although gender and age

also constitute important bases of stratification, income and race strike us as central to spatial sorting. Their prominence is evident in research on residential segregation, which represents the most obvious "face" of sorting (Iceland 2009; Reardon and Bischoff 2011). As typically conceptualized and measured, segregation reveals the degree to which groups live apart—in separate residential environments—rather than together. It thus provides a snapshot of the racial and socioeconomic lay of the land. However, it says little about the processes directly responsible for that pattern.

The other three forms of sorting featured in this volume tend to be more processual or dynamic in nature. Like segregation, locational attainment, residential mobility, and neighborhood change have each inspired their own distinct research literatures. Nevertheless, all four manifestations of sorting remain tightly bundled in the real world. Segregation, for example, largely reflects the accumulation of household moves and the resulting shifts in neighborhood composition. (Differential fertility and mortality can affect composition as well.) In locational attainment research, the focus is on variation in the ability of group members to achieve favorable housing, community, and socioeconomic outcomes as they navigate the life course and, in the case of immigrants, become more acculturated. Constrained by the magnitude of segregation, these outcomes can be improved or diminished when conditions (e.g., upturns or downturns in local job and housing markets and hurricanes or other environmental shocks) prompt people to move. Residential mobility also serves as the proximate mechanism underlying neighborhood change. Whether a neighborhood undergoes transition or remains stable depends on the balance of characteristics among households that move in, stay put, or move out. But homeseekers' decisions to enter the neighborhood may in turn be influenced by the neighborhood's current composition or perceived trajectory. Mobility and neighborhood change exhibit similar reciprocal associations with segregation.

Such associations recommend that the four types of spatial sorting be examined in combination with one another or at least side by side. We have selected the articles for the volume with this objective in mind. As an illustration, several authors take into account mobility decision-making and behavior to better understand how neighborhoods evolve and why certain kinds of ethnoracial segregation persist at a high level. Our choice of articles was further guided by the authors' ability to draw connections between sorting and life chances (in employment, education, health, and the like). In short, what the articles have in common is thoughtful attention to the entwined nature of segregation, locational attainment, mobility, and neighborhood change, and to the roles played by these sorting phenomena as causes, correlates, or consequences of other, nonresidential aspects of social stratification.

Contexts

Woven into the volume is an awareness of contexts beyond the obvious residential ones. At the most general level of abstraction, sorting occurs in time as well as space. With a single exception, the articles employ data that include the post-2000 era or fall entirely within it, and roughly half incorporate at least some evidence collected during the past five years. Moreover, a majority of authors explore changes in sorting manifestations or processes, although the periods covered range from a few years to multiple decades. The lead article by John Logan and his colleagues is the only piece that does not use contemporary data. By offering new insight into patterns of black segregation and locational attainment from the late nineteenth through the mid-twentieth century, the article encourages the remaining articles to be viewed through a comparative historical lens.

New sociodemographic and economic realities represent another type of context germane to residential inequality. For instance, the dramatic rise in immigrant-driven ethnoracial diversity—marked by whites' shrinking slice of the U.S. population "pie" and the rapid expansion of the Hispanic and Asian slices—has attracted substantial attention as a societal master trend. While the magnitude and pervasiveness of the trend are well documented, debate continues over the implications of diversity for segregation, neighborhood change, and minority residential and status attainment (Lee, Iceland, and Farrell 2014; Parisi, Lichter, and Taquino 2015). These issues are addressed by several of the articles in this volume. Other contributions examine aspects of spatial sorting against the backdrop of the Great Recession, the housing crisis, and the widening gap between rich and poor. Given the heightened economic uncertainty that such events have fueled, one can readily imagine how preexisting racial and class differences in neighborhood quality and shelter security might be exacerbated, perhaps more so in some regions or metropolitan areas than others.

Because of the hardships accompanying them, major social and economic changes often elicit or require a response from government. This fact points to public policy as a third, cross-cutting context within which patterns of residential inequality must be understood. The time span emphasized in the volume aligns with a variety of federal and local efforts to deconcentrate poverty through the demolition of public housing projects, an increased reliance on housing voucher and tax credit programs, and the creation of mixed-income communities (Schwartz 2010). Similarly, fair housing laws designed to combat racially discriminatory practices by real estate agents, landlords, and other actors have remained in effect during the 2000s. Despite inconsistent enforcement, the laws appear to be slowly modifying agent behavior in the intended direction. People of color, however, still receive less favorable treatment than whites during the search for housing, based on recent monitoring evidence (Turner et al. 2013). Of broader concern is whether incremental legislative steps of any kind are strong enough to significantly reduce the inequalities in life chances associated with spatial sorting. A number of the volume's authors speak to this question about policy efficacy.

Volume Overview

We have organized the seventeen papers from the conference into five thematic sections devoted to (1) racial segregation, (2) the income divide, (3) locational attainment and housing insecurity, (4) residential mobility, and (5) neighborhood change. Four of the sections are loosely anchored by a particular form of spatial sorting. The third section is more hybrid in nature, with income inequality serving as the central theme for analyses that cover multiple types of spatial sorting. But the close interrelations among these sorting mechanisms mean that all sections of the volume have a somewhat mixed flavor.

In the first section on *racial residential segregation*, John Logan and his collaborators, Weiwei Zhang, Richard Turner, and Allison Shertzer, use 1880 through 1940 census data geocoded at a variety of spatial scales to evaluate arguments about the timing of ghettoization and potential class differences in the degree of residential isolation within the African American population. Their results highlight the key role played by the color line in both early ghetto formation and the entrenchment of black segregation to the present day. Daniel Lichter, Domenico Parisi, and Michael Taquino pursue geographic rather than historical comparisons, examining Hispanic segregation trends since 1990 in nonmetropolitan communities as well as in major cities and their suburbs. While higher Hispanic income is inversely related to segregation from whites, declines have been more rapid in Hispanic-black than Hispanic-white segregation, hinting at segmented assimilation.

The last article in the section, by Justin Steil, Jorge De la Roca, and Ingrid Gould Ellen, demonstrates how metropolitan segregation levels negatively affect educational, labor market, and social outcomes for young Hispanic adults much the same as they do for African Americans, thus restricting life chances and perpetuating social inequality. This analysis blends micro data from the 2007–11 American Community Survey (ACS) with information from varied sources on school test scores, neighborhood crime, local business density, and other variables to evaluate possible mechanisms through which segregation may operate.

Though hardly ignoring race, the next set of articles gives priority to residential expressions of the *income divide* in the United States. For Sean Reardon, Lindsay Fox, and Joseph Townsend, the task is to describe the average neighborhood income and racial environments that metropolitan dwellers inhabit. Applying a novel method to two decades of census and ACS data, they identify striking disparities in neighborhood contexts by household income, race, and metropolitan area of residence. Ann Owens addresses the income divide via an important policy issue: whether, as intended, the geographic deconcentration of assisted housing units within metropolitan areas (through housing vouchers and construction subsidies for affordable units) has reduced income segregation among neighborhoods. She finds only weak support for the realization of this goal since 1980, in part due to the small number of residents who live in assisted units.

Using unique panel data for specific housing units in low-income neighborhoods of ten cities, Brett Theodos, Claudia Coulton, and Rob Pitingolo link two spatial sorting mechanisms—residential mobility and neighborhood change—as they investigate the degree to which neighborhood socioeconomic composition shifts in response to housing turnover. Even with controls for a variety of relevant factors, neighborhood poverty proves to be quite sticky over time, even at the housing-unit level: vacancies created by the departure of poor households are filled by other poor people. Drawing on case studies of three mixed-income developments in Chicago, Robert Chaskin and Mark Joseph detail how new urbanism design principles, regulatory practices, and other aspects of these developments have marginalized one class of targeted beneficiaries—former public housing occupants—who experience "incorporated exclusion" rather than engagement with their higher-status neighbors.

Locational attainment processes, broadly construed, are the topic of the next four articles. Robert Sampson, Robert Mare, and Kristin Perkins lead with a study of neighborhood income mobility, emphasizing exposure to naturally occurring "mixed middle-income" neighborhoods distinguished by diverse income compositions. The authors show the fragility of these neighborhoods not only in Chicago but nationally (compared with the persistence of extremely affluent or poor contexts), then reveal that few Chicago adolescents experience mixed-income neighborhoods during their transition to young adulthood despite the popularity of policies intended to create such settings. Chenoa Flippen and Eunbi Kim shift the focus from neighborhood to metropolitan contexts, using 2009–11 ACS data to assess the degree to which Asian socioeconomic attainments (in income, occupational status, and homeownership) differ across new immigrant destinations and traditional gateways. Their results indicate some divergence in socioeconomic outcomes by metropolitan housing and labor market conditions, gender, and Asian origin group, with Chinese, Japanese, Indians, and Filipinos—the advantaged groups—benefitting the most from residence outside of gateway areas.

The two other articles in the section examine mortgage foreclosure, a form of residential "disattainment," or downward housing mobility. In the first article, Rachel Dwyer and Lora Phillips Lassus treat metro-level variation in recent foreclosure rates as a function of residential segregation and measures of labor and housing market precarity, each of which is hypothesized to increase the risks associated with homeownership. They also make a compelling case for framing policy debates about urban inequality in terms of economic insecurity. Their analysis is nicely complemented by Matthew Hall, Kyle Crowder, and Amy Spring, who describe differences in the timing and depth of the foreclosure crisis by region, metro area, place, and neighborhood, with race as a prominent factor at all geographic scales. Combining RealtyTrac information on 9.5 million foreclosures between 2005 and 2012 with census and Panel Study of Income Dynamics data, Hall and colleagues highlight the overrepresentation of foreclosures in minority and racially diverse neighborhoods and document the greater exposure of black and especially Hispanic households to foreclosure—either their own foreclosures or those of their neighbors.

The fourth section of the volume, which comprises three articles, is devoted to understanding residential mobility, a foundational type of spatial sorting tightly linked with racial and income inequality. The initial article by Lincoln Quillian criticizes traditional mobility and locational attainment models for their failure to adequately capture the complex decision-making of homeseekers when weighing alternative destinations. Quillian identifies discrete-choice models as a promising alternative and, informed by empirical comparisons, discusses when a particular model is most appropriate. Community perceptions, an important component of the housing search process and ultimately of segregation, are explored by Michael Bader and Maria Krysan. Analyzing Chicago survey data from 2004 to 2005, the authors demonstrate the impact that the ethnoracial composition of a community has on homeseekers' perceptions of its desirability, particularly among whites.

S. Michael Gaddis and Raj Ghoshal inquire whether discriminatory treatment constrains the residential mobility of Arab Americans in Detroit, Houston, Los Angeles, and New York. Their innovative audit study of responses to roommate-wanted ads posted on Craigslist finds that the likelihood of discrimination against people with Arab-origin names is increased in one or more of the four metro areas by a housing unit's proximity to a mosque and by the concentration of mosques and the percentage of Arab Americans in the surrounding neighborhood. This pattern of results offers more support for ethnic competition logic than for the contact hypothesis.

A fifth set of articles investigates new forms of neighborhood change that, at first glance, would seem to have the potential to reduce inequality. Lance Freeman and Tiancheng Cai employ data from the decennial census and noncensus sources both to describe the significant increase between 2000 and 2010 in the number of black neighborhoods experiencing white population growth and to identify the distinctive features of these areas. The correlates of white "invasion" that Freeman and Cai uncover are consistent with emergent gentrification and thus speak to the future of segregation and neighborhood diversity. Jackelyn Hwang explores the relationship between immigrant settlement and gentrification with measures from an array of sources (including Google Street View). She first examines the immigration-gentrification linkage across twenty-three cities then strategically focuses her attention on Seattle and Chicago, two very different metropolitan contexts with respect to segregation and immigrant incorporation. Her overall conclusion is that immigration and gentrification combine to sustain rather than diminish ethnoracial residential inequality.

Last but not least, Michael Friedson and Patrick Sharkey probe local manifestations of the national crime decline during the past two decades, asking if violent crime rates have decreased uniformly or unevenly across neighborhoods in Chicago, Cleveland, Denver, Philadelphia, St. Petersburg, and Seattle. The authors find the greatest drops in crime in the most dangerous neighborhoods, suggestive of weakened linkages among poverty, segregation, and crime. Nevertheless, the pervasive nature of the declines have left the distribution of violence relatively stable in each of the six cities: the most disadvantaged neighborhoods rank as the most violent at both the beginning and end of the period under investigation.

The Friedson and Sharkey article and the sixteen preceding it inspired lively debate at the Penn State conference where they were first presented. We hope that they will add further momentum to scholarship on the connections between spatial sorting and stratification. Such a line of inquiry strikes us as critical in a period when American communities are encountering new demographic, economic, and political conditions that will shape their inhabitants' life chances. Of course, this is hardly the first time social science research has addressed issues facing the nation's cities and neighborhoods. Over the past half century, volumes of *The ANNALS* have examined urban crime; economic restructuring; housing quality and affordability; inner-city disadvantage; community development strategy; and local challenges posed by globalization, immigration, and sprawl. While each of these topics speaks to racial and income disparities directly or indirectly, one *ANNALS* volume about "Race and Residence" (Roof 1979) published more than 35 years ago provides an appropriate bookend to the set of articles presented here. We revisit that volume in the conclusion, using it to frame a summary of significant results and policy lessons from current research on residential inequality.

References

Iceland, John. 2009. *Where we live now: Immigration and race in the United States*. Berkeley, CA: University of California Press.

Krysan, Maria, and Michael D. M. Bader. 2009. Racial blind spots: Black-white-Latino differences in community knowledge. *Social Problems* 56:677–701.

Lee, Barrett A., John Iceland, and Chad R. Farrell. 2014. Is ethnoracial integration on the rise? Evidence from metropolitan and micropolitan America since 1980. In *Diversity and disparities: America enters a new century*, ed. John R. Logan, 415–56. New York, NY: Russell Sage Foundation.

Logan, John R., and Harvey L. Molotch. 1987. *Urban fortunes: The political economy of place*. Berkeley, CA: University of California Press.

Newberger, Harriet B., Eugenie L. Birch, and Susan M. Wachter, eds. 2011. *Neighborhood and life chances: How place matters in modern America*. Philadelphia, PA: University of Pennsylvania Press.

Parisi, Domenico, Daniel T. Lichter, and Michael C. Taquino. 2015. The buffering hypothesis: Growing diversity and declining black-white segregation in America's cities, suburbs, and small towns? *Sociological Science* 2:125–57.

Reardon, Sean F., and Kendra Bischoff. 2011. Income inequality and income segregation. *American Journal of Sociology* 116:1092–153.

Roof, Wade Clark, ed. 1979. Race and residence in American cities. *The ANNALS of the American Academy of Political and Social Science* 441.

Sampson, Robert J. 2012. *Great American city: Chicago and the enduring neighborhood effect*. Chicago, IL: University of Chicago Press.

Schwartz, Alex F. 2010. *Housing policy in the United States*. 2nd ed. New York, NY: Routledge.

Sharkey, Patrick. 2013. *Stuck in place: Urban neighborhoods and the end of progress toward racial equality*. Chicago, IL: University of Chicago Press.

Turner, Margery Austin, Robert Santos, Diane K. Levy, Doug Wissoker, Claudia Aranda, and Rob Pitingolo. 2013. *Housing discrimination against racial and ethnic minorities 2012*. Washington, DC: U.S. Department of Housing and Urban Development.

Racial and Ethnic Segregation

Creating the Black Ghetto: Black Residential Patterns before and during the Great Migration

By
JOHN R. LOGAN,
WEIWEI ZHANG,
RICHARD TURNER,
and
ALLISON SHERTZER

Were black ghettos a product of white reaction to the Great Migration in the 1920s and 1930s, or did the ghettoization process have earlier roots? This article takes advantage of recently available data on black and white residential patterns in several major northern cities in the period 1880–1940. Using geographic areas smaller than contemporary census tracts, we trace the growth of black populations in each city and trends in the level of isolation and segregation. In addition we analyze the determinants of location: which blacks lived in neighborhoods with higher black concentrations, and what does this tell us about the ghettoization process? We find that the development of ghettos in an embryonic form was well underway in 1880, that segregation became intense prior to the Great Migration, and that in this whole period blacks were segregated based on race rather than class or southern origin.

Keywords: ghettoization; residential segregation; Great Migration

While much attention is being given to the persistence and slow decline of black-white residential segregation in the United States since the 1960s, much less is known about its origins in the late nineteenth and early twentieth centuries. A standard account holds that segregation was modest in northern cities in the decades following the Civil War: "No matter what other disadvantages urban blacks suffered in the aftermath of the Civil War, they

John R. Logan is a professor of sociology at Brown University and director of the Initiative in Spatial Structures in the Social Sciences (S4).

Weiwei Zhang is a postdoctoral research associate in the Population Studies and Training Center (PSTC) at Brown University.

Richard Turner is a postdoctoral research associate at the National Strategic Planning & Analysis Research Center (nSPARC) at Mississippi State University.

Allison Shertzer is an assistant professor of economics at the University of Pittsburgh.

DOI: 10.1177/0002716215572993

were not residentially segregated from whites" (Massey and Denton 1993, 17). Then, following the initial wave of the Great Migration of blacks from the South to the North during the First World War, whites responded by erecting new barriers that sharply restricted where blacks could live. At the risk of oversimplification, let us refer to this as the "threatening presence" account (echoing Blalock 1956) because it attributes so much to the impact of rapid black population growth (for similar interpretations see Lemann 1991; Lieberson 1980).

We argue here that this account is incomplete for two reasons. First, it is based on analyses of data for city wards, which are too large a geographic unit to capture the segregation of small population groups. Second, it does not take into account information about the processes that underlie racial separation, especially the extent to which the black ghetto trapped all blacks regardless of their social class or other attributes. We use newly available data here to address both concerns. We map and measure segregation at the scale of enumeration districts (EDs), areas as small as a few city blocks. And we examine the association between the racial composition of the ED that blacks lived in and their own background characteristics in a multilevel model. For the first time we use fine-area data to study racial residential patterns for many cities over several decades in the early twentieth century. Our results support an alternative narrative about the origins of segregation, placing the era of high segregation and the entrapping ghetto well before the Great Migration. The massive movement of blacks from the South to new destinations in the North certainly affected the geographic scale and the racial homogeneity of the areas where blacks clustered. In addition, residential segregation became more complete from decade to decade. Yet segregation was undeniably high even in 1880 in most cities that we study, and the rise in segregation was apparent even by 1900 or 1910, prior to the Great Migration. The locational process for blacks never gave blacks of higher social class much chance of living in a more racially mixed neighborhood, and blacks born in the North lived in very similar neighborhoods to those of migrants from the South. The main characteristics, then, that we have come to associate with black ghettoization were already in place in the late nineteenth century, and we demonstrate that in this contribution. We agree with the standard account that black neighborhoods grew larger and more homogeneous in the next several decades.

The Problem of Geographic Scale

The importance of geographic scale is underlined by reviewing the ward-level evidence that supports the "threatening presence" account. Figure 1 presents

NOTE: This research was supported by research grants from National Science Foundation (SES-1355693) and National Institutes of Health (1R01HD075785-01A1) and by the staff of the research initiative on Spatial Structures in the Social Sciences at Brown University. The Population Studies and Training Center at Brown University (R24HD041020) provided general support. The authors have full responsibility for the findings and interpretations reported here.

FIGURE 1
Segregation (D), Isolation (P°bb) and Black Population Share, Ward Level Data
(Weighted Average for 10 northern Cities)

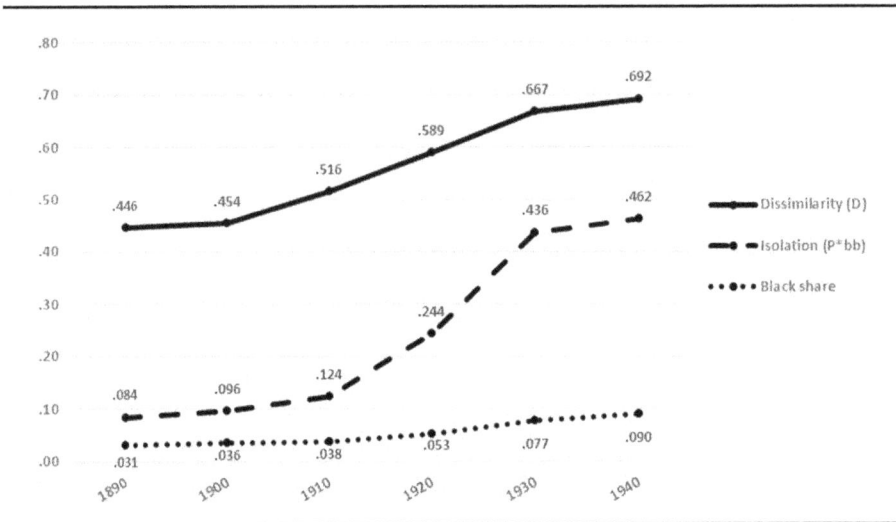

such data for the ten cities that we study here for 1890–1940, showing the black population share and two standard measures of segregation (the isolation index and the index of dissimilarity).[1] These ward data show that black isolation was quite low through 1910, when the average black lived in a city that was only 3.8 percent black and lived in a ward that was overwhelmingly white (only 12.4 percent black). Black isolation rose appreciably by 1920 (when the average black person's ward was 24.4 percent black), but greatly accelerated after that time, reaching 43.6 percent in 1930 and 46.2 percent in 1940. This timing is crucial—blacks' neighborhoods did not approach being majority black until the 1930s.

The index of dissimilarity followed a similar trajectory. It was moderate in 1890 and 1900, rising slightly by 1910. By 1930 it reached the level of .60 that most analysts consider "high." On the basis of similar data for all large cities, Cutler, Glaeser, and Vigdor (1999, 456) conclude: "Where only one city had a ghetto by our definition in 1890 (Norfolk, Va.), 55 cities had a ghetto by 1940." Segregation came late to northern cities.

One historian with access to data at a finer spatial scale, Philpott (1978, 120–21), complained that ward data were misleading in the case of Chicago (for comparison see the case studies of Chicago and Harlem by Spear [1967] and Osofsky [1963]). The 1900 ward map for Chicago, he said, "shows blacks scattered over all of the Southwest Side, most of the South Side, and much of the West Side as well." In fact, he argued, "the residential confinement of the blacks was nearly complete at the turn of the century." More recently Logan, Zhang, and Chunyu (2015) have analyzed newly available data at the ED level for Chicago and New York, showing that segregation was already high in 1880 in these two cities and was rising steadily prior to the Great Migration. In this study we expand that analysis to ten large northern cities. We analyze census data at the level of EDs, which averaged fewer than 1,500 people, somewhat smaller than contemporary census tracts (which have about 4,000 residents). In contrast, wards are much larger. For example, in 1880 Chicago—with a population of more than 500,000—had only 18 wards, an average of nearly 30,000 people per ward.

The Segregating Process

Equally important in our analysis is the question of how black residents were consigned to black neighborhoods. Was it based on racial exclusion, or was the effect of race tempered by differences in social class and geographic origin? It would be considered "normal" for immigrants in this period to live in segregated ethnic neighborhoods, in large part because of the disadvantages associated with their relatively low initial class position and newcomer (i.e., "un-acculturated") status. But suppose the key source of segregation in their case is not being Italian or Russian, but rather another attribute that happens to be associated with those ethnicities. Suppose, as assumed by the spatial assimilation perspective (Massey 1985) that their residential enclaves are left behind as individual families achieve better jobs, the second generation becomes more American, and they learn how to better navigate the city. This would contrast with the ghetto, which is thought of as an absorbing state (Logan, Alba, and Zhang 2002). Contemporary studies show that blacks are less likely than comparable whites to escape poor neighborhoods, even across generations (South and Crowder 1997; Sharkey 2013).

When did race become so consequential for where people lived? Research on the 1930s and 1940s (see, for example Frazier [1937] on Harlem and Duncan and Duncan [1957, 237–98] on Chicago) makes clear that there was class variation within the ghetto at that time, but that both middle-class and working-class blacks were unable to escape its grasp. Massey and Denton (1993, 30) argue that entrapment was a new development, and that previously "well-to-do African Americans" had been more able to find housing commensurate with their social status. This conclusion is contested by early studies in Chicago. Comstock (1912, 255) observed that "[t]he strong prejudice among the white people against having colored people living on white residence streets ... confines the opportunities for residence open to colored people of all positions in life to relatively small and well-defined areas" (see also Breckinridge 1913).

Research Design

When did northern blacks become highly segregated, and what was the process behind their residential separation? We address these questions with information on ten cities for the period 1880–1940: New York City, Brooklyn, Chicago, Boston, Philadelphia, Cincinnati, Cleveland, St. Louis, Pittsburgh, and Detroit.[2] These cities included nearly 18 million residents in 1940.

We draw on data at the individual level and at the level of EDs and cities. In models where the unit of analysis is the individual we rely on microdata made available by the Minnesota Population Center (MPC), including a 100 percent population sample for 1880, a 5 percent sample for 1900, and 1 percent samples for 1910–1940. In these models, we draw a subsample comprising either the household head or spouse (selected randomly) and all unrelated adults (age 15 and above, including only those with a coded occupation) in the household.

ED-level counts have been aggregated from individual-level records from three sources. One source is Ancestry.com, which has transcribed portions of all the individual records from pre-1950 censuses. Allison Shertzer (University of Pittsburgh) obtained permission to assemble these data from Ancestry's webpage for the four census years between 1900 and 1930 for the northern cities. We have cleaned the individual records and aggregated data on racial composition to the ED level. In addition we have created historically accurate GIS maps of the EDs in each decade 1900–1930.[3]

The source for 1880 is the Urban Transition HGIS (Logan et al. 2011), which is based on the 100 percent population microdata distributed by MPC. ED-level data were aggregated and joined to historically accurate GIS maps. For 1940 we have aggregated 100 percent population microdata to the ED level based on an early release from MPC. ED maps are not yet available, and we rely here on a mapping of 1940 census tracts completed by the National Historical GIS Project at MPC.

The ED data are used to calculate indices of segregation. Two segregation measures are used for comparison across cities and over time. These are the index of dissimilarity (D_{bw}) and index of isolation (P^*_{bb}, see Lieberson and Carter 1982). As used here, dissimilarity measures the degree to which blacks and whites were unevenly distributed across EDs in a city. The more blacks are clustered in some EDs and whites in others, the higher the value of D, with a maximum of 1. There is a consensus among demographers that a value of .60 or above is very high. The average value of D in metropolitan regions in 2010 was close to this level (.591). If there were no segregation, D would reach its minimum value of 0. The isolation index measures the exposure of a group to itself. For example, a P^*_{bb} value of .50 indicates that the average black person lived in an ED that was 50 percent black. Even if segregation (D) remains the same over time, growth in a minority population will tend to leave it more isolated; that is, leave group members in neighborhoods where they are a larger share of the population. As we will see, D began at fairly high levels in 1880 and continued to rise at the same time that the black share of the population was also rising. Consequently black isolation reached extremely high levels by the end of our study period.

We also combine ED data with sample data for individual black residents to estimate locational attainment models. In 1880 we have a 100 percent sample; in 1900 it is 5 percent; in other years it is a 1 percent sample (Ruggles et al. 2010). In these models we analyze how people's residential outcome (the percent black in their neighborhood) is associated with their individual-level characteristics. An example with contemporary data is provided in Logan, Alba, and Zhang (2002), who predict living in ethnic neighborhoods for Hispanic and Asian groups in New

York and Los Angeles. The predictors used here are gender, age, marital status (single, married, and divorced/widowed), household composition (living alone or with relatives vs. living with only nonrelatives), southern birth,[4] and occupational standing. Occupation is the only available social class measure prior to 1940. It is typically included in analyses as an interval scale socioeconomic index (SEI) based on rankings of occupations' income, education, and prestige in 1950. Sobek (1996) has demonstrated that it provides a reliable ranking of occupations as far back as the late nineteenth century. We operationalize it as the highest SEI of any family member in the household.

Finally, we have constructed additional city-level variables for each year to use as predictors of segregation and isolation. The city percent black of the total population was calculated from our 100 percent population samples in 1880 and 1940 and from the ED counts in 1900–1930. The measure of relative class position of black residents is the ratio of the mean value of SEI for blacks to the mean SEI for whites. The measure of southern origin is the percentage of black adults (age 18 and above) who were born in the South, as defined above. The latter two indicators are based on calculations from the various microdata samples.

Results

City trends in segregation and isolation, 1880–1940

Blacks were present in small numbers in northern cities throughout the nineteenth century. Again citing Massey and Denton (1993, 17): "There was a time, before 1900, when … in the north, a small native black population was scattered widely throughout white neighborhoods." Flamming's (2006) study of black Los Angeles describes the trend from a historian's perspective: The "quieter" migration of the better educated and more ambitious African Americans during 1890–1915 "filtered into small, loosely knit communities that were, in large part, middle class …There was some racial segregation, but there were no black ghettos to speak of" (2006, 45). But following World War I, provoked by the first wave of the Great Migration, whites panicked: "They erected residential boundaries, through violence and law … thereby penning the migrants into black-only districts that proved to be embryonic ghettos" (Flamming 2006, 46).

These authors agree on several points: the black population was initially small but grew appreciably after World War I; the black population in the North was initially mainly of northern origin but later included much larger shares of southern-born migrants; and ghettos were only "embryonic" prior to World War I. After World War I, segregation and racial isolation spiked in response to the Great Migration. Our analyses offer some new insights into and in some ways contradict these conclusions.

Because black population growth plays such an important role in this account, we begin by outlining the trend in the black population share of the population in our ten cities (Figure 2).[5] In the years through 1910 these values are tightly bunched together in a range between 1 percent and 5 percent (St. Louis is an

FIGURE 2
Black Proportion in 10 Northern Cities, 1880–1990

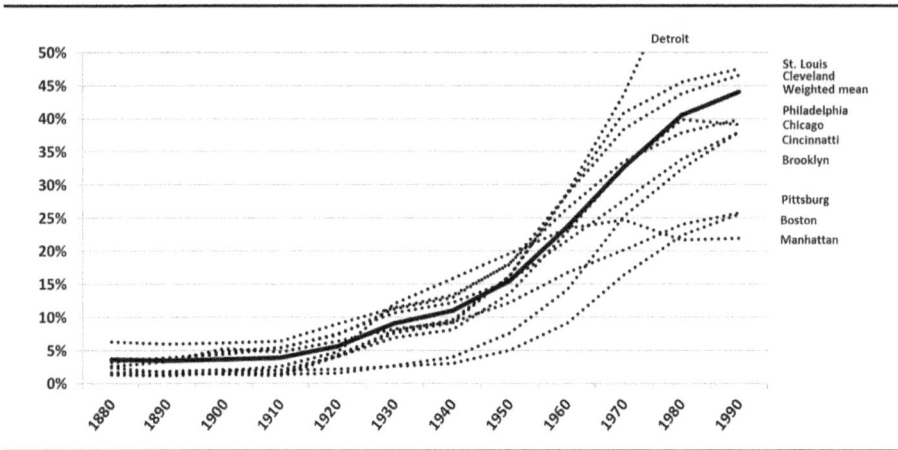

outlier, already 6.4 percent black in 1910; see Table A1 in the online appendix). The average (weighting by the size of the black population in the city) was 3.7 percent in 1880, rising to just 3.9 percent in 1910. At this point the cities' paths began to diverge, with little change in Brooklyn but an increase of around 2 percent in other cities, and reaching an average of 5.6 percent in 1920. It is surprising that the sharp rise in segregation over the 1880 to 1940 period was accompanied by such a modest increase in black share, particularly compared with the substantial growth in the black population in the post–World War II era. The greater impacts occurred from 1950 (when the black share averaged 15.5 percent) to 1960 (23.7 percent), 1970 (32.7 percent), and 1980 (40.5 percent). This timeline raises a question: if there had been little segregation prior to 1920, and if the black population had risen only from 5.6 percent in 1920 to 11.0 percent in 1940, what drove this sudden ghettoization?

Possibly whites responded not to the actual change in the black population but to perceptions of racial change, fueled by rumor and media coverage. We are unable to test this hypothesis. Another plausible answer is that it corresponded with the influx of southern migrants into cities that had better accommodated the local black population. Although Tolnay (2003, 218) points to evidence that southern blacks were positively selected and not particularly disadvantaged in some ways, his review of the literature acknowledges this point of view: "In virtually all destinations, the southern migrants were greeted with suspicion and hostility by black and white northerners alike. With generally minuscule black populations before the Great Migration, northern and western cities had achieved a relatively stable state of race relations, albeit one characterized by distinct racial inequality. That situation began to change, however, as waves of migrants from the South produced extraordinary growth in local black populations." An early study of sixteen cities by Woofter (1928, 97) concluded that southern migrants' neighborhoods "did not measure up" to those of blacks raised in the North. It should be pointed out, however, that throughout this period

FIGURE 3
Trend in Index of Dissimilarity, 1880–1940

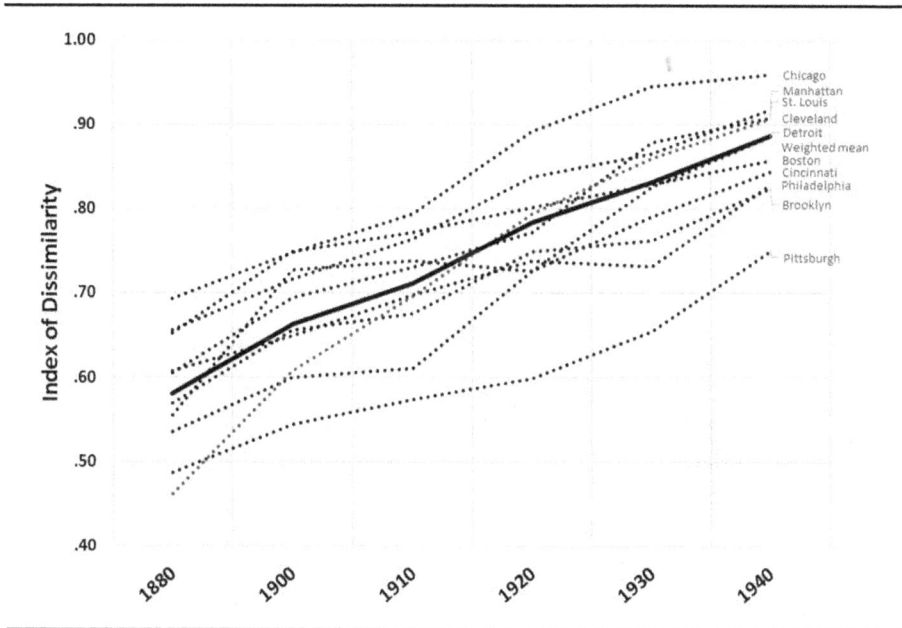

northern-born blacks remained about half of the black population. In the cities studied here, the southern-born share of the black population as early as 1880 ranged from 29 percent in Brooklyn to 60 percent in Cincinnati. By 1940 the range was from 30 percent in Boston to 64 percent in Detroit. If the influx from the South is what made the difference, it must have been mostly because of sheer population increase, not dependent on the source of the increase.

We can also draw inferences from analysis of the decade-by-decade trends in two measures of residential patterns: segregation (measured by D) and isolation (measured by P°). Figure 3 reports trends in D. Note that the average black person lived in a city with an ED-based value of D close to 60 as early as 1880. In 1910 it was above 70, and it approached 90 in 1940. By this measure segregation of blacks was always high, even when less than 5 percent of city residents were black. This is a very different conclusion than has been reached in prior studies using ward data. There was variation among cities, to be sure. Segregation was consistently lower in Pittsburgh than in the other cities, and most extreme in Chicago. However, of these two cities Pittsburgh always had a larger share of black residents.

Figure 4 reports trends in racial isolation. Not surprisingly isolation was initially quite low because it was limited by the overall black population share. The average black person across these cities lived in an ED that was 14.5 percent black in 1880 and still only 22.1 percent black by 1900. After 1900 it rose rapidly, but it was always in the range of 7 to 8 times the black share. In three of these ten cities the average black lived in a majority black ED by 1920; six cities by 1930; and in all but Brooklyn by 1940.

FIGURE 4
Trend in Isolation (P^*_{bb}), 1880–1940

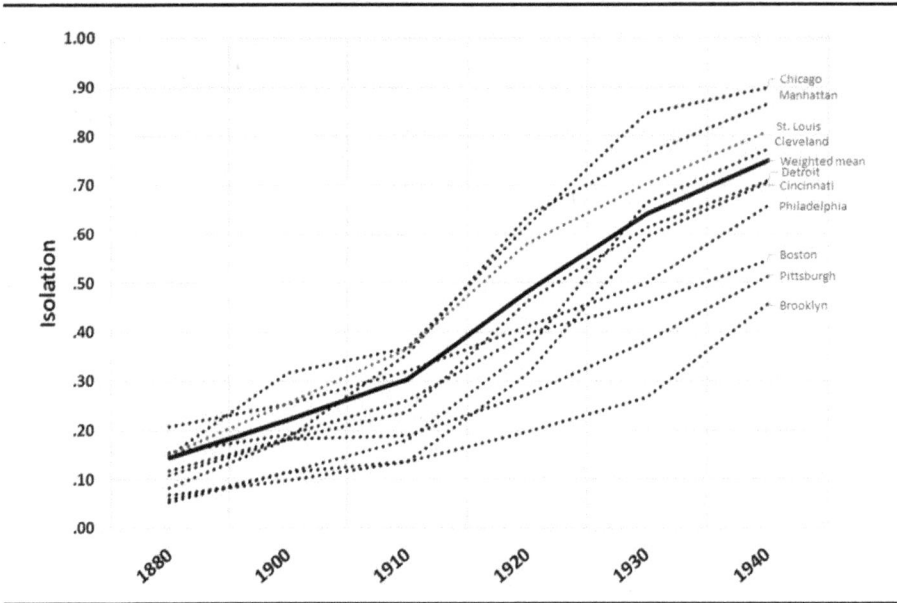

Another view of these trends is provided in Figure 5, where we have pooled all of the city data for D and P^* in a single scatterplot. The hollow circles represent cities in 1880 and 1900, the gray circles represent 1910 and 1920, and the black circles represent 1930 and 1940. One can see the progression over time toward higher levels of both segregation and isolation.

There was also a clear evolution in the spatial pattern of predominantly black areas, as displayed in the online supplemental Figures 1–10,[6] which map the black settlement pattern in all of these cities. Maps are shown for EDs in 1880–1930 and for census tracts in 1940. These maps have a common feature: the main areas of black settlement expand over time. For example, in St. Louis in 1880 when D was below .50, P^* was only .151, and the black population was less than 6 percent; segregation was visible in the contrast between EDs with virtually no black residents and those at 10 percent or 20 percent black. In 1900 several EDs that were more than 70 percent black appear, and the metaphor of "embryonic ghetto" seems useful to describe this case. In each subsequent decade this predominantly black zone expands and slowly spreads, to the point that by 1940 a majority of census tracts are nearly all white, while the zone of black settlement has clearly solidified.

Predicting variation across cities

Though we see similar trends across all ten cities, there is also considerable variation among them. We turn now to prediction models, seeking further clues to the sources of growing separation. In these models the dependent variables

FIGURE 5

Index of Dissimilarity (D) by Isolation (P^*_{bb}) for Ten Cities, 1880–1940

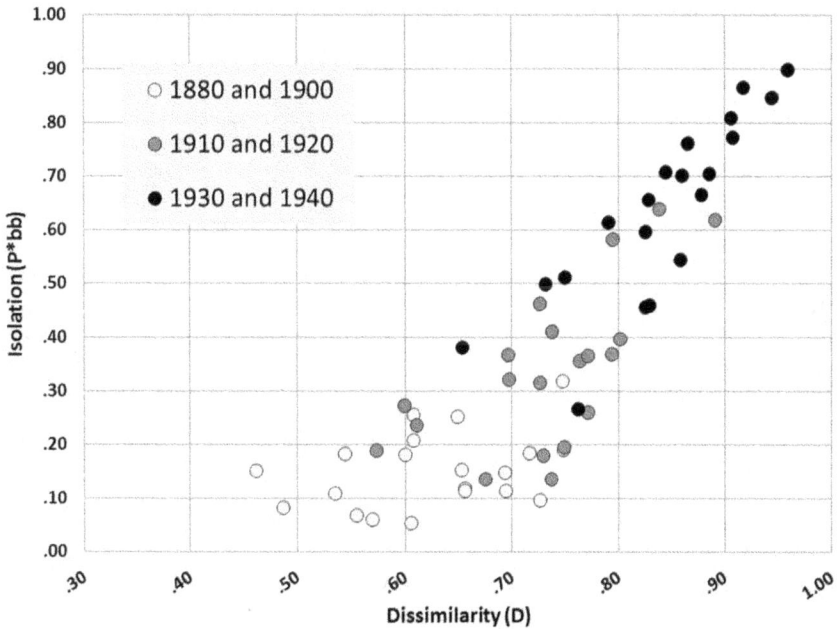

are the city's level of dissimilarity and isolation in a given year (so there are 60 cases, six time points for each of the 10 cities). Time itself is a key predictor, given the evident time trend in the data. Other predictors are characteristics of the city in that year that have been pointed to as reasons for segregation to be higher or lower. One of these is the black share of the population, prominent in theories of the "threat" of a minority population (it is only a control variable in predicting isolation, since percent black in the city is an integral part of the definition of isolation). Another characteristic is the relative class position of whites and blacks. From a market perspective and also from the theory of spatial assimilation, one would expect greater black separation in cities where blacks' average occupational standing (measured here as the mean SEI of employed blacks) is especially low in comparison to whites.[7] Finally we include a measure of southern origin: the percentage of adult blacks (age 18 and above) who were born in southern states. To the extent that socioeconomic and cultural boundaries restricted the incorporation of southern blacks in these cities, this predictor would be positively associated with residential separation.

Results are reported in Table 1. The time trend is evident in the effects of dummy variables for year, which are responsible for most of the explained variance in the models. Controlling for the three city characteristics (for example, taking into account variation in the black population share), D rose by thirty points and P^* by thirty-eight points between 1880 and 1940. Dissimilarity was not

TABLE 1
OLS Models Predicting Segregation (D_{bw}) and Isolation (P^*_{bb}), Pooled Cross-Sections for
1880–1940

	Dissimilarity		Isolation	
	b	SE	b	SE
Percent black	.10	(.43)	2.84	(.67) °°°
Black to white SEI ratio	.56	(.18) °°	.57	(.28) °
Percent adult black born in South	.10	(.11)	.27	(.18)
Year (ref=1880)				
1900	.00	(.04)	−.03	(.06)
1910	.04	(.04)	.02	(.06)
1920	.14	(.04) °°	.17	(.06) °°
1930	.21	(.04) °°°	.26	(.06) °°°
1940	.30	(.04) °°°	.38	(.06) °°°
Constant	.26	(.12) °	−.35	(.18)
R^2, Adjusted	.68		.81	
N	60		60	

°$p < .05.$ °°$p < .01.$ °°°$p < .001.$

affected by variation in black population share. Neither indicator of racial separation was affected by variation in the share of southerners in the city's black population. And although the ratio of black to white average SEI had a highly
significant effect on both measures, it was in the opposite direction of the prediction from market and assimilation theory. Where blacks in a city were closer to
whites in SEI, they were more separated.

Bayer, Fang, and McMillan (2011) noticed this latter result with current census data, and they argue that it represents a "re-sorting" of blacks, as newly
middle-class black residents upgrade into homogeneous new black neighborhoods. This was probably not the mechanism in the early period when the black
middle class was much smaller and less residentially mobile than it is now. An
alternative explanation is that perhaps the causal order of the relationship is different—that blacks have greater occupational opportunities in cities where they
are more segregated. A longstanding understanding of the black ghetto is that it
created a market for services to the black community, ranging from professional
occupations such as ministers, teachers, and medical practitioners to people
engaged in personal services. Semyonov (1988) identified this phenomenon in
Israel, where Arabs working in (mono-ethnic) Arab communities are occupation-
ally advantaged in comparison with Arabs working in labor markets where they
must compete with Jews. Segregation can increase opportunities if the minority
community "reaches a critical mass and is large enough to develop independent,
mono-ethnic, labor markets" (1988, p. 257). Lieberson (1980, 297–98) made the
same point with respect to historical black neighborhoods: "If the black

population base is large enough, there will be support for black doctors, black clergy, and so on, even if they remain totally unacceptable to others."

How individual background translates into locational outcomes

Aside from city variations, we can learn more about the processes underlying segregation by investigating which black residents lived in more or less segregated neighborhoods (that is, neighborhoods with a higher share of black neighbors). We tackle this question in a series of multilevel regression models where we have individual-level data on where black people lived. Models of this sort are referred to as locational attainment models (Alba and Logan 1992). Similar models have been estimated with historical data for American cities by Tolnay, Crowder, and Adelman (2002) and Logan and Zhang (2012). We include several demographic characteristics as control variables: gender, age, marital status, and household composition (living with nonrelatives vs. living with family members). Southern birth is included to probe further into the possible disadvantages of migrants. Other predictors identify aspects of people's class position, on the assumption that those with more human capital would be more likely to escape predominantly black zones. These include literacy (a dummy variable in the census), the SEI of the highest status person in the respondent's family (or the person's own SEI in the case of unrelated adults), and whether the housing unit is rented or owned by the head of household (unrelated persons are always coded as renters). One final variable, whether the person is a live-in servant in another family's household, is especially relevant in the early decades, when a large share of employed blacks lived with their white employers.

This is a multilevel random effects model with two variables at the city level: the index of dissimilarity in a given year and the city's percent of black residents. The mathematical form of the model is shown in the following equation:

$$y_{ij} = \alpha + \beta_1 X_{ij} + \beta_2 X_j + u_j + e_{ij}$$

where X_{ij} are individual-level covariates; X_j are city-level covariates; u_j is the city-specific error term; e_{ij} is the individual-specific error. Unexplained variation at the city level in the racial composition of people's neighborhoods depends only on average differences between cities themselves. The coefficients of city-level variables contribute to city-specific slopes. The effects of individual covariates are constant across cities.

Table 2 shows that the city-level predictors are the dominant factors in the nature of black people's neighborhoods.[8] Most of the variance explained by the models is between cities, and a smaller share is within cities. If one lives in a city with a larger black population that is highly segregated, one will live in a neighborhood with a larger share of black neighbors.

At the individual level, southern birth has a significant negative effect in 1880, and it is not a significant factor in 1900, 1910, or 1920. But it emerges in both 1930 and 1940 as a significant predictor. In these decades, blacks born in the

TABLE 2

Multilevel Regression Predicting Percent Black in the ED Where a Person Lives, 1880–1940 (Black persons age 15 and above)

	1880		1900		1910		1920		1930		1940	
	b	SE	b	SE	b	SE	b	SE	b	SE	b	SE
Female	-.33	(.13)*	-2.41	(.56)***	-3.30	(1.25)	-1.27	(1.11)	-.88	(.87)	-1.67	(.75)*
Age	.01	(.01)	-.01	(.03)	.02	(.06)	-.13	(.05)*	.00	(.04)	-.02	(.03)
Southern born	-.34	(.12)*	.02	(.56)	2.43	(1.25)	.47	(1.15)	5.05	(.95)***	3.36	(.83)***
Marital status (REF=single)												
Married	-.36	(.21)	1.99	(.80)*	6.24	(1.79)***	2.74	(1.63)	-.23	(1.28)	4.33	(1.01)***
Divorced/widowed	.96	(.23)***	4.43	(.98)***	5.93	(2.03)**	5.89	(2.02)**	-.84	(1.51)	3.73	(1.33)**
Lives with nonrelatives only	-1.00	(.20)***	1.15	(.74)	2.64	(1.63)	2.70	(1.48)	1.28	(1.16)	.62	(.91)
Literate			1.18	(.83)	5.06	(2.26)*	2.59	(2.31)	4.67	(2.32)*	6.22	(2.02)**
Highest family member's SEI	-.03	(.00)***	.12	(.02)***	-.03	(.04)	.05	(.04)	.13	(.03)***	.04	(.02)
Owner	-12.49	(.97)***										
Live-in servant	-9.32	(.18)***	-12.49	(.97)***	-16.82	(2.33)***	-19.95	(1.96)***	-8.39	(1.33)***	-8.94	(1.41)***
City-level D$_{bw}$	.85	(.01)***	.91	(.07)***	1.26	(.12)***	1.94	(.08)***	1.88	(.05)***	2.07	(.07)***
City-level % black	4.42	(.06)***	4.09	(.26)***	3.40	(.40)***	3.07	(.28)***	2.25	(.16)***	1.45	(.10)***
Constant	-46.63	(.95)***	-55.31	(5.89)***	-82.46	(10.07)***	-121.35	(7.62)***	-120.58	(5.86)***	-135.35	(6.37)***
R^2: overall	.18		.08		.12		.21		.24		.20	
R^2: within	.09		.05		.05		.04		.02		.01	
R^2: between	.82		.57		.85		.89		.91		.91	
N	45,863		5,801		1,678		3,062		4,923		6,139	

*$p < .05$. **$p < .01$. ***$p < .001$.

South lived in neighborhoods that were 3 to 5 percent higher in black share. We cannot explain why this effect varies over time. It does not seem to be associated with a spike in the southern share. There are only two cities (Cleveland and Detroit) where the southern share was quite modest in early decades (20–40 percent in 1880–1910) and much higher subsequently (60–75 percent in 1920–1940, though actually declining between 1930 and 1940). Possibly there was a shift in the composition of black migrants that is not associated with other variables in the model.

Almost by definition, live-in domestic servants in 1880 lived on average in neighborhoods with fewer blacks. Black owners (around 10 percent of the total) did also. However some other human capital measures worked in the opposite direction: literacy (in 1910, 1930, and 1940) and occupational SEI (in 1900 and 1930) were positively related to living in areas with larger black shares. We advise caution in interpreting the SEI effect because it appears in only two years. However it is parallel to the effect of the black/white SEI ratio on both dissimilarity and isolation in Table 1. Possibly blacks with certain kinds of higher status occupations who served black clients were especially likely to live in black neighborhoods.

Discussion and Conclusion

The chief conclusion from this study is that black separation from whites in northern cities was much greater and appeared much earlier than has previously been documented. If the Blalock hypothesis that segregation was imposed on blacks because they posed a threat to whites stands, then they must have already been threatening when they were only 2 percent or 3 percent of the population. The index of dissimilarity was in the high range (above 60) in several cities studied here in 1880, and the average value was above 60 by 1900. Black isolation was much higher at the ED level than at the ward level throughout this period. Whether one interprets the actual values as "high" is a matter of interpretation. Our view is that if the average city was only 3 percent black in 1880 but the average black person lived in a neighborhood that was 15 percent black, then their neighborhood isolation is greatly out of proportion and should be considered high. Isolation is a function of both segregation (appropriately measured by D) and the size of the black population. As both of these factors rose over time, isolation skyrocketed. For those who consider the central fact of ghettoization to be the creation of zones of the city that are predominantly black, it is unlikely that ghettos could have formed to such a degree except for the Great Migration. But if segregation had not also been high and rising, black population growth could not have created the ghetto.

We also argue that the processes underlying segregation are key to the concept of ghettoization. This is similar to the view of Marcuse (1997), who notes that there are several forms of racial separation. What distinguishes the ghetto is not its size or homogeneity but rather the process of race-based exclusion. In the multivariate models we present above we examined the correlates of segregation to find clues about this process. The analysis is limited by the small number of cities, but it is strengthened by the fact that this number includes many of the key cities in the northern ghetto belt and that data follow these cities over several decades. We find little evidence that racial separation was due to human capital deficiencies of black residents in terms of either low class standing or southern migrant origin. In the city-level analysis the share of southern blacks proved to be unrelated to dissimilarity and isolation, and cities where blacks' occupational level was closer to that of whites had higher separation. In the individual-level analysis the results are mixed. There is no evidence of a southern migrant disadvantage through 1920, but in both 1930 and 1940 southerners lived in neighborhoods with higher black shares. The relatively few blacks who owned their own homes lived in neighborhoods with smaller black shares, but literacy and higher SEI were associated with living in neighborhoods with larger shares of black neighbors.

These findings for individual locational outcomes merit further investigation. Why did results change over time, and why did literacy and occupational standing have opposite effects to that of home ownership? Yet these within-city effects should not be overstated. In 1930 and 1940 the explained variance within cities is no greater than .01 and .02. Almost all of the variation in outcomes is between cities, dependent on the level of segregation and the size of the black population. For this reason we draw the more general conclusion that individual variation among blacks had minimal impact on where they lived. Blacks lived in black neighborhoods because of their race, and this was already the case in 1880.

This conclusion is relevant because it dates northern residential segregation to a time soon after the Civil War. This timing is consistent with what has already been well known about the strict limits on black peoples' opportunities in the labor market, education, and other spheres at that time, and in fact it would be surprising if a society that was so divided by race in these other ways had not also been divided at the neighborhood level. Scholars have pointed to a number of conditions specific to the period after World War I as the causes of segregation: the wave of bombings in Chicago in the 1920s, the creation of racial covenants in housing, redlining by federal officials, and exclusion of blacks from most early housing subdivisions outside the urban core. Our results suggest that the roots of the ghetto can be found much earlier. These mechanisms did not originate the ghetto, rather they supplemented the strong boundaries that were already in place by 1880 or 1900. They facilitated and accentuated segregation, making possible the extreme form of the ghetto that existed in 1940 and beyond.

These results also have some implications for contemporary patterns. First, they may help to explain why entrenched black neighborhoods remain so persistent in the current time, even when many of the mechanisms that promoted segregation have been outlawed. The ghetto does not depend on restrictive covenants or redlining or openly discriminatory real estate practices. It depends more fundamentally on the existence of strong social boundaries based on race, and these were in place in the late nineteenth century and, we believe, are in place today. Second, our observations about the spatial scale at which segregation should be studied are especially relevant to smaller racial/ethnic groups or groups in cities with small minority shares today. When a particular group—Chinese, Central American, Afro-Caribbean—is present in small numbers, it may be highly segregated at a fine geographic scale without being detectable in units as large as census tracts. Third, our emphasis on segregation not only as a level of separation but also as a process of residential mobility has great relevance for distinguishing among group experiences at the current time. Prior research has noted especially that some Asian national origin groups are fairly highly segregated, but that living in a more ethnic neighborhood may be characteristic of more advantaged group members, and therefore more likely a positive choice (Logan, Alba, and Zhang 2002). In contrast, the residential separation of Hispanics is strongly tied to their relatively low socioeconomic achievement and English-language skills (see Lichter, Parisi, and Taquino, this volume). As we interpret contemporary residential patterns, we have to be aware that the ghetto, the ethnic community, and the immigrant enclave have very different sources and consequences for minority residents.

Notes

1. The figure is reproduced from Logan, Zhang and Chunyu (2015). We draw on the ward data gathered and disseminated by Cutler, Glaeser, and Vigdor (1999). Their analyses use a variation of the isolation index that seeks to standardize for the effect of the black population share in a city. We present the more familiar index, which can be interpreted as the proportion of black residents in the neighborhood where the average black person lived.

2. Brooklyn was a separate city in 1880, and we treat Kings County as a separate city for the whole period. The Shertzer project collected data only for the Borough of Manhattan (not including the Bronx, which was also part of New York City in 1880, or Queens and Staten Island, which were incorporated into the city in 1898). Therefore our results for New York City cover only Manhattan.

3. See online supplement, Figures 1–10; http://ann.sagepub.com/supplemental.

4. "Southern" in this study is defined to include sixteen southern states and the District of Columbia: South Atlantic (DC, Florida, Georgia, Maryland, North Carolina, South Carolina, Virginia, West Virginia, and Delaware), East South Central (Alabama, Kentucky, Mississippi, and Tennessee), and West South Central (Arkansas, Louisiana, Oklahoma, and Texas).

5. Because our concern is with residential segregation and because children do not have a large role in location decisions, we include only the population age 18 and above.

6. See http://ann.sagepub.com/supplemental.

7. Using the occupational SEI probably understates the class inequality between blacks and whites, because blacks very likely earned less than whites in the same occupation. However this "compression" of measured inequality should be similar across these ten northern industrial cities.

8. Our findings diverge in some ways from the published research that most closely parallels it, a study of 103 northern and western cities in 1920, where the dependent variable was the percent native white in the ward where the person lived (Tolnay, Crowder, and Adelman 2002). These authors also found that city-level effects were dominant. After controlling for city characteristics, there was no difference in locational outcome for northern- vs. southern-born blacks. Literacy, SEI, and homeownership were all associated with living in neighborhoods with a higher share of native white residents. However these latter effects were based on pooling data for both whites and blacks.

References

Alba, Richard D., and John R. Logan. 1992. Analyzing locational attainments: Constructing individual-level regression models using aggregate data. *Sociological Methods and Research* 20:367–97.

Bayer, Patrick J., Haming Fang, and Robert McMillan. 2011. Separate when equal? Racial inequality and residential segregation. Economic Research Initiatives at Duke (ERID) Working Paper No. 100. Available from http://papers.ssrn.com.

Blalock, Hubert M., Jr. 1956. Economic discrimination and Negro increase. *American Sociological Review* 21:584–88.

Breckinridge, Sophonisba P. 1913. The color line in the housing problem. *The Survey* 40:575–76.

Comstock, Alzada P. 1912. Chicago housing conditions, VI: The problem of the Negro. *American Journal of Sociology* 18:241–57.

Cutler, David M., Edward L. Glaeser, and Jacob L. Vigdor. 1999. The rise and decline of the American ghetto. *Journal of Political Economy* 107:455–506.

Duncan, Otis D., and Beverly Duncan. 1957. *The Negro population of Chicago: A study of residential succession.* Chicago, IL: University of Chicago Press.

Flamming, Douglas. 2006. *Bound for freedom: Black Los Angeles in Jim Crow America.* Berkeley, CA: University of California Press.

Frazier, E. Franklin. 1937. Negro Harlem: An ecological study. *American Journal of Sociology* 43:72–88.

Lemann, Nicholas. 1991. *The Promised Land: The Great Black Migration and how it changed America.* New York, NY: Knopf.

Lichter, Daniel T., Domenico Parisi, and Michael C. Taquino. 2015. Spatial assimilation in U.S. cities and communities? Emerging patterns of Hispanic segregation from blacks and whites. *The ANNALS of the American Academy of Political and Social Science* (this volume).

Lieberson, Stanley. 1980. *A piece of the pie: Blacks and white immigrants since 1880.* Berkeley, CA: University of California Press.

Lieberson, Stanley, and Donna K. Carter. 1982. Temporal changes and urban differences in residential segregation: A reconsideration. *American Journal of Sociology* 88:296–310.

Logan, John R., Richard D. Alba, and Wenquan Zhang. 2002. Immigrant enclaves and ethnic communities in New York and Los Angeles. *American Sociological Review* 67:299–322.

Logan, John R., Jason Jindrich, Hyoungjin Shin, and Weiwei Zhang. 2011. Mapping America in 1880: The Urban Transition Historical GIS Project. *Historical Methods* 44:49–60.

Logan, John R., and Weiwei Zhang. 2012. White ethnic residential segregation in historical perspective: US cities in 1880. *Social Science Research* 41:1292–306.

Logan, John R., Weiwei Zhang, and Miao Chunyu. 2015. Emergent ghettos: Black neighborhoods in New York and Chicago, 1880–1940. *American Journal of Sociology.*

Marcuse, Peter. 1997. The enclave, the citadel, and the ghetto: What has changed in the postfordist U.S. city? *Urban Affairs Review* 33:228–64.

Massey, Douglas S. 1985. Ethnic residential segregation: A theoretical synthesis and empirical review. *Sociology and Social Research* 69:315–50.

Massey, Douglas S., and Nancy A. Denton. 1993. *American apartheid: Segregation and the making of the underclass.* Cambridge, MA: Harvard University Press.

Osofsky, Gilbert. 1963. *Harlem: The making of a ghetto. Negro New York, 1890–1930.* New York, NY: Harper and Row.

Philpott, Thomas Lee. 1978. *The slum and the ghetto: Neighborhood deterioration and middle-class reform, Chicago, 1880–1930.* New York, NY: Oxford University Press.

Ruggles, Steven, J. Trent Alexander, Katie Genadek, Ronald Goeken, Matthew B. Schroeder, and Matthew Sobek. 2010. *Integrated Public Use Microdata Series: Version 5.0* [Machine-readable database]. Minneapolis, MN: University of Minnesota.

Semyonov, Moshe. 1988. Bi-ethnic labor markets, mono-ethnic labor markets, and socioeconomic inequality. *American Sociological Review* 53:256–66.

Sharkey, Patrick. 2013. *Stuck in place: Urban neighborhoods and the end of progress toward racial equality.* Chicago, IL: University of Chicago Press.

Sobek, Matthew. 1996. Work, status, and income: Men in the American occupational structure since the late nineteenth century. *Social Science History* 20:169–207.

South, Scott J., and Kyle D. Crowder. 1997. Escaping distressed neighborhood: Individual, community, and metropolitan influences. *American Journal of Sociology* 4:1040–84.

Spear, Allan H. 1967. *Black Chicago: The making of a Negro ghetto 1890–1920.* Chicago, IL: University of Chicago Press.

Tolnay, Stewart E. 2003. The African American "Great Migration" and beyond. *Annual Review of Sociology* 29:209–32.

Tolnay, Stewart E., Kyle D. Crowder, and Robert M. Adelman. 2002. Race, regional origin, and residence in northern cities at the beginning of the Great Migration. *American Sociological Review* 67:456–75.

Woofter, T. J., Jr. 1928. *Negro problems in cities: A study.* New York, NY: Doubleday, Doran, and Company.

Spatial Assimilation in U.S. Cities and Communities? Emerging Patterns of Hispanic Segregation from Blacks and Whites

DANIEL T. LICHTER,
DOMENICO PARISI,
and
MICHAEL C. TAQUINO

This article provides a geographically inclusive empirical framework for studying changing U.S. patterns of Hispanic segregation. Whether Hispanics have joined the American mainstream depends in part on whether they translate upward mobility into residence patterns that mirror the rest of the nation. Based on block and place data from the 1990–2010 decennial censuses, our results provide evidence of increasing spatial assimilation among Hispanics, both nationally and in new immigrant destinations. Segregation from whites declined across the urban size-of-place hierarchy and in new destinations. Hispanics are also less segregated from whites than from blacks, but declines in Hispanic-black segregation have exceeded declines in Hispanic-white segregation. This result is consistent with the notion of U.S. Hispanics as a racialized population—one in which members sometimes lack the freedom to join whites in better communities. Hispanic income was significantly associated with less segregation from whites, but income inequality alone does not explain overall Hispanic segregation, which remains high. The segmented assimilation of Hispanics that we observe supports two seemingly contradictory theories: both the idea that spatial assimilation can come from economic and cultural assimilation and the notion that economic mobility is no guarantee of residential integration.

Keywords: segregation; assimilation; Hispanics; race relations; racial boundaries; new destinations; rural communities

The recent Hispanic diaspora—the movement from immigrant gateways to new destinations—has raised concerns about

Daniel T. Lichter is Ferris Family Professor in the Department of Policy Analysis and Management, professor of sociology, and director of the Cornell Population Center, all at Cornell University.

Domenico "Mimmo" Parisi is a professor of sociology and director of the National Strategic Planning & Analysis Research Center (nSPARC) at Mississippi State University.

Michael C. Taquino is deputy director and an associate research professor at nSPARC at Mississippi State University.

DOI: 10.1177/0002716215572995

36

ANNALS, *AAPSS*, 660, July 2015

patterns of racial and ethnic change and residential segregation in America's cities and communities (Johnson and Lichter 2010; Lee, Iceland, and Farrell 2014; Tienda and Fuentes 2014). Estimates of neighborhood segregation from the 2010 census indicate that Hispanic-white segregation is at a standstill or even increasing in some cities (Logan and Stults 2011; Rugh and Massey 2014); metropolitan segregation levels, as measured by the index of dissimilarity, have been fixed at moderately high levels for at least 30 years—at roughly 50.[1] Some of the most heavily populated and segregated Hispanic metropolitan areas also have experienced increasing segregation. For example, the Los Angeles metropolitan area has experienced increases in Hispanic segregation every decade since 1980, rising from 57.3 in 1980 to 63.4 in 2010 (Logan and Stults 2011). In metropolitan areas, the typical Hispanic person today lives in a neighborhood that is 46 percent Hispanic, 35 percent white, and 11 percent black. Mapping changing patterns of Hispanic residential segregation at other spatial scales, however, has been surprisingly limited (Fossett et al. 2014; Lee, Iceland, and Farrell 2014; Parisi, Lichter, and Taquino 2011, 2015), even as America's Hispanic population has become more spatially dispersed. Hispanics have increasingly bypassed traditional gateway states and cities, moving down the urban hierarchy into smaller metropolitan cities, suburbs, and rural communities.

Our fundamental goal here is to provide a geographically inclusive analytic framework for studying racial residential segregation, one that acknowledges that metropolitan (metro) and nonmetropolitan (nonmetro) areas represent systems of places—cities, suburbs, and small towns—that shape the ethnic and racial profile of neighborhoods. As Hispanics become "racialized" (Rumbaut 2009), cultural and economic boundaries and social distance from other racial groups will likely be revealed in shifting patterns of residential segregation. Tienda and Fuentes (2014, 500) now claim that a burgeoning and diverse Hispanic population has "reconfigured the ethnic and geographic landscape" of the United States. Indeed, much of the Hispanic growth has occurred in new immigrant destinations, including suburban and rural "boomtowns" (Lichter 2012; Massey 2008). Hispanics will redefine America's shifting racial and ethnic boundaries and residence patterns over the foreseeable future.

This article builds on previous empirical studies of racial and ethnic segregation by providing up-to-date national estimates of Hispanic residential segregation in 2010, along with comparative estimates for 1990 and 2000 (Iceland, Weinberg, and Hughes 2014; Logan and Stults 2011). Our analyses are inclusive of metro central cities, suburban communities, and nonmetro small towns. Second, we highlight recent patterns of Hispanic segregation from both whites and African Americans. Rapid Hispanic population growth makes the usual black-white paradigm of racial and ethnic relations seem increasingly outmoded

NOTE: The authors acknowledge the helpful comments of the volume coeditors and critical suggestions of the two reviewers, as well as support from the Cornell Population Center and the National Strategic Planning & Analysis Research Center (nSPARC) at Mississippi State University. The authors' previous collaborative studies of place-based or multiscale segregation, including rural segregation, appear in *Social Forces, Rural Sociology, Demography, Sociological Science, Social Problems,* and *Social Science Research,* among other publications.

in today's multicultural, multiracial society (Waters, Kasinitz, and Asad 2014). Third, we update recent estimates of Hispanic segregation in so-called new destinations. We evaluate whether the recent movement from traditional urban gateways to new Hispanic destinations has upended conventional theoretical interpretations of immigrant integration and cultural and economic incorporation (also see Flippen and Kim, this volume). Fourth, our analytical approach explicitly recognizes that segregation occurs at different geographic scales (Lee et al. 2008; Reardon et al. 2008), i.e., neighborhoods are nested within places, which in turn are located within metro (and nonmetro) areas (Lichter, Parisi, and Taquino 2012; Parisi, Lichter, and Taquino 2011). Here we estimate hierarchical linear models to examine the extent to which place-to-place differences in segregation account for the overall differences in Hispanic neighborhoods in metro (nonmetro) areas.

Recent settlement patterns raise an obvious question: Are America's Hispanics "moving toward" whites or blacks, as measured by conventional segregation indices (i.e., an indirect indicator of social distance)? For Hispanics, has integration been achieved by distancing themselves from blacks? Are Hispanics "whitening," as indicated by shifts in segregation along the black-white continuum of spatial exclusion? Or, instead, are (some) Hispanics being "racialized," i.e., integrating spatially with America's historically disadvantaged black population?

Spatial Assimilation: Hispanic Segregation in the New Century

Recent studies of Hispanic segregation have been framed largely by dueling theoretical perspectives that emphasize either spatial assimilation or place stratification (Iceland and Nelson 2008; Kim and White 2010). *Spatial assimilation theory* argues that spatial integration is a product of increasing cultural and economic integration. That is, segregation of immigrants from the majority population declines with upward social and economic mobility. U.S. immigrants typically first settle in gateway communities (e.g., enclaves where cultural and institutional support is ensured) but "move up and out" over time or across generations as they learn English, acquire new job skills, and become "Americanized." Spatial mobility reflects and reinforces socioeconomic mobility among America's new immigrant populations, including Hispanics from Mexico and other Latin American countries. That family income among Hispanics is positively associated with neighborhood quality gives empirical credence to the spatial assimilation model (Alba et al. 2014).

Indeed, Iceland and Scopilliti (2008), using restricted U.S. census data for metro areas, found that native-born Hispanics ($D = 48.1$) were less segregated than foreign-born Hispanics (59.9) from native-born whites in 2000. Native/foreign-born differences in segregation disappeared when sociodemographic characteristics of Hispanics (e.g., household income) and metro characteristics were controlled in the multivariate analysis. Although these results are consistent with spatial assimilation theory, native-born Hispanics have remained highly segregated from native-born whites because they have not been fully incorporated into American society on many important dimensions, such as education, income, and English language usage. Among the foreign-born population, Hispanic segregation declined only modestly with length of time in the country (Iceland and Scopilliti 2008; Iceland and Nelson 2008). In new immigrant destinations, Hispanic segregation also is often higher than in established gateways (Hall 2013; Lichter et al. 2010; Fossett et al. 2014; but see Park and Iceland 2011). This pattern seemingly reflects low-wage, low-skill Hispanic immigrants increasingly bypassing traditional gateways for new destinations (Lichter and Johnson 2009). Traditionally, spatial and social mobility went hand in hand. Today, however, the historical link between social and spatial assimilation is perhaps less easily established, especially in smaller cities, suburban communities, and rural places that represent sites of first settlement for Hispanics.

The widespread spatial dispersion of Hispanics seemingly suggests greater spatial integration with the native population (Massey, Rothwell, and Domina 2009), but the continuing concentration and neighborhood segregation of Hispanics in new destination communities also suggest something less benign. At a minimum, new patterns of Hispanic population redistribution call into question the inevitability of spatial assimilation and suggest that Hispanic origin is becoming a racial marker that is reflected in persistent segregation from whites (Tienda and Fuentes 2014). This view is probably accommodated best by the *place stratification* perspective. This conceptual framework has many variations, but the essential point is that upward social and economic mobility, which often accompanies assimilation, is no guarantee of spatial integration with the majority population. The racialization of Hispanics, in particular, means that they sometimes lack the freedom of movement to join whites in better communities or neighborhoods, where job opportunities, quality schooling, and safety may be more abundant (Alba et al. 2014). Hispanics instead are often locked in place by low incomes and housing discrimination (Iceland and Scopilliti 2008); segregation from whites persists but perhaps is observed in a different spatial form (e.g., in new immigrant enclaves and destinations) from the past (Lee, Iceland, and Farrell 2014; Lichter 2013).

Persistent Hispanic segregation also is reinforced by a new "political economy of place," one in which local communities mobilize politically to exclude racial and immigrant minorities. Exclusionary zoning is on the upswing, and discriminatory or exploitive lending practices prey disproportionately on historically disadvantaged populations (Rothwell and Massey 2009; Rugh and Massey 2010). At the same time, places with rapidly growing immigrant populations, including some new immigrant destinations, face the prospect of "white flight" (Crowder, Hall, and Tolnay 2011), which may contribute to a new kind of segregation in which entire communities (rather than just neighborhoods) shift from majority white to majority minority. Under the circumstances, it is perhaps unsurprising that past declines in Hispanic-white segregation have slowed or even reversed. A spatially inclusive approach—one that fully recognizes that Hispanic segregation might occur simultaneously at different scales of geography—may yield complementary insights and new conclusions that build on a large metro-centric empirical literature on residential segregation.

Although the spatial assimilation and place stratification perspectives are sometimes represented in the literature as competitors, they might better be viewed as complementary conceptual and analytical frameworks. It is an empirical fact that some Hispanic subgroups are becoming more residentially integrated over time with whites; racial boundaries of all kinds are being "crossed" with Hispanic gains in education, upward socioeconomic mobility, and more intermarriage. Other Hispanic populations—those distinguished by national origin (e.g., Mexicans versus others), phenotype (e.g., skin color or stature), or legal status—may be self-segregating or may face discrimination or other barriers that prevent them from relocating into white neighborhoods (Iceland, Weinberg, and Hughes 2014). A recent study by Hall and Stringfield (2014), for example, found that local metro shares of undocumented immigrants were associated with more neighborhood segregation from whites and less segregation from blacks. One interpretation is that some economically and culturally disadvantaged Hispanic groups are assimilating into a black underclass, sealed away in neighborhoods and places that provide few opportunities for upward mobility. Iceland and Nelson (2008) also found that foreign-born black Hispanics (mostly from the West Indies) were far more likely than foreign-born white Hispanics to be segregated from whites (D = 79 vs. 58 in 2000). Hispanic assimilation, if measured by residential segregation patterns, is highly segmented.

Current Study

Our article makes several specific contributions to the literature on Hispanic settlement patterns. First, we provide an up-to-date and geographically inclusive empirical approach (using block and place data from the 2010 and earlier decennial censuses) that acknowledges the new geographic dispersal of Hispanics over the past 20 years. We document emerging patterns of segregation in cities, suburban areas, and small towns, and, using hierarchical linear modeling (HLM), the extent to which demographic and economic differences between places contribute to the overall differences in residential segregation across metro (nonmetro) areas.

Second, we document residential segregation of Hispanics from both whites and blacks. Although roughly 50 percent of Hispanics self-identify as white, the percentages of Hispanics reporting as such vary widely over geographic space (Rumbaut 2009). This suggests that racial identification—and its implications for Hispanic-white segregation—are situational or contextual. Hispanics in the South, for example, are far more likely to identify as nonwhite than they are in other parts of the country. Increasingly, Hispanicity is a racial marker that is both imposed and internalized, which will be reflected in changing patterns of Hispanic-white and Hispanic-black segregation.

Third, we provide a baseline assessment of trends in Hispanic residential segregation and update several previous studies of segregation in new Hispanic destinations (Hall 2013; Lichter et al. 2010) based on 2000 or earlier data. Rather than being viewed as an indicator of growing social and economic integration, the emergence of new destinations for Hispanics may represent a new kind of racial and spatial balkanization in America. Neighborhood segregation may be giving way to new forms of segregation, where place-oriented social, economic, and political processes have become ascendant in the new century (Lichter, Parisi, Taquino 2015).

Methods

Data and measurement

Most previous studies of segregation use metro areas, central (principal) cities, or urbanized areas as units of analyses. They typically emphasize changing patterns of segregation across metro neighborhoods, as proxied by census tracts; the entire metro area is treated as a single housing or labor market that sorts different population groups into different neighborhoods in the metro region. The new "political economy of place" suggests, however, that communities rather than neighborhoods also represent important but often unrecognized actors that include or exclude desirable and undesirable population groups. Indeed, Farrell (2008, 467) claims that "[u]rban and suburban municipalities are replacing neighborhoods as the central organizing units of metro segregation." As organized collective actors, places can include or exclude specific populations while either promoting or discouraging ethnoracial diversity and intergroup relations.

Racial segregation across neighborhoods may therefore start first with the sorting of individuals across cities and communities, which are perceived by movers as possible residential destinations or not (Krysan and Bader 2009; Sharkey 2012). For the purposes of our study, we use places as defined by the U.S. Census Bureau in 1990, 2000, and 2010. Places include all incorporated cities, towns, and villages, as well as unincorporated communities and housing developments (called "census designated places" by the Census Bureau) that lack municipal governments. Places are located both in metro and nonmetro areas, which are defined using a fixed 2013 definition throughout the 1990–2010 period of study.

For our analysis, we focus on all places with at least 200 Hispanics residents in 1990, 2000, or 2010. We identified 6,806 cities, suburbs, and rural communities. In 2010, these places were home to 83.3 percent of the U.S. Hispanic population. Of these places, 5.4 percent were classified as metro central cities, 72.3 percent suburban cities and communities, and 22.3 percent as nonmetro small towns. Our selection criterion—one based on population size—reflects a floating definition from decade to decade. This means that changes in Hispanic segregation levels may reflect changes in the universe of places that meet our selection criterion. Our approach nevertheless arguably represents the reality of changing Hispanic segregation in America over a period of widespread growth and dispersion in the Hispanic population. Eliminating "new places" or destinations (i.e., those places now meeting but previously failing to meet the 200-person criterion) obviously would misrepresent a major dimension of changing Hispanic segregation in America.

We also classified some places as new Hispanic destinations. These are places with comparatively small Hispanic populations initially but that subsequently experienced unusually rapid Hispanic growth (Lichter et al. 2010). Identifying new destinations involved three specific steps. First, as a baseline, we selected all places in 1990 with Hispanic percentages of less than the U.S. Hispanic percentage in 1990 (9 percent). Second, from these places, we identified those that grew by at least 200 Hispanics between 1990 and 2000 or 2000 and 2010, which served to eliminate fast-growing places that added only a small number of Hispanics. Third, new destinations—those with exceptional Hispanic growth—were defined as places with rates exceeding the average Hispanic growth rate by at least one standard deviation. These thresholds were set separately for metro central cities, suburban areas, and nonmetro places. Our measurement approach yielded a total of 446 new destinations. Of these, 257 were classified as new destinations during the 1990s, and an additional 189 places were defined as new destinations for the 2000s. More than 11 percent of new destinations were central cities, 67.5 percent were suburban places, and the remaining 21.1 percent were nonmetro places.

We use blocks rather than census tracts (neighborhoods) as accounting units to calculate Hispanic segregation (e.g., Lee et al. 2008; Lichter, Parisi, and Taquino 2012; Parisi, Lichter, and Taquino 2015). Blocks are ideal for multiscale analyses of place-to-place differences in residential segregation in metro and nonmetro areas. For small towns and many suburban places, census tracts are often too large to effectively measure the changing distribution of Hispanics from other populations. Some small suburban or rural towns also may be represented by only one or two census tracts, which sometimes do not represent legal boundaries. Blocks can be uniquely nested within place boundaries and provide the spatial granularity needed to measure the within-place spatial distribution of different racial and ethnic groups. Blocks, however, should not be viewed as neighborhoods, nor can segregation within small towns—especially rural communities—be treated as conceptually similar to segregation in big cities. In rural communities, highly segregated populations may nevertheless share the same streets and sidewalks, shopping centers, local public schools, and playgrounds and parks.

We measure place segregation using the index of dissimilarity (D). D_t is defined as

$$D_t = \frac{1}{2} \sum_{i=1}^{k} |h_{it} - w_{it}|$$

where h_{it} and w_{it} are the respective percentages of the Hispanic population and whites residing in census block i at time t. This index is based on pair-wise comparisons and varies from 0 (no segregation) to 100 (complete segregation). D indicates the percentage of Hispanics that would have to move to other city or community blocks to achieve parity between Hispanics and whites in their percentage distributions across all blocks.

Analytical strategy

The analysis begins with a simple description of levels and trends in Hispanic segregation between 1990 and 2010 for the total United States and, separately, for central cities, suburbs, and nonmetro places. We then estimate multilevel models of Hispanic segregation in 2010 to examine (1) the extent to which place characteristics influence neighborhood segregation and (2) the extent to which place-to-place differences account for overall metro (nonmetro) segregation. Specifically, we estimate models in which place coefficients are allowed to vary by metro and nometro areas (i.e., varying intercept model). The former are typically but not exclusively composed of multiple metro counties, and the latter are distinguished either as micropolitan areas, which may include one or more nonmetro counties, or as noncore counties. We also estimate the intraclass correlation coefficient (ICC) to highlight the contribution of metro-to-metro and place-to-place differences to the overall differences in neighborhood segregation. Finally, for the same time period, we examine the level and trends of Hispanic segregation in "new" Hispanic destinations and similarly model the within- and between-place components of segregation using HLM with 2010 data.

Results

Hispanic segregation patterns, 1990–2010

Hispanic segregation from the rest of America. Table 1 (top panel) provides unweighted and weighted segregation indices between Hispanics and non-Hispanics during 1990–2010. Weights are based on the relative size of the Hispanic population, i.e., we give greater weight to places with larger Hispanic populations. The weighted value indicates the level of segregation experienced by the average Hispanic person rather than the average segregation of places (based on the unweighted D).

These initial benchmark estimates indicate the level of residential segregation of Hispanics from the "rest of America"; as such, D indicates the broad spatial integration of Hispanics in American society. These data provide several specific insights. For starters, Hispanic/non-Hispanic segregation in America's cities and communities is moderately high nationally (first column), indicating the slow spatial integration of Hispanics with the rest of American society. Between 1990 and 2010, Hispanic/non-Hispanic segregation—those weighted by the Hispanic population—nevertheless declined from 48.5 to 44.0, roughly a 10 percent decrease. Similar but smaller declines in segregation were observed using the unweighted Ds (i.e., average decline from 45.9 to 43.3 over the 20-year period). Such declines are consistent with spatial assimilation theory (Iceland and Scopilliti 2008), but these results are perhaps surprising in light of the fact that much of recent Hispanic population growth is due to immigration (and the second-order effects of high rates of natural decrease; Johnson and Lichter 2008). Immigrants are typically more segregated than the native born (Hall and Stringfield 2014; Iceland and Scopilliti 2008).

TABLE 1

Average Segregation of Hispanics (H) from Non-Hispanics (NH), Non-Hispanic Whites (NHW), and Non-Hispanic Blacks (NHB), 1990–2010

| | United States | | Metro Places | | | | | | | | Nonmetro Places | |
| | | | Metro Central Cities | | Metro Suburban Places | | Total Metro Places | | | | | |
	Unweighted	Weighted*	Unweighted	Weighted*	Unweighted	Weighted*	Unweighted	Weighted*			Unweighted	Weighted*
H-NH Segregation												
1990	45.9	48.5	56.1	55.2	42.8	41.1	44.2	48.3			53.6	53.3
2000	45.8	47.1	52.5	53.8	43.1	40.6	43.9	46.9			53.0	50.8
2010	43.3	44.0	48.0	50.8	40.4	38.4	40.9	43.9			51.4	48.3
H-NHW Segregation												
1990	47.1	52.8	57.4	61.4	44.1	43.6	45.5	52.7			54.6	55.0
2000	47.7	52.0	55.4	61.0	45.1	43.9	46.0	51.9			54.5	52.8
2010	45.5	49.3	51.6	58.5	42.7	42.1	43.3	49.3			52.9	50.4
H-NHB Segregation												
1990	67.0	68.5	68.5	74.5	63.7	61.1	64.2	67.9			80.6	81.2
2000	65.2	65.1	63.9	71.3	62.0	58.6	62.2	64.6			77.2	76.1
2010	60.0	54.9	56.3	57.4	56.0	51.4	56.0	54.1			73.6	71.4

*Weighted by number of Hispanics.

Declines in Hispanic/non-Hispanic segregation also were observed across the urban-rural continuum (Table 1). Our estimates clearly show that segregation is not strictly a metro or big-city phenomenon. In fact, overall segregation among nonmetro Hispanics was 48.3 in 2010, compared with 43.9 in metro areas overall and 50.8 in central cities. Hispanic/non-Hispanic segregation has been decidedly lower in suburban communities (e.g., 38.4 in 2010) than elsewhere, but recent declines in Hispanic segregation have also been more modest in suburban areas over the past two decades. For Hispanics, suburbanization clearly is associated with spatial integration.

Hispanic segregation from whites and blacks. Overall segregation of the Hispanic population from the non-Hispanic population reflects reinforcing and offsetting patterns and shifts in segregation from both whites and blacks. The bottom two panels of Table 1 provide parallel estimates of Hispanic-white and Hispanic-black segregation since 1990. We address the question of whether Hispanics are more segregated from blacks or from whites, and whether increases or decreases in segregation over time suggest more spatial assimilation with whites or blacks.[2]

These analyses reveal several conclusions. For example, the Ds indicate high but declining Hispanic-white segregation, on average, for all places (that met our selection criteria). The overall Hispanic-white D was 49.3 for all places in 2010, a slight decline from 52.8 in 1990. (These patterns are similar when D is weighted by the white population.) In comparison, the Hispanic-black D was higher in 2010 than the Hispanic-white D (55.0 vs. 49.3), but the former has experienced significantly larger declines since 1990. Hispanic-black segregation declined by 20 percent between 1990 and 2010, from 68.5 to 55.0; while Hispanic-white segregation declined by roughly 10 percent over the same period. Spatial integration of Hispanics into American society is occurring much more slowly with whites than with blacks. Although Hispanics are still more spatially integrated with whites than with blacks, current trends suggest that this may not be true in the future.

These recent declines in Hispanic-white and Hispanic-black segregation are also broadly observed across the urban hierarchy. That is, declines in D have occurred in metro central cities and suburbs and in nonmetro communities. Still, large spatial differences in the segregation of Hispanics from whites and blacks persist across different types of places. For example, these data reveal large city-suburb differences in Hispanic-white segregation but not in Hispanic-black segregation. In cities, Hispanics are segregated at similarly high levels from whites and blacks (Ds are 58.5 and 57.4, respectively). The big differences in segregation are in the suburbs of metro areas, where Hispanics are more segregated from blacks than from whites but where segregation from blacks has declined most rapidly since 1990 (from 61.1 to 51.4 between 1990 and 2010). Hispanic segregation from whites is lower in the suburbs, with Ds fixed over time in the low 40s. Declines in Hispanic-black segregation are now being reinforced by the growing racial diversity in (at least some of) America's suburbs.

TABLE 2
Descriptive Statistics, 2010

	All		New Destinations	
	Mean	SD	Mean	SD
Demographics				
Total population	28,418	133,843	26,729	60,257
Percent black	9.8	15.2	15.3	16.8
Percent Hispanic	19.7	21.3	15.7	13.4
Percent foreign born	12.1	11.1	11.2	8.7
Economic structure				
Percent new housing, 1990–2000	24.0	74.6	73.5	219.5
Percent in poverty	14.5	9.6	16.7	9.5
Industry				
Percent in manufacturing	11.0	7.3	15.5	9.9
Percent in service	53.5	10.3	50.2	10.4
Percent in government	5.3	4.5	4.5	3.4
Economic well-being				
Hispanic mean household income	58,259	31,867	50,326	22,571
Total number of places	6,806		446	

In nonmetro communities, Hispanic-black segregation is very high (over 70) but has also declined very rapidly since 1990; indeed, Hispanic-black declines have been much more rapid than Hispanic-white segregation declines. The main point is clear: Everywhere, Hispanics are assimilating more rapidly with blacks than with whites, a trend that is likely to manifest itself in lower Hispanic-black than Hispanic-white segregation in the future. Still, over the foreseeable future, Hispanics are almost certain to remain highly segregated from other U.S. racial and ethnic groups, including both whites and blacks. Our results clearly show that Hispanics are highly segregated spatially.

Multilevel models of Hispanic segregation, 2010. Our descriptive analyses of recent trends in Hispanic segregation raises an obvious question: What accounts for the current high levels of neighborhood residential segregation between Hispanics and both whites and blacks? Here, we fit multilevel models of Hispanic segregation nationally but restrict our analyses to the 2010 period. These analyses link our place-based segregation estimates drawn from the 100 percent count of the 2010 decennial census with the 2008–2012 American Community Survey. Variable definitions and descriptive statistics, which are identified from previous studies (e.g., Iceland 2004; Lichter et al. 2010), are reported for all places and for new destinations in Table 2.

The HLM estimates in Table 3, at a minimum, reinforce the conclusions based on the bivariate segregation results previously reported. For example, they indicate that Hispanic segregation, nationally, is significantly higher in new destinations than in established gateways in the 1990s (b = 1.95) and in nonmetro places (b = 4.89). Suburban places of all sizes and places in the West had lower levels of Hispanic segregation, net of the other variables in the model. Hispanic-white segregation is also positively associated with population size (b = 0.56) and size of the black population (b = 0.13). One interpretation is that Hispanic segregation from whites is greatest in places with a large black presence, perhaps because the housing stock has been highly differentiated along racial lines over a long period of time or because large shares of African Americans exacerbate the racial "threat" associated with rapid Hispanic population growth. Yet our results also show that increases in the share of Hispanics and the foreign-born population are statistically unrelated to changes in Hispanic-white segregation, a finding that argues against the "minority group threat" hypothesis (see also Iceland and Scopilliti 2008).

Spatial patterns of Hispanic-white segregation are also associated with the economic or industrial structure of a community. As expected, Hispanic-white segregation increases with overall poverty rates but declines with growth in the housing stock (i.e., growth tends to occur at the periphery of many communities; see Lichter et al. 2010). Hispanic-white segregation is lowest in places with disproportionate shares of employment in the government sector, but Hispanic segregation does not vary by the percentage of workers employed in the service sector or in manufacturing (which comprises nondurable manufacturing, such as food processing).

Last, we included a measure of Hispanic economic well-being, which we defined as the mean household income of Hispanics. As expected, Hispanic segregation from whites is lower in places with higher Hispanic incomes (b = –0.34; Table 3). Moreover, in some additional analyses (not shown), using the same model specification but focused on Hispanic-black segregation, we discovered that Hispanic segregation from blacks increased significantly with increases in Hispanic household income (b = 0.04). Higher incomes clearly "move" some Hispanics toward whites and away from blacks on the racial continuum, which is consistent with our hypothesis of segmented assimilation. Income "whitens," seemingly separating high-income Hispanics from their less affluent counterparts and from blacks. Although such results are consistent with a spatial assimilation perspective, they also indicate a highly segmented process based on socioeconomic status.

In a final model (column 2, Table 3), we included an interaction term between Hispanic income and percent Hispanic to address the question of a "Hispanic threat." The expectation is that any threat associated with growing shares of Hispanics would dissipate with increasing Hispanic income. Indeed, the significant positive effect of percent Hispanic (b = –18) is offset by the significant and negative interaction term (b = –0.01). The results are consistent with expectations that Hispanics are seen as less threatening—and segregation is reduced—as Hispanics "whiten" with increasing household income. Additional analyses are required before making strong conclusions about the substantive significance of Hispanic income vis-à-vis other key predictors of Hispanic-white segregation.

TABLE 3
Multilevel Models of Hispanic Segregation, 2010

	White				Black			
	Model 1		Model 2		Model 3		Model 4	
	b	SE	b	SE	b	SE	b	SE
Intercept	44.54***	1.92	43.14***	1.90	113.59***	2.62	111.74***	2.60
Destination place (other place as reference)								
New destination in 1990s	1.95**	0.72	2.02**	0.71	−0.34	0.97	−0.26	0.96
New destination in 2000s	0.60	0.76	0.50	0.75	0.04	1.03	−0.08	1.03
Place type (central city as reference)								
Suburban place	−2.31***	0.62	−2.50***	0.61	−6.60***	0.84	−6.88***	0.83
Nonmetro place	4.89***	0.73	4.71***	0.72	3.54***	0.97	3.28***	0.96
Region (South as reference)								
Northeast	−1.02	0.90	−0.87	0.89	−6.86***	1.14	−6.67***	1.12
Midwest	1.35*	0.62	1.39*	0.61	4.47***	0.80	4.56***	0.79
West	−6.69***	0.62	−6.56***	0.61	7.39***	0.79	7.58***	0.78
Demographics								
Log population size	0.56***	0.13	0.67***	0.13	−3.25***	0.18	−3.12***	0.17
Percent black	0.13***	0.01	0.14***	0.01	−0.16***	0.01	−0.15***	0.01
Percent Hispanic	−0.01	0.01	0.18***	0.02	0.10***	0.01	0.36***	0.03
Percent foreign born	−0.03	0.02	−0.01	0.02	−0.33***	0.03	−0.31***	0.03
Economic structure								
Percent new housing, 1990–2000	−0.01***	0.00	−0.01***	0.00	−0.01***	0.00	−0.01***	0.00
Percent in poverty	0.12***	0.02	0.05*	0.02	0.13***	0.03	0.04	0.03
Industry								
Percent in manufacturing	−0.02	0.03	−0.01	0.03	−0.16***	0.04	−0.14***	0.04
Percent in service	0.03^	0.02	0.03^	0.02	−0.27***	0.03	−0.27***	0.03
Percent in government	−0.45***	0.04	−0.44***	0.04	−0.84***	0.05	−0.83***	0.05
Economic well-being								
Hispanic mean household income (in 1,000)	−0.04***	0.00	−0.02**	0.01	0.04***	0.01	0.07***	0.01
Percent Hispanic × Hispanic mean household income			−0.01***	0.00			−0.01***	0.00
RANDOM EFFECTS								
Unconditional variance	70.28		70.28		157.37		157.37	
Model residual variance	27.64		26.40		37.31		35.32	
% of variance explained	60.67		62.44		76.29		77.56	
Intraclass correlation coefficient (Unconditional Model)		0.42				0.45		

$^\wedge p < .10$. $^* p < .05$. $^{**} p < .01$. $^{***} p < .001$ (two-tailed test).

For comparative purposes, we also provide similar HLM estimates of Hispanic-black segregation in 2010. Unlike the Hispanic-white segregation results, Hispanic-black segregation was lowest in the largest cities ($b = -3.25$) and in places with relatively large black populations ($b = 0.16$). This provides some evidence that Hispanic populations are merging with black populations in the neighborhoods of the nation's largest cities and communities with sizable black populations and where employment is concentrated in manufacturing, services, and government. Finally, unlike the negative effects of Hispanic income on Hispanic-white segregation, Hispanic income is positively and significantly associated with Hispanic-black segregation ($b = 0.04$). Money whitens. Indeed, the lack of money tends to bring Hispanics and blacks closer but tends to set Hispanics apart from whites.

The results reported in the bottom of Table 3 indicate the value of our place-based theoretical and empirical approach. The intraclass correlation indicates that 42 percent of the overall segregation of Hispanics from whites is explained by metro/nonmetro-to-metro/nometro differences in neighborhood segregation, while 58 percent is explained by place-to-place differences within metro or non-metro areas. To be sure, places are useful analytic units, as they effectively sort Hispanics into neighborhoods with different mixes of whites and minority populations.

Hispanic segregation in new destinations

Segregation in the 1990s and 2000s. Much of the recent attention on Hispanic redistribution patterns has centered on new immigrant destinations (Crowley and Ebert 2014). In our earlier work (Lichter et al. 2010), Hispanic segregation from whites was found to be greater in new destinations than in established gateways, based on census data from 1990 and 2000. The implication is that the 1990s introduced a new form of spatial and racial segregation, an emerging macroseg-regation, as Hispanics became increasingly concentrated in specific new immigrant destinations (Lichter 2012; Tienda and Fuentes 2014). With newly released data from the 2010 decennial census, ongoing trajectories of Hispanic spatial assimilation can now be tracked by trends in segregation of new Hispanic destinations over the subsequent 10-year period, from 2000 to 2010.

The weighted segregation indices reported in Table 4 (first column) clearly indicate continuing high levels of Hispanic segregation in new destinations (as defined in the methods section)—higher than among all places (see Table 1). Yet in new destinations (defined by 1990s growth), Hispanic-white segregation is 51.7 in 2010, down significantly from 58.7 in 2000. Segregation levels in 2010 among the "newest" new destinations (49.5) are still lower. These data provide clear empirical evidence of (slowly) increasing spatial assimilation of Hispanics between 2000 and 2010 in new Hispanic destinations.

Some caveats nevertheless are warranted. As found in previous studies (Hall 2013; Lichter et al. 2010), Hispanic-white segregation remains slightly higher in the 1990s among new destinations (51.9) than for the country overall (49.3; see Table 2). Moreover, for the first wave of new destinations, Hispanic-white

TABLE 4
Average Segregation of Hispanics (H) from Non-Hispanic Whites (NHW) and Non-Hispanic Blacks (NHB) in New Destinations, 1990–2010

| | United States | | Metro Places | | | | | | Nonmetro Places | |
| | | | Metro Central Cities | | Metro Suburban Places | | Total Metro Places | | | |
	Unweighted	Weighted*	Unweighted	Weighted	Unweighted	Weighted	Unweighted	Weighted	Unweighted	Weighted
H-NHW Segregation										
1990–2000 New Destinations										
2000	59.4	58.7	65.0	66.2	57.9	53.4	58.5	58.1	62.1	62.1
2010	51.8	51.7	58.9	60.6	49.6	45.9	50.4	51.1	55.9	56.0
2000–2010 New Destinations										
2010	50.4	49.5	54.8	53.3	47.1	44.4	48.7	49.1	58.4	59.9
H-NHB Segregation										
1990–2000 New Destinations										
2000	64.8	63.6	61.2	64.3	61.4	60.1	61.4	61.7	75.2	75.8
2010	58.8	52.8	53.8	50.6	55.0	50.6	54.9	50.6	71.2	68.8
2000–2010 New Destinations										
2010	56.5	54.0	58.9	58.7	54.3	47.7	55.2	53.5	63.1	67.2

*Weighted by number of Hispanics.

declines in segregation paled in comparison to declines in Hispanic-black segregation, which dropped from 63.6 to 52.8 during the 2000s (see Table 4, panel B). Spatial integration of Hispanics is occurring more rapidly with blacks than with whites in new destinations. Hispanic segregation from blacks, however, was very similar nationally (55.0) to segregation patterns in new destinations that emerged both in the 1990s (D = 53.8) and 2000s (D = 54.0). But the evidence nevertheless supports a singular conclusion: Hispanics in new destinations are assimilating more rapidly with blacks than with whites.

Recent declines in Hispanic segregation have proceeded more rapidly in new destinations compared to all places, whether in central cities, suburbs, or non-metro places. For example, Hispanic-white segregation declined only slightly overall in metro areas during the 2000s (from 52.7.0 to 49.3) but dropped more rapidly in the 1990s' new destinations (from 58.1 to 51.1). This pattern of differential declines in Hispanic-white segregation was also apparent in suburban and nonmetro places. Still, in each instance, Hispanic-white segregation in 2010 was higher in new destinations than in all places.

Unlike the case with whites, Hispanic segregation from blacks in new destinations was slightly lower (rather than higher) than the estimates nationally (Table 1). For example, Hispanic-black segregation was 52.8 and 54.0 in the 1990s' and 2000s' new destinations, respectively, compared with 54.9 nationally. As is the case nationally, Hispanic-black segregation in nonmetro places is exceptionally high in new destinations (68.8 and 67.2; see Table 4). The current widespread redistribution of America's Hispanic population arguably is associated with greater spatial integration with America's black population, even as segregation from whites has declined.

Multilevel models of Hispanic segregation in new destinations. An important general lesson from the multilevel models (not shown) is that conventional predictors of segregation are much less useful for understanding segregation in fast-growing Hispanic communities. Compared with the results for all places (Table 3), far fewer independent variables are statistically significant. We therefore limit our discussion to some empirical commonalities that new destinations share with all places (reported earlier in Table 3). The full results are available in our online supplemental tables.[3]

For example, Hispanic-white segregation in new destinations is lowest in the West (b = –4.94) but also rises as the black percentage increases (b = 0.20). These associations in new destinations are generally similar in direction and significance to those estimated for all places (see Table 3). One clear exception is population size. In the national sample, Hispanic-white segregation increases with population size (model 1, Table 3), but in new destinations segregation is statistically unrelated with population size. Measured by the relative sizes of the income coefficient, the association with Hispanic income and Hispanic-white segregation is stronger in new destinations than it is nationally (b = –0.15 vs. –0.04). Upward income mobility seems to be a clear route to greater spatial integration with whites in new destinations. Economic incorporation is tied to spatial assimilation among Hispanics. This finding arguably provides some support for the spatial assimilation model, but it might also indicate that some low-income Hispanic communities are "left behind." The experience of incorporation—spatially or otherwise—is highly segmented in new destinations.

The results indicate that Hispanics are least segregated from blacks in the suburbs, which suggests that the new destinations in the suburbs are disproportionately black in racial composition. Hispanic-black segregation also declines with increasing population size ($b = -3.19$) but increases with larger shares of Hispanics in the community ($b = 0.21$). This positive effect, however, is offset by growing shares of the foreign-born population ($b = -0.30$), a result that suggests that foreign-born Hispanics rather than native-born Hispanics are concentrating more rapidly in black neighborhoods in these new destinations. Finally, more Hispanic income in new destinations is associated with less segregation from blacks ($b = -0.10$), as it is with whites ($b = -0.15$). This pattern is different from the national results reported in Table 3. It seems that more Hispanic income contributes to greater spatial assimilation with the local population—whether it is white or black.

Finally, the interclass correlations from these supplementary analyses of new destinations indicate that 41 percent of Hispanic-white segregation is explained by metro/nonmetro-to-metro/nometro differences in within-place neighborhood segregation. The remaining 59 percent of Hispanic-white segregation is explained by place-to-place differences within metro or nonmetro areas. Clearly, even in new destinations, segregation patterns among Hispanics arguably must be contextualized by demographic, economic, and political processes occurring at the community or place level (Waters and Jiménez 2005).

Discussion and Conclusion

The new growth of Hispanics in smaller metro cities, in rapidly diversifying suburban communities, and in new rural destinations has upended conventional interpretations of spatial assimilation and intergroup relations among America's minority and immigrant populations (Parisi, Lichter, and Taquino 2015; Tienda and Fuentes 2014; Waters, Kasinitz, and Asad 2014). In this article, we argue for a new scholarly commitment to spatial inclusion in the study of racial and ethnic segregation, one that acknowledges the growing significance and impact of a highly diverse and spatially dispersing Hispanic population. The singular focus on neighborhood segregation in the largest metro areas is clearly important, but it arguably misses part of an evolving story of racial change and differentiation in the U.S. settlement system (Lee, Iceland, and Farrell 2014; Lichter, Parisi, and Taquino 2015).

Whether Hispanics are joining the majority mainstream depends, at least in part, on whether they are able to translate upward mobility into residence patterns that mirror those of the rest of the nation. Based on our results, we view spatial assimilation theory and the place stratification model as complementary rather than strictly competitive perspectives. Indeed, our empirical results provide evidence of spatial assimilation among Hispanics, both nationally and in new immigrant destinations. From a place rather than big-city or metro-level perspective, Hispanic segregation from whites has declined significantly over the past 20 years. Hispanic income is also associated with less segregation from whites. At the same time, our results also raise the prospect of growing Hispanic spatial assimilation with blacks; declines in Hispanic-black segregation have proceeded more rapidly than declines in Hispanic-white segregation. The segmented assimilation of Hispanics therefore seems to accommodate spatial assimilation theory for some groups (e.g., highly educated Hispanics) but a place stratification perspective for the most disadvantaged Hispanic populations, who are integrating most rapidly with blacks. Our results raise the question of spatial assimilation "with whom."

Our analyses provide an empirical baseline for additional research, especially on the apparently divergent residential trajectories of different segments of the Hispanic population. Our results showing average declines in Hispanic segregation from both whites and blacks over the past 20 years suggest the need to consider multigroup segregation or even the new blending of residence patterns among different racial and ethnic groups in an increasingly diverse society (Logan and Zhang 2010). The presence of new Hispanic populations is sometimes viewed as serving as a spatial buffer between blacks and whites (Iceland 2004; Parisi, Lichter, and Taquino 2015), perhaps even contributing to declines in black-white segregation. Our results give little indication that the black population plays a similar role between Hispanics and whites. Quite the opposite is true: Hispanic segregation from whites increased (rather than decreased) with the growth in the relative size of the black population. One obvious interpretation is that Hispanics—or some segment of the Hispanic population—are now joining blacks in highly segregated minority and immigrant neighborhoods. In fact, this finding may account for the rapid declines in Hispanic-black segregation over the study period, even as Hispanic-white segregation declined more slowly.

Our study is not without some limitations. We have neither considered racial differences in Hispanic segregation, nor differences among Hispanic national origin groups. To be sure, Mexicans, Puerto Ricans, and Cubans are highly differentiated spatially, in part because of cultural differences and economic disparities. Iceland and colleagues' (2014) recent analyses of 1980–2010 data showed widely uneven patterns of segregation among Hispanic national origin groups. Mexicans were less segregated ($D = 50.3$ in 2010) than other Hispanic groups (e.g., El Salvadorans) from whites and blacks. Moreover, a recent metro-based study by Kim and White (2010) reported very high rates of segregation among national origin groups, even among those that shared the same pan-ethnic identification (e.g., Asians). Treating Hispanics as a monolithic may misrepresent the experiences of different Hispanic national origin groups.

Our study has also not identified the specific demographic mechanisms that have produced declining Hispanic segregation from whites and blacks. For example, a growing Hispanic population can either replace or displace white or black populations in local communities, or, stated differently, declining Hispanic segregation can reflect the movement of whites to better neighborhoods (leaving behind less expensive housing for other minorities to fill). Or, instead, whites can leave predominately white neighborhoods ("white flight") that are undergoing a new minority population infusion. Neither scenario suggests that declining segregation means that Hispanics are increasingly residing in "good" communities or better neighborhoods than in the past.

In conclusion, our study has placed the spotlight squarely on recent trends in Hispanic segregation—both from whites and blacks—across many different levels of geography. The results clearly provide an empirical basis for both optimism and pessimism. The hope of improving race relations implied by declining Hispanic segregation and greater ethnoracial inclusion is counterbalanced by persistently high levels of residential segregation among America's disadvantaged Hispanic and other minority populations. Ongoing trends in Hispanic population redistribution and residential patterns are perhaps the linchpin to shifting patterns of racial residential segregation in an increasingly diverse and multicultural American society.

Notes

1. This means that 50 percent of Hispanics would need to move to white neighborhoods before the distribution of Hispanics and whites would become equalized across all metropolitan neighborhoods.

2. It is not especially surprising that Hispanics—even today—are more segregated from blacks than from whites. After all, according to the 2010 decennial census, more than one half (53 percent) of Hispanics self-identified as white, and only a very small percentage (2.5 percent) identified as black (Ennis, Ríos-Vargas, and Albert 2011). Some additional analyses (not shown) revealed that segregation of white Hispanics from black Hispanics was 66.1 in 2010, down from 73.6 in 1990. Still, white Hispanics are much more likely to be segregated from black Hispanics than are all Hispanics segregated generally from blacks (66.1 vs. 55.0). White-Hispanic/white segregation in 2010 was 48.5, which is lower still. Race clearly trumps ethnicity in the residential sorting process.

3. See http://ann.sagepub.com/supplemental.

References

Alba, Richard D., Glenn Deane, Nancy Denton, Ilir Disha, Brian McKenzie, and Jeffrey Napierala. 2014. The role of immigrant enclaves for Latino residential inequalities. *Journal of Ethnic and Migration Studies* 40:1–20.

Crowder, Kyle, Matthew Hall, and Stewart E. Tolnay. 2011. Neighborhood immigration and native out-migration. *American Sociological Review* 76:25–47.

Crowley, Martha, and Kim Ebert. 2014. New rural immigrant destinations: Research for the 2010s. In *Rural America in a globalizing world: Problems and prospects for the 2010s*, eds. Conner Bailey, Leif Jensen, and Elizabeth Ransom, 401–8. Morgantown, WV: West Virginia University Press.

Ennis, Sharon R., Merarys Ríos-Vargas, and Nora G. Albert. 2011. *The Hispanic population: 2010*. Washington, DC: U.S. Department of Commerce, Economics and Statistics Administration, U.S. Census Bureau.

Farrell, Chad R. 2008. Bifurcation, fragmentation, or integration? The racial and geographic structure of U.S. metropolitan segregation, 1990–2000. *Urban Studies* 45:467–99.

Flippen, Chenoa, and Eunbi Kim. 2015. Immigrant context and opportunity: New destinations and socio-economic attainment among Asians in the United States. *The ANNALS of the American Academy of Political and Social Science* (this volume).

Fossett, Mark, Amber R. Fox, Rogelio Saenz, and Wenquan Zhang. 2014. White-Latino residential segregation in new destinations: Trends for metropolitan, micropolitan, and non-micropolitan areas, 1990–2010. Paper presented at the annual meeting of the Population Association of America, Boston, MA.

Hall, Matthew. 2013. Residential integration on the new frontier: Immigrant segregation in established and new destinations. *Demography* 50:1873–96.

Hall, Matthew, and Jonathan Stringfield. 2014. Undocumented migration and the residential segregation of Mexicans in new destinations. *Social Science Research* 47:61–78.

Iceland, John. 2004. Beyond black and white: Metropolitan residential segregation in multi-ethnic America. *Social Science Research* 33:248–71.

Iceland, John, and Kyle Anne Nelson. 2008. Hispanic segregation in metropolitan America: Exploring the multiple forms of spatial assimilation. *American Sociological Review* 73:741–65.

Iceland, John, and Melissa Scopilliti. 2008. Immigrant residential segregation in U.S. metropolitan areas, 1990–2000. *Demography* 45:79–94.

Iceland, John, Daniel Weinberg, and Lauren Hughes. 2014. The residential segregation of detailed Hispanic and Asian groups in the United States: 1980–2010. *Demographic Research* 31:593–624.

Johnson, Kenneth M., and Daniel T. Lichter. 2008. Natural increase: A new source of population growth in emerging Hispanic destinations in the United States. *Population and Development Review* 34:327–46.

Johnson, Kenneth M., and Daniel T. Lichter. 2010. Growing diversity among America's children and youth: Spatial and temporal dimensions. *Population and Development Review* 36:151–76.

Kim, Ann H., and Michael J. White. 2010. Panethnicity, ethnic diversity, and residential segregation. *American Journal of Sociology* 115:1558–96.

Krysan, Maria, and Michael D. M. Bader. 2009. Racial blind spots: Black-white-Latino differences in community knowledge. *Social Problems* 56:677–701.

Lee, Barrett A., John Iceland, and Chad R. Farrell. 2014. Is ethnoracial residential integration on the rise? Evidence from metropolitan and micropolitan America since 1980. In *Diversity and disparities: America enters a new century*, ed. John Logan, 415–56. New York, NY: Russell Sage Foundation.

Lee, Barrett A., Sean F. Reardon, Glenn Firebaugh, Chad R. Farrell, Stephen A Matthews, and David O'Sullivan. 2008. Beyond the census tract: Patterns and determinants of racial segregation at multiple geographic scales. *American Sociological Review* 73:766–91.

Lichter, Daniel T. 2012. Immigration and the new racial diversity in rural America. *Rural Sociology* 77:3–35.

Lichter, Daniel T. 2013. Integration or fragmentation? Racial diversity and the American future. *Demography* 50:359–91.

Lichter, Daniel T., and Kenneth M. Johnson. 2009. Immigrant gateways and Hispanic migration to new destinations. *International Migration Review* 43:496–518.

Lichter, Daniel T., Domenico Parisi, and Michael C. Taquino. 2015. Toward a new macro-segregation? Decomposing segregation within and between metropolitan cities and suburbs. *American Sociological Review*, forthcoming.

Lichter, Daniel T., Domenico Parisi, and Michael C. Taquino. 2012. The geography of exclusion: Race, segregation, and concentrated poverty. *Social Problems* 59:364–88.

Lichter, Daniel T., Domenico Parisi, Michael C. Taquino, and Steven Michael Grice. 2010. Residential segregation in new Hispanic destinations: Cities, suburbs, and rural communities compared. *Social Science Research* 39:215–30.

Logan, John R., and Brian J. Stults. 2011. *The persistence of segregation in the metropolis: New findings from the 2010 census*. New York, NY: Russell Sage Foundation and Brown University.

Logan, John R., and Charles Zhang. 2010. Global neighborhoods: New pathways to diversity and separation. *American Journal of Sociology* 115:1069–1109.

Massey, Douglas S., ed. 2008. *New faces in new places: The changing geography of American immigration*. New York, NY: Russell Sage Foundation.

Massey, Douglas S., Jonathan Rothwell, and Thurston Domina. 2009. The changing bases of segregation in the United States. *The ANNALS of the American Academy of Political and Social Science* 626:74–90.

Parisi, Domenico, Daniel T. Lichter, and Michael C. Taquino. 2011. Multi-scale residential segregation: Black exceptionalism and America's changing color line. *Social Forces* 89:829–52.

Parisi, Domenico, Daniel T. Lichter, and Michael C. Taquino. 2015. The buffering hypothesis: Growing diversity and declining black-white segregation in America's cities, suburbs, and small towns? *Sociological Science*, forthcoming.

Park, Julie, and John Iceland. 2011. Residential segregation in metropolitan established immigrant gateways and new destinations, 1990–2000. *Social Science Research* 40:811–21.

Reardon, Sean F., Stephen A. Matthews, David O'Sullivan, Barrett A. Lee, Glenn Firebaugh, Chad R. Farrell, and Kendra Bischoff. 2008. The geographic scale of metropolitan segregation. *Demography* 45:489–514.

Rothwell, Jonathan, and Douglas S. Massey. 2009. The effect of density zoning on racial segregation in U.S. urban areas. *Urban Affairs Review* 44:779–806.

Rugh, Jacob S., and Douglas S. Massey. 2010. Racial segregation and the American foreclosure crisis. *American Sociological Review* 75:629–51.

Rugh, Jacob S., and Douglas S. Massey. 2014. Segregation in post-civil rights America: Stalled integration or end of the segregated century? *Du Bois Review* 11:205–32.

Rumbaut, Rubén G. 2009. Pigments of our imagination: On the racialization and racial identities of "Hispanics" and "Latinos." In *How the U.S. racializes Latinos: White hegemony and its consequences*, eds. José A. Cobas, Jorge Duany, and Joe R. Feagin, 15–36. St. Paul, MN: Paradigm Publishers.

Sharkey, Patrick. 2012. Residential mobility and the reproduction of unequal neighborhoods. *Cityscape: A Journal of Policy Development and Research* 14:9–31.

Tienda, Marta, and Norma Fuentes. 2014. Hispanics in metropolitan America: New realities and old debates. *Annual Review of Sociology* 40:499–520.

Waters, Mary C., and Tomás R. Jiménez. 2005. Assessing immigrant assimilation: New empirical and theoretical challenges. *Annual Review of Sociology* 31:105–25.

Waters, Mary C., Philip Kasinitz, and Asad L. Asad. 2014. Immigrants and African Americans. *Annual Review of Sociology* 40:369–90.

Desvinculado y Desigual: Is Segregation Harmful to Latinos?

By
JUSTIN STEIL,
JORGE DE LA ROCA,
and
INGRID GOULD ELLEN

Despite the high levels of metropolitan-area segregation that Latinos experience, there is a lack of research examining the effects of segregation on Latino socioeconomic outcomes and whether those effects differ from the negative effects documented for African Americans. We find that segregation is consistently associated with lower levels of educational attainment and labor market success for both African American and Latino young adults compared with whites, with associations of similar magnitudes for both groups. One mechanism through which segregation may influence outcomes is the difference in the levels of neighborhood human capital to which whites, Latinos, and African Americans are exposed. We find that higher levels of segregation are associated with lower black and Latino neighborhood exposure to residents with college degrees, relative to whites. We also find support for other commonly discussed mechanisms, such as exposure to neighborhood violent crime and the relative proficiency of the closest public school.

Keywords: segregation; inequality; stratification; racial disparities; ethnic disparities

In the twentieth century, the Great Migration transformed the United States as millions of African Americans left the rural South and remade the nation's cities. The growth of the U.S. Latino population is provoking a similar transformation in the twenty-first century. Between 1970 and 2010, the Latino population

Justin Steil is a legal research fellow at NYU's Furman Center and a doctoral candidate in Urban Planning at Columbia University. His recent research investigates diverging municipal responses to immigration. He is a coeditor of Searching for the Just City: Debates in Urban Theory and Practice *(Routledge 2009).*

Jorge De la Roca is an economics research fellow at the Furman Center. His fields of interest include urban economics, economic geography, and labor economics, with a particular focus on the study of urban migration and workers' sorting and learning across cities of different sizes.

DOI: 10.1177/0002716215576092

grew from 8 million to more than 45 million, and most of this population lives in the nation's largest metropolitan areas.

During the Great Migration, both de jure and de facto segregation policies severely constrained the options of African American migrants, steering them into segregated housing and labor markets and contributing to what the Kerner Commission in 1968 described as a nation "moving toward two societies, one black, one white—separate and unequal" (National Advisory Commission on Civil Disorders 1968, 1). Although levels of black-white residential segregation have decreased from their 1968 levels, they remain high, and evidence suggests that segregation continues to produce separate and unequal access to resources, such as schools or jobs, and exposure to hazards, such as violence or environmental risks. As the Latino population continues to grow, Latinos seem to be inheriting the segregated urban structures experienced by African Americans. As metropolitan-area levels of segregation for Latinos increase toward the levels observed for African Americans, to what extent are the effects of segregation similar or different for the two groups?

Previous studies of black-white segregation have found that higher levels of segregation lead to worse health, educational, and socioeconomic outcomes for African Americans (Cutler and Glaeser 1997; Ellen 2000; Card and Rothstein 2007), but we know much less about the effects that residential segregation has on socioeconomic outcomes for Latinos. On one hand, there are some reasons to think that segregation could have comparable or even worse effects on Latinos. For instance, if we assume that segregation reduces neighbors' human capital, then the low levels of mean educational attainment and occupational status in predominantly Latino neighborhoods could significantly undermine the life chances of Latinos. On the other hand, largely Latino neighborhoods boast higher levels of employment than largely black neighborhoods and may offer enclave economies that help coethnic workers to find jobs and build skills and experience (Portes and Shafer 2007).[1] Consistent with higher employment rates and potential enclave effects, Latino neighborhoods may not have suffered the same level of disinvestment as have largely black neighborhoods (Small and McDermott 2006; W. J. Wilson 1991). Finally, largely Latino neighborhoods tend to have lower levels of violent crime than largely black neighborhoods (De la Roca, Ellen, and O'Regan 2014).

Compared with the extensive literature on changing patterns (Park and Iceland 2011; Logan and Turner 2013; Tienda and Fuentes 2014) and causes of Latino segregation (Bayer, McMillan, and Rueben 2004; Iceland and Nelson

Ingrid Gould Ellen is the Paulette Goddard Professor of Urban Policy and Planning at NYU's Robert F. Wagner Graduate School of Public Service and faculty director of the Furman Center. She is the author of Sharing America's Neighborhoods: The Prospects for Stable Racial Integration *(Harvard University Press 2000) and many articles on segregation.*

NOTE: "Desvinculado y Desigual" can be translated as "Separate and Unequal." We thank Gerard Torrats and Justin Tyndall for their exceptional research assistance as well as the anonymous reviewers for their very helpful comments. We also thank the organizers of and participants in the 2014 Penn State Stratification Conference on Residential Inequality in American Neighborhoods.

2008; Rugh and Massey 2014), there is far less research on how metropolitan-area segregation levels affect Latinos, and how those effects differ from those for African Americans. We address this gap by exploring how levels of metropolitan-area segregation relate to the socioeconomic outcomes of young, native-born Latinos and how those associations differ from those for African Americans. We also examine the relevance of several key mechanisms that may drive the relationship between residential segregation and individual outcomes.

Changing Dynamics of Segregation

Although African Americans have historically been far more segregated than other minority groups, Latino-white and black-white segregation levels began to converge between 1980 and 2010. Black-white dissimilarity declined consistently between 1980 and 2010, while Latino-white dissimilarity remained relatively steady.[2] By 2010, black-white segregation still surpassed Latino-white segregation, but the difference was far smaller than it had been three decades earlier (De la Roca, Ellen, and O'Regan 2014). Although Latino isolation (that is, the share of Latino residents in a neighborhood where the average Latino lives) has risen less rapidly than Latinos' quickly rising share of the population, average levels of Latino isolation have still risen substantially and matched average levels of African American isolation in 2010.[3] Over this time period, African American isolation declined from 0.61 to 0.46 while Latino isolation rose from 0.38 to 0.46 (De la Roca, Ellen, and O'Regan 2014).

How Does Segregation Matter?

Segregation can affect the outcomes of individuals by constraining residential options and shaping the characteristics of the population and the quality of the resources and services available in the neighborhoods in which individuals live. To be clear, ethnic concentration is not inherently harmful; the effects of segregation may vary significantly by the political and socioeconomic context of concentration. Below we explore several potential avenues through which segregation may affect individual outcomes.

Human capital

Residential segregation can lead to large disparities in levels of neighborhood human capital, which may be critical to youth outcomes. William Julius Wilson (1997) argued that the educational and lifetime experiences of adult residents in a neighborhood can powerfully affect the outcomes of youth by shaping their access to conventional role models and the mainstream social networks that facilitate social and economic advancement. Recent research on black-white segregation supports Wilson's theory about neighborhood human capital and social isolation, finding that increases in the proportion of college-educated African American adults in a metropolitan area significantly reduce the negative effects of segregation on black youths' educational attainment (Bayer, Fang, and McMillan 2014).

Similar dynamics have been identified in the experiences of immigrants and their children. The educational attainment of the children of immigrants has been found to depend not only on their parents' educational attainment but also on the mean attainment of coethnics in the parents' generation, and particularly the mean educational attainment of coethnics residing in the same neighborhood (Borjas 1995). Recent research on the labor market experiences of the foreign born comes to similar conclusions, finding that segregation among a highly educated ethnic group is economically beneficial, while segregation among an ethnic group with below-average educational attainment leads to lower employment rates and earnings (Cutler, Glaeser, and Vigdor 2008).

Ethnic enclaves

Some scholars have emphasized that spatial clustering among coethnics can create supportive enclaves, especially in largely immigrant neighborhoods where the shared experience of immigrant origins reinforces social ties (K. Wilson and Portes 1980; Light and Bonacich 1988). For entrepreneurs, ethnic enclaves can simultaneously create a market for ethnic goods, a readily available pool of committed labor, and access to coethnic sources of capital (Portes and Sensenbrenner 1993). Group members with low levels of human or financial capital can in theory find employment more easily in enclaves despite low levels of English proficiency or formal education, and those with higher levels of education may be able to find jobs that are more commensurate with their skills or more easily access capital to start their own business than in mainstream labor and capital markets.

Scholars have critiqued the enclave thesis on various grounds, including pointing out that while ethnic enclaves may benefit entrepreneurs and business owners who can access capital and coethnic labor, enclaves may not benefit workers who end up exploited by coethnic employers and trapped in low-paying jobs (Sanders and Nee 1992; see also W. J. Wilson 1997; Logan, Alba, and Stults 2003). Still, existing research provides some support for the notion that residence in an enclave can improve labor market outcomes, even for unskilled immigrants (Edin, Fredriksson, and Åslund 2003; Portes and Shafer 2007).

Public services

The quality of life in a neighborhood is profoundly shaped by the availability and the quality of public services that are delivered, from schooling to public safety to sanitation. Residential segregation may affect individual outcomes by contributing to unequal access to crucial municipal services. In the education context, poorly performing local schools reduce the likelihood that local children will have the foundation necessary to maximize their educational attainment or labor market potential.

Institutional density

Segregation may also shape outcomes through effects on institutional density. Recent scholarship on urban inequality has highlighted the importance of local institutions and the way that they structure life in disadvantaged neighborhoods, potentially fostering access to local and extra-local resources and creating neighborhood-level collective efficacy (Marwell 2007; Small 2009; Sampson 2012). Neighborhood institutions, such as community organizations or childcare centers, serve as resource brokers, with networks to private employers and public agencies that enable these institutions to connect residents to schools and to jobs and provide resources that can support educational attainment, employment, and parenting (Small 2009). The neighborhood benefits of institutions can extend to private businesses as well, through their contribution to neighborhood vitality and informal social control (Small and McDermott 2006).

Violence

Another way in which segregation may have an effect on individuals is through shaping exposure to neighborhood violence. Sharkey et al. (2014) found that exposure to neighborhood violence affects children's academic performance. Indeed, Harding (2009) estimated that neighborhood violent crime rates account for half of the association between neighborhood disadvantage and high school graduation.

Weighing the mechanisms together

These mechanisms may not disadvantage Latinos and African Americans equally. For example, if segregation affects the socioeconomic outcomes of youth through shaping exposure to neighborhood levels of human capital, then given the lower levels of human capital in predominantly Latino neighborhoods we would expect segregation to have a larger negative effect on Latino socioeconomic outcomes. Although there is significant variation among Latinos in mean human capital levels by ancestry group, recent high levels of immigration by those with comparatively low educational levels mean that predominantly Latino neighborhoods on average have lower levels of educational attainment than largely black neighborhoods.[4]

Other mechanisms, however, suggest greater disadvantage for predominantly black neighborhoods. To the extent that enclave effects may reduce the harms of segregation or even create benefits for individual outcomes, they may then be more likely to benefit Latinos who live among a higher proportion of foreign-born residents and in neighborhoods with higher employment rates.[5]

To the extent that segregation matters through its relationship with the quality of local public services, the long history of unequal public investment in predominantly black neighborhoods suggests that segregation may thus have a larger effect on African Americans than on Latinos. Further, W. J. Wilson (1991, 654–55) has suggested that disinvestment from predominantly African American neighborhoods has "sapped the vitality of local businesses and other institutions" and thus isolated neighborhood residents further. Although there is little research on institutional density in Latino neighborhoods, if we assume that predominantly Latino neighborhoods have not experienced the same disinvestment and isolation as predominantly black neighborhoods, then any effect that segregation has on individual outcomes through exposure to local institutional density would be greater for African Americans than Latinos. Finally, to the extent that segregation affects individual outcomes through its relationship with exposure to violence, we would expect segregation to have a larger effect on African Americans than on Latinos because of the greater levels of violence to which black residents are exposed in segregated metropolitan areas.

Data and Empirical Approach

To examine individual socioeconomic outcomes and relate them to levels of segregation in a metropolitan area, we utilize public-use microdata gathered by the U.S. Census Bureau and provided by Integrated Public Use Microdata Series (IPUMS-USA) of the University of Minnesota Population Center. We focus our analysis on data from the American Community Survey five-year estimate (2007–2011) to study the relationship between residential segregation and socioeconomic outcomes of native-born individuals between the ages of 20 and 30. We consider an array of educational (probability of high school and college graduation), labor market (earnings and likelihood of being idle), and social (likelihood of being a single mother) outcomes. We exclude the foreign born because the data do not provide precise information on their year of arrival and, hence, we cannot tell how long they have experienced segregation. We base our empirical analysis on data for individuals from 199 Core Based Statistical Areas (CBSA) across the United States with a total population greater than 100,000 residents. In addition, because small black or Latino populations can lead to misleading segregation scores, a metropolitan area must have at least 5,000 Latino residents to be included in the white-Latino models and at least 5,000 African American residents to be included in the white-black models. Throughout the study, we use both the dissimilarity and isolation indices to measure segregation.[6] Across the sample, the mean Latino-white dissimilarity index score is 0.468, with a standard deviation of 0.108, and the mean black-white dissimilarity index score is 0.579, with a standard deviation of 0.122.[7]

To describe the characteristics of neighborhood residents, we rely on the Neighborhood Change Database developed by GeoLytics and the Urban Institute, which provides rich data at the census-tract level. We complement this dataset with data from the U.S. Department of Education to describe the test scores of the local elementary schools nearest to each census tract for the 2008–2009 school year, relative to other schools in the metropolitan area.[8] For data on crime, we rely on the National Neighborhood Crime Study, a nationally representative sample of crime data for ninety-one U.S. cities, collected by Peterson and Krivo (2000) between 1999 and 2001. To capture the neighborhood density of business establishments and nonprofit organizations, we employ data from Esri's Business Analyst, which relies on information for more than 12 million licensed businesses assembled by Dun and Bradstreet in 2010. Establishment-level data are geocoded, which allows us to calculate counts of businesses by census tract.

To investigate how individual outcomes relate to metropolitan-area levels of segregation, we estimate regressions to test if any relationship between segregation and the outcome of interest persists after controlling for a wide array of metropolitan-area variables that could partially account for such a relationship. Specifically, in Table 2, we regress an individual outcome, such as the probability of college graduation, on measures of black-white and Latino-white segregation in 2010. We estimate separate models for blacks and Latinos. Following Cutler and Glaeser (1997), we let the coefficient on segregation differ for whites and for blacks or Latinos. Therefore, we test whether segregation has a differential effect on blacks and Latinos relative to its effect on whites.

We include several individual variables as controls, including age, gender, a black indicator variable in specifications for blacks, and a set of indicator variables for Latino groups of different origin (Mexicans, Puerto Ricans, Dominicans, Cubans, Central Americans, South Americans, and other Latinos) in specifications for Latinos. These census-defined "Hispanic origin" groups exhibit substantial differences in levels of educational attainment, earnings, and potentially unobserved traits that could explain differences in outcomes within the Latino population. By including these ancestry-group indicator variables, we can capture a share of the variance in outcomes that can be attributed to the fact that Latinos of specific subgroups, who may be concentrated in different metropolitan areas, bring different backgrounds. We also control for a large set of metropolitan-area variables and analyze them with a black or Latino indicator variable to let the effects of city characteristics differ for blacks and Latinos as compared with whites. We calculate robust standard errors clustered at the metropolitan-area level to account for the fact that all individuals in a given metropolitan area share the same values for metropolitan-area controls.

It is worth emphasizing that we measure segregation at the level of the metropolitan area rather than at the level of the neighborhood. We do so in part for theoretical reasons, as we believe that metropolitan area–level segregation may restrict choices and opportunities even for minorities who live in integrated neighborhoods. But studying segregation at the level of the metropolitan area also offers empirical advantages, as individuals are less likely to select into a particular metropolitan area than into a neighborhood based on their tastes, preferences, and unobserved resources. Although the research design focusing on metropolitan areas eliminates error that could be introduced by more highly educated or economically successful individuals choosing to reside in certain neighborhoods within a metropolitan area, selective migration among metropolitan areas may still occur, with less successful minority adults sorting into more segregated cities. We have limited ability to control for sorting among metropolitan areas, and we do not know the exact length of time an individual in the sample has been exposed to a particular level of segregation. We try to address these concerns in several ways. First, we restrict the analysis to native-born individuals between 20 and 30 years old, given that their metropolitan area of residence is more likely to be predetermined by parental location choices. Second, we exclude those individuals who moved across state lines in the previous year to eliminate individuals who are likely to have moved from another CBSA and, hence, were not exposed to the assigned level of segregation.[9] Still, we do not interpret our results as indicating causal relationships.

After examining outcomes, we explore potential mechanisms by analyzing differences in the typical neighborhoods lived in by blacks and Latinos in more and less segregated metropolitan areas. Specifically, we first stratify metropolitan areas into segregation quartiles according to their dissimilarity index: "very low," "low," "high," and "very high" (the group of metropolitan areas in each quartile of levels of segregation between blacks and whites is different from the group in each quartile of levels of segregation between Latinos and whites). We then calculate weighted averages of neighborhood attributes, such as the share of college-educated residents or the test scores of the local school, to characterize the average neighborhoods lived in by African Americans, Latinos, and whites in a given set of metropolitan areas. The weights in each of these calculations are the number of residents of a given race or ethnicity in a neighborhood divided by the total number of people of that race or ethnicity in the metropolitan areas in that segregation quartile. These exposure rates indicate the degree to which the average person of a particular group is exposed to a given neighborhood characteristic. Through this exercise based on raw associations, we assess the extent to which differences in exposure to neighborhood conditions between whites and blacks or whites and Latinos widen in more segregated metropolitan areas.

Results

Ordinary least squares results on the relation between segregation and individual outcomes

Table 1 presents raw correlations between metropolitan-area segregation and each of the individual outcomes for young adults. The upper panel shows segregation quartiles based on the 2010 Latino-white dissimilarity index and the lower panel shows quartiles constructed using the 2010 black-white dissimilarity index. Whites exhibit better outcomes than blacks and Latinos across the board: they are more likely to graduate from high school and college, to be employed, and to have higher earnings. Further, higher levels of segregation are consistently associated with larger gaps in outcomes between whites and blacks and between whites and Latinos. For African Americans, differences in outcomes relative to whites are consistently larger in "very high" segregation metropolitan areas than in than in "very low" segregation metropolitan areas, but the pattern is weaker for the middle quartiles. Differences between white and Latino outcomes are monotonic, systematically increasing with the level of segregation for every outcome except single motherhood.

TABLE 1
Raw Correlations between Metropolitan Segregation and Outcomes

	High school graduation	College graduation	Idleness	Log earnings	Single motherhood
	(1)	(2)	(3)	(4)	(5)
A. Whites					
Very low segregation	94.0%	34.6%	11.1%	10.00	13.4%
Low segregation	93.9%	39.9%	9.8%	10.07	12.1%
High segregation	94.4%	40.2%	9.2%	10.10	11.6%
Very high segregation	96.1%	50.9%	7.9%	10.26	8.8%
B. Latinos					
Very low segregation	88.5%	18.0%	12.6%	9.83	21.2%
Low segregation	86.1%	20.5%	13.2%	9.85	24.4%
High segregation	85.9%	18.2%	13.1%	9.85	23.3%
Very high segregation	85.7%	21.8%	12.2%	9.95	23.4%
White-Latino gap (A-B)					
Very low segregation	5.5%	16.7%	−1.5%	0.17	−7.8%
Low segregation	7.9%	19.4%	−3.5%	0.23	−12.3%
High segregation	8.5%	21.9%	−3.9%	0.24	−11.7%
Very high segregation	10.4%	29.1%	−4.4%	0.30	−14.6%
C. Whites					
Very low segregation	94.7%	36.5%	10.5%	10.07	11.8%
Low segregation	94.3%	39.2%	9.5%	10.11	11.4%
High segregation	94.0%	40.4%	9.5%	10.09	12.3%
Very high segregation	95.8%	50.4%	7.9%	10.23	9.5%
D. Blacks					
Very low segregation	88.4%	19.3%	12.6%	9.76	39.8%
Low segregation	88.1%	18.9%	13.0%	9.71	40.5%
High segregation	87.4%	21.8%	12.5%	9.71	41.7%
Very high segregation	88.3%	21.5%	13.4%	9.74	40.7%
White-black gap (C-D)					
Very low segregation	6.3%	17.2%	−2.1%	0.31	−28.0%
Low segregation	6.2%	20.3%	−3.5%	0.40	−29.1%
High segregation	6.6%	18.5%	−3.0%	0.38	−29.4%
Very high segregation	7.5%	28.9%	−5.5%	0.49	−31.2%

NOTE: Metropolitan areas are classified into segregation quartiles (very low, low, high, and very high) based on their 2010 Latino-white dissimilarity index in the upper panel and on their 2010 black-white dissimilarity index in the lower panel. The data are obtained from the five-year estimates from the American Community Survey 2007–2011. Samples are restricted to native-born blacks, Latinos, and whites between 25 and 30 years old living in metropolitan areas with more than 100,000 residents and 5,000 Latinos or blacks in each panel, accordingly. Individuals who lived in a different state in the previous year are excluded. Idleness is defined as not working and not enrolled in school. The sample for earnings is people who are working, not enrolled in school, and have nonnegative earnings. Single motherhood includes only women who are unmarried.

Table 2 presents results from ordinary least squares regressions of each individual outcome on metropolitan-area levels of segregation. The first row in each subpanel reports the coefficient on the Latino-white dissimilarity index and the second row in each subpanel reports the interaction between the dissimilarity index and a Latino indicator variable. The sample consists of only whites and Latinos, so the coefficient on the dissimilarity index can be interpreted as the association between Latino-white segregation and white outcomes, while the coefficient on the interaction between the dissimilarity index and the Latino indicator variable shows any difference in the association between segregation and outcomes for Latinos as compared to whites. The next two rows in each subpanel report analogous coefficients on black-white segregation for whites and blacks. We estimate our regressions separately for native-born young adults aged 20–24 in column (1) and aged 25–30 in column (2). We include a large set of metropolitan area–level controls, specifically population and median household income; the fraction of the population that is Asian, Latino, black, over 65 years, under 15 years, and unemployed; and the share of workers employed in the manufacturing sector, the share of residents with a college degree, and the share of residents in poverty. We interact all of these metropolitan area controls with a Latino or black indicator variable to let effects differ for Latinos and blacks as compared to whites. To conserve space, we focus discussion on results for 25–30 year olds. Results are generally similar for the younger group, though somewhat weaker for outcomes such as college graduation and earnings given the larger proportion of the younger sample that is still in college.

Results reveal significant associations between metropolitan-area segregation levels and individual outcomes for Latinos and African Americans. Starting with the probability of having completed high school, we find that segregation has no significant association with the probability of completing high school for whites, but the interaction coefficients for African Americans and Latinos are negative and significant, indicating that in more segregated metropolitan areas African American and Latino young adults are relatively less likely to complete high school. Indeed, a one standard deviation increase in the Latino-white dissimilarity index is associated with a decline in the probability of finishing high school of 3 percentage points for Latinos relative to whites (with an overall difference in means between whites and Latinos aged 25–30 of 9.1 percentage points). For African Americans, a one standard deviation increase in the black-white dissimilarity index is associated with a decline in the probability of completing high school of 1.4 percentage points relative to whites (with an overall difference in means between whites and blacks aged 25–30 of 6.9 percentage points). To be more concrete, for African Americans, a move from Phoenix, with a black-white dissimilarity score of 0.413, to New Orleans, with a black-white dissimilarity score of 0.633, would be associated with a decreased likelihood of high school graduation compared with whites of roughly 2.4 percent. For Latinos, a move from Las Vegas, with a Latino-white dissimilarity score of 0.420, to Los Angeles, with a Latino-white dissimilarity score of 0.622, would be associated with a decreased likelihood of high school graduation compared with whites of roughly 5.4 percent. In sum, both native-born Latinos and native-born African Americans

TABLE 2
Ordinary Least Squares Results on the Relation between Segregation
and Individual Outcomes

	Age 20–24	Age 25–30
	(1)	(2)
High school graduation		
Latino-white dissimilarity index	−.016	−.010
	(.024)	(.023)
Latino-white dissimilarity index × Latino	−.237	−.277
	(.042)°°°	(.045)°°°
Black-white dissimilarity index	−.005	−.003
	(.020)	(.016)
Black-white dissimilarity index × black	−.176	−.114
	(.044)°°°	(.038)°°°
College graduation		
Latino-white dissimilarity index	.010	.106
	(.041)	(.076)
Latino-white dissimilarity index × Latino	−.155	−.428
	(.038)°°°	(.076)°°°
Black-white dissimilarity index	.069	.149
	(.036)°	(.065)°°
Black-white dissimilarity index × black	−.249	−.397
	(.040)°°°	(.095)°°°
Idleness		
Latino-white dissimilarity index	−.046	−.075
	(.018)°°	(.018)°°°
Latino-white dissimilarity index × Latino	.161	.153
	(.030)°°°	(.031)°°°
Black-white dissimilarity index	.001	−.014
	(.016)	(.017)
Black-white dissimilarity index × black	.160	.133
	(.056)°°°	(.048)°°°
Log earnings		
Latino-white dissimilarity index	.149	.170
	(.086)°	(.067)°°
Latino-white dissimilarity index × Latino	.007	−.711
	(.119)	(.117)°°°
Black-white dissimilarity index	−.164	−.017
	(.084)°	(.065)
Black-white dissimilarity index × black	−.223	−.690
	(.133)°	(.126)°°°
Single motherhood		
Latino-white dissimilarity index	.005	−.034
	(.023)	(.028)

(continued)

TABLE 2 (CONTINUED)

	Age 20–24	Age 25–30
	(1)	(2)
Latino-white dissimilarity index × Latino	.228	.391
	(.058)°°°	(.079)°°°
Black-white dissimilarity index	.040	.010
	(.019)°°	(.023)
Black-white dissimilarity index × black	.127	.286
	(.059)°°	(.057)°°°

NOTE: All specifications include a constant term, census region, age indicators, and an indicator variable for females. Specifications for blacks include a black indicator variable, and those for Latinos include seven ancestry-group indicator variables (Mexicans, Puerto Ricans, Cubans, Central Americans, Dominicans, South Americans, and other Latinos). Coefficients are reported with robust standard errors in parenthesis, which are clustered by metropolitan area. Additional controls included for metropolitan areas are log population, log median household income, and shares of population that are black, Latino, Asian, over 65 years, under 15 years, unemployed, working in manufacturing, in poverty status, and with college degree. These metropolitan area controls are also interacted with a black or Latino indicator variable in each specification accordingly. Samples of metropolitan areas have to have at least 5,000 residents of a given minority group to be included in the specifications for that group. Individuals who lived in a different state in the previous year are excluded.
°°°$p < .01$. °°$p < .05$. °$p < .10$.

are significantly less likely compared with whites to graduate from high school in more segregated metropolitan areas, and the magnitude of the association between segregation and the likelihood of completing high school is greater for Latinos than for African Americans.

Higher levels of metropolitan-area segregation are also associated with a reduced likelihood of college completion for Latinos and African Americans. An increase of one standard deviation in segregation lowers the odds of completing college by 4.8 percentage points for blacks and 4.6 percentage points for Latinos relative to white graduation rates (with an overall difference in mean college graduation rates compared with whites of 23.1 percentage points for blacks and 23.7 for Latinos). For African Americans, the move from Phoenix to New Orleans would be associated with an 12 percentage-point widening in the gap with white college graduation rates. For Latinos, the move from Las Vegas to Los Angeles would be associated with an 10.8 percentage-point widening in the gap with white college graduation rates.

With regard to the likelihood of being simultaneously out of school and out of work, frequently referred to as "idleness," a one standard deviation increase in segregation is associated with a 1.6 percentage-point increase in idleness for African Americans relative to whites and a 1.7 percentage-point increase in idleness for Latinos relative to whites. The increase in black-white dissimilarity from

Phoenix to New Orleans then is associated with a 3.2 percentage-point increase in the likelihood of idleness for black 25–30 year olds relative to whites, while the increase in Latino-white dissimilarity from Las Vegas to Los Angeles is associated with a 4.6 percentage-point increase in the likelihood of idleness for Latino 25–30 year olds relative to whites.

The association with segregation is perhaps most dramatic for earnings. A one standard deviation increase in segregation is associated with a 7.7 percent decline in earnings for Latinos relative to whites and an 8.4 percent decline in earnings for African Americans relative to whites. Thus the increase in black-white segregation from Phoenix to New Orleans is associated with a 14.8 percent decline in black earnings relative to whites, whereas the increase in Latino-white segregation from Las Vegas to Los Angeles is associated with a 17.7 percent decline in Latino earnings relative to whites.

In terms of single motherhood, a one standard deviation increase in the segregation index raises the likelihood of being a single mother by 3.5 percentage points for African Americans and by 4.2 percentage points for Latinas relative to whites. The move from Phoenix to New Orleans is thus associated with a 6.1 percentage-point increase in the likelihood of single motherhood for African American women between the ages of 25 and 30, while the move from Las Vegas to Los Angeles is associated with a 8.6 percentage-point increase in single-motherhood for Latinas between the ages of 25 and 30.

In alternative estimations we used the isolation index as our measure of metropolitan-area segregation and obtained very similar results. We also assessed a potential lag in the effects of segregation by regressing outcomes in 2010 on 2000 segregation levels.[10] The results again are strikingly similar in both significance and magnitude. In sum, using nationwide individual-level data for 2010, our findings indicate that segregation has consistent negative correlations with socioeconomic outcomes for both African American and Latino young adults.

Potential mechanisms: Exposure to neighborhood conditions and services

The results above suggest that metropolitan-area segregation levels continue to be associated with reductions in educational attainment and labor market success for African Americans, and that segregation is associated with diminished outcomes for Latinos that are generally as large as or larger than those for blacks. In this section, we explore potential mechanisms that could explain these patterns.

Neighborhood human capital. Table 3 highlights the strong negative relationship between Latino segregation and exposure to college-educated neighbors. For African Americans, the gap between white and black exposure to college educated neighbors in low segregation metropolitan areas is 5.7 percentage points, but it rises to 13.3 percentage points in high segregation areas. For Latinos, the gap is smaller in low segregation metropolitan areas, at 3.1 percentage points, and nearly quintuples to 15.1 percentage points in high segregation areas. The consistent relationship between segregation and exposure to neighborhood human capital suggests that the significant association between segregation and outcomes for both blacks and Latinos could be attributable in large part to a neighborhood human capital channel.

Enclave effects. To explore the role of enclave effects in explaining the effects of segregation, we examine the degree to which segregation is related to exposure to employed, coethnic neighbors. Specifically, we calculate the share of neighbors who are employed coethnics for each racial or ethnic group. As expected, as segregation increases the share of employed coethnic residents to which the average African American or Latino resident is exposed increases consistently. It is possible that this greater exposure to employed coethnics in more segregated areas mitigates the harms of segregation operating through other channels, but given overall negative associations, these mitigating benefits do not appear to be large. The results should be interpreted with caution, however, as the measure of enclaves is a very rough approximation.

Public services. Public schools are a critical, and typically neighborhood-based, public service. As a proxy for the quality of neighborhood public services, we explore differences in the relative test scores of elementary schools to which children from different backgrounds have access in metropolitan areas with different levels of segregation. Table 3 reveals that the gaps between whites and African Americans in the exposure to neighborhood school proficiency are large, starting at an index score of 4.7 in very low segregation areas. The gap increases by more than a factor of five to 25.6 in very high segregation areas. The white-Latino gaps in school proficiency also increase consistently and by roughly a factor of five, from 3.9 in very low segregation areas to 19.9 in very high segregation areas. The results suggest that school quality may also be an important mechanism through which segregation operates.

Institutional density. To approximate the level of neighborhood institutional density we compute census tract counts of several types of for-profit and not-for profit establishments. Drawing on Small and McDermott (2006), Table 3 shows the gap in the average black and Latino metropolitan-area resident's exposure to the listed private business establishments relative to the average white resident's exposure. We also add the density of not-for-profit establishments such as business or civic associations and political or religious organizations.

When examining individual private business categories, we find some consistent patterns. Most notably, as levels of metropolitan-area segregation increase, the black-white and Latino-white gaps in exposure to banks widen dramatically, implying that both blacks and Latinos are exposed to fewer bank branches. With black-white segregation, pharmacies exhibit a similar pattern and with Latino-white segregation, childcare centers show the same pattern of greater gaps in exposure in more segregated areas. Grocery stores, on the other hand, display a pattern in the opposite direction than banks—as levels of metropolitan area segregation increase, African Americans and Latinos are exposed to more grocery stores. These counts do not control for the size or quality of the grocery stores,

TABLE 3
Racial Gaps in Exposure to Neighborhood Conditions and Services

	Very low	Low	High	Very high
Neighborhood residents with B.A.				
White	29.7%	28.3%	30.7%	35.7%
Black	24.0%	20.9%	22.5%	22.4%
White-black gap	5.7%	7.5%	8.2%	13.3%
White	26.3%	28.9%	30.3%	36.5%
Latino	23.2%	21.2%	19.4%	21.4%
White-Latino gap	3.1%	7.7%	10.9%	15.1%
Neighborhood coethnic workers				
White	65.8%	57.1%	65.0%	63.8%
Black	16.6%	17.3%	24.6%	33.0%
White-black gap	49.2%	39.8%	40.4%	30.8%
White	65.1%	65.8%	64.5%	63.0%
Latino	8.8%	14.0%	32.0%	34.8%
White-Latino gap	56.3%	51.8%	32.6%	28.3%
School proficiency ranking (0–100)				
White	57.0	57.4	57.9	59.0
Black	52.3	46.4	40.6	33.5
White-black gap	4.7	11.0	17.3	25.6
White	56.9	56.8	57.6	59.4
Latino	53.0	49.0	47.2	39.4
White-Latino gap	3.9	7.8	10.3	19.9
Density of for-profit establishments				
White-black gap: Banks	0.26	0.14	0.65	1.57
White-black gap: Pharmacies	−0.04	0.03	0.05	0.23
White-black gap: Childcare centers	−2.01	−1.07	−1.74	−1.78
White-black gap: Grocery stores	−0.35	−0.61	−0.88	−1.34
White-Latino gap: Banks	0.14	1.25	1.36	1.58
White-Latino gap: Pharmacies	−0.12	0.44	0.38	0.16
White-Latino gap: Childcare centers	−0.49	0.10	0.41	0.22
White-Latino gap: Grocery stores	−0.10	−0.54	−0.34	−1.55
Density of not-for-profit institutions				
White	15	16	18	14
Black	24	22	25	19
White-black gap	−9	−6	−7	−5
White	19	18	16	14
Latino	17	14	11	12
White-Latino gap	2	4	5	2
Neighborhood violent crime ranking (0–100)				
White		44.1	40.9	32.3

(continued)

TABLE 3 (CONTINUED)

	Very low	Low	High	Very high
Black		62.6	68.4	66.2
White-black gap		–18.5	–27.5	–33.9
White		41.3	42.4	34.3
Latino		54.7	55.8	58.6
White-Latino gap		–13.5	–13.5	–24.2

NOTE: Units of analysis are neighborhoods or census tracts in 2010. The samples in the first five panels include metropolitan areas (CBSAs) with more than 1,000,000 residents and 5,000 black or Latino residents. The first and second panel use data from the Neighborhood Change Database. The third panel uses data from U.S. Department of Education for school year 2008–2009. The fourth panel uses data from Esri's Business Analyst for 2010. Density is the count of establishments or institutions in each tract per 10,000 residents. In the last panel, the sample of crime exposure includes ninety-one cities in sixty metropolitan areas for 2000. We have pooled categories for "very low" and "low" levels of segregation given the small number of metropolitan areas in "very low" segregation quartile.

however, and it is possible that in more segregated metropolitan areas, blacks and Latinos have access to a greater number of small, lower quality stores. Therefore, although levels of segregation are not consistently related to a greater or reduced overall presence of business establishments, the composition of the establishments varies with segregation, and for some categories segregation appears to have different effects for African Americans and Latinos (e.g., pharmacies and childcare centers).

In terms of nonprofits, African Americans are exposed to a greater density of nonprofit organizations than either whites or Latinos. The white-black gap in exposure to not-for-profit institutions generally declines with segregation while no clear pattern is observed for the white-Latino gap. Thus, in less segregated metropolitan areas, African Americans presumably have relatively greater access to a wide array of civic, political, fraternal, and religious organizations than African Americans in more segregated metropolitan areas.

Given the mixed patterns in the relationship between metropolitan-area levels of segregation and available measures of the density of common institutions, it seems that institutional density broadly conceived may not account for much of segregation's impact on individual outcomes. Institutional presence certainly may still matter, but further exploration is needed to study the characteristics of the different institutions present in different neighborhoods.

Violence. The relationship between segregation and exposure to violent crime presents a particularly striking pattern. For both blacks and Latinos, the gap with whites in exposure to violent crime increases relatively consistently with levels of segregation. Most of that gap, however, is driven by the dramatic reduction in white exposure to neighborhood violent crime as both white-black and white-Latino segregation increases. It appears that segregation may enable whites to cluster in neighborhoods that are insulated from violence, perhaps through pub-lic or private security investments. Black and Latino exposure to neighborhood violent crime, by contrast, remains relatively similar even as segregation increases. The white-black gap in neighborhood exposure to violent crime is large and increases substantially from an 18.5-point gap to a 33.9-point gap as the level of segregation increases (see Table 3). Both the white-black gap and the differ-ence between the gaps in low- and high-segregation metropolitan areas are larger for African Americans than for Latinos, suggesting that, to the extent that segre-gation affects individual outcomes through its correlation with exposure to vio-lence, segregation is likely to have a greater impact on African Americans than Latinos. The results should be considered with some caution, however, as they are based on a smaller sample of cities for which neighborhood-level crime data were available and on crime data from 1999–2001, several years prior to the IPUMS sample used here, which together provide less confidence in their gen-eralizability.

Discussion and Conclusion

This research finds that segregation continues to be associated with significant reductions in educational attainment and labor market success for African Americans, and that the associations between segregation and outcomes for Latinos are at least as large as those for African Americans. For native-born African American and Latino young adults between the ages of 20 and 30, increases in metropolitan-area segregation are associated with significant reduc-tions in the likelihood of high school and college graduation, with lower earnings and employment rates, and with an increase in single motherhood.

These findings are somewhat unanticipated given the long history of intense black-white segregation and the systematic disinvestment in black neighborhoods through much of the last century, when compared to the historically more mod-erate levels of Latino-white segregation (Marable 1983; Massey and Denton 1993; Telles and Ortiz 2008). These findings raise the question of which mecha-nisms may be at play to generate these differences.

One crucial mechanism seems to be the levels of neighborhood human capital to which whites, Latinos, and African Americans are exposed; they are consistent with the negative associations for both blacks and Latinos and with the differ-ences in the magnitude of the association between them. The white-Latino gap in neighborhood exposure to human capital increases dramatically as levels of segregation increase.

The significance of neighborhood levels of human capital is consistent with existing research on the effects of segregation for African Americans and for immigrants (Borjas 1995; Cutler, Glaeser, and Vigdor 2008; Edin, Fredriksson, and Åslund 2003). The crucial questions for further research that emerge relate to the mechanism driving this association. Do more highly educated neighbors improve the outcomes of young neighborhood residents by setting high-achieving norms of educational and occupational attainment (W. J. Wilson 1991)? Does the educational attainment of neighbors act directly by actually connecting residents to networks and resources that facilitate greater levels of schooling and more remunerative employment? Or does neighborhood human capital act more indirectly, by contributing to a context of greater collective efficacy in which there is both neighborhood "social cohesion and the ability to realize shared expectations for neighborhood control" (Sampson 2012, 152)?

This research also finds support for other mechanisms connecting segregation to individual outcomes, particularly disparities in access to the quality of public services, as measured by local school proficiency and exposure to violence. The dramatic increase in gaps between the average level of school proficiency to which whites as compared to blacks and Latinos are exposed as segregation levels increase is also consistent with the significant association with individual outcomes. Although the available measures on which we rely here are not ideal, the widening disparities in black and Latino exposure to violence as compared to whites are also consistent with the associations between segregation and outcomes that we identify; they certainly merit further research.

Our findings, however, suggest that several commonly discussed mechanisms may be less powerful than previously thought in explaining segregation's effects. For instance, with the exception of banks, the densities of the establishments we studied do not vary consistently with levels of metropolitan-area segregation. This finding complements Small's (2008) work, pointing out that, contrary to prevailing notions, organizational density varies widely among low-income, predominantly black neighborhoods. Admittedly, we do not have detailed information on the size and quality of the establishments, and further research is needed. Similarly, the limited data we have suggest that ethnic enclaves do not seem to play a significant role in limiting harms of segregation for native-born young adults in most predominantly Latino neighborhoods. Several other studies have also questioned the benefits of enclaves, especially for employees and for groups with lower levels of human capital (e.g., Sanders and Nee 1992; Logan, Alba, and Stults 2003; Borjas 2000). Any inferences about enclaves here should take into account both that we were able to use only a very rough approximation of a residence-based ethnic enclave and that young native-born workers are not the prototypical beneficiaries of an ethnic enclave.

We hope this article spawns further research on how segregation is shaping, and constraining, the social and economic mobility of African Americans and Latinos. The research reported here suggests that segregation may have as negative effects for Latinos as it does for African Americans and that persistent Latino-white segregation is of serious concern as the nation's metropolitan areas continue to become more diverse.

Notes

1. See online appendix Table A1 for further data about characteristics of predominantly white, black, and Latino neighborhoods. See http://ann.sagepub.com/supplemental.

2. The dissimilarity index measures the evenness with which two different groups are distributed across neighborhoods within a metropolitan area. The index computes the proportion of one group that would need to exchange neighborhoods to achieve a uniform distribution of the groups across the city and thus provides a sense of how spatially concentrated one population group is in relation to the other.

3. The isolation index captures the proportion of the neighborhood population that belongs to a given group. It can be conceptualized as a measure of the extent to which the average member of a group is likely to be exposed to members of that same group within his or her neighborhood.

4. See online appendix Table A1.

5. See online appendix Table A1.

6. We use the metropolitan area–level segregation indices provided by US2010, a joint project between the Russell Sage Foundation and Brown University. See http://www.s4.brown.edu/us2010/About/History .htm; see also Logan and Stults (2011).

7. Further data about the distribution of the segregation indices are available in online appendix Table A3. See http://ann.sagepub.com/supplemental.

8. Although the matching is based on the nearest distance from each public elementary school to the centroid of each census block group and not on actual schooling attendance zones, Ellen and Horn (2011) show that the nearest school is also the zoned school in the overwhelming majority of cases.

9. We also experimented with excluding individuals who had moved across Public Use Microdata Areas in the previous year and obtained similar results.

10. Results using the isolation index as a measure of metropolitan-area segregation are available in online appendix Table A4, and results of the lagged effects of segregation are available in online appendix Table A5. See http://ann.sagepub.com/supplemental.

References

Bayer, Patrick, Hanming Fang, and Robert McMillan. 2014. Separate when equal? Racial inequality and residential segregation. *Journal of Urban Economics* 82:32–48.

Bayer, Patrick, Robert McMillan, and Kim S. Rueben. 2004. What drives racial segregation? New evidence using census microdata. *Journal of Urban Economics* 56 (3): 514–35.

Borjas, George. 1995. Ethnicity, neighborhoods, and human capital externalities. *American Economic Review* 85 (3): 365–90.

Borjas, George. 2000. Ethnic enclaves and assimilation. *Swedish Economic Policy Review* 7:89–122.

Card, David, and Jesse Rothstein. 2007. Racial segregation and the black-white test score gap. *Journal of Public Economics* 91 (11–12): 2158–84.

Cutler, David M., and Edward L. Glaeser. 1997. Are ghettos good or bad? *Quarterly Journal of Economics* 112 (3): 827–72.

Cutler, David M., Edward L. Glaeser, and Jacob L. Vigdor. 2008. When are ghettos bad? Lessons from immigrant segregation in the United States. *Journal of Urban Economics* 63 (3): 759–74.

De la Roca, Jorge, Ingrid Gould Ellen, and Katherine M. O'Regan. 2014. Race and neighborhoods in the 21st century: What does segregation mean today? *Regional Science and Urban Economics* 47:138–51.

Edin, Per-Anders, Peter Fredriksson, and Olof Åslund. 2003. Ethnic enclaves and the economic success of immigrants: Evidence from a natural experiment. *Quarterly Journal of Economics* 118 (1): 329–57.

Ellen, Ingrid Gould. 2000. Is segregation bad for your health? The case of low birth weight. Brookings-Wharton Papers on Urban Affairs 2000:203–38.

Ellen, Ingrid Gould, and Keren Horn. 2011. Do households with housing assistance have access to high quality public schools? Evidence from New York City. In *Finding common ground: Coordinating housing and education policy to promote integration*, ed. Philip Tegeler, 9–14. Washington, DC: Poverty & Race Research Action Council.

Harding, David J. 2009. Collateral consequences of violence in disadvantaged neighborhoods. *Social Forces* 88 (2): 757–84.

Iceland, John, and Kyle A. Nelson. 2008. Hispanic segregation in metropolitan America: Exploring the multiple forms of spatial assimilation. *American Sociological Review* 73 (5): 741–65.

Light, Ivan, and Edna Bonacich. 1988. *Immigrant entrepreneurs: Koreans in Los Angeles 1965–1982.* Berkeley, CA: University of California Press.

Logan, John R., Richard D. Alba, and Brian J. Stults. 2003. Enclaves and entrepreneurs: Assessing the payoff for immigrants and minorities. *International Migration Review* 37 (2): 344–88.

Logan, John R., and Brian J. Stults. 2011. *The persistence of segregation in the metropolis: New findings from the 2010 census.* Report, US2010 Project. Providence, RI, and New York, NY: Brown University and Russell Sage Foundation.

Logan, John R., and Richard N. Turner. 2013. *Hispanics in the United States: Not only Mexicans.* Report, US2010 Project. Providence, RI, and New York, NY: Brown University and Russell Sage Foundation.

Marable, Manning. 1983. *How capitalism underdeveloped black America: Problems in race, political economy, and society.* Boston, MA: South End Press.

Marwell, Nicole P. 2007. *Bargaining for Brooklyn: Community organizations in the entrepreneurial city.* Chicago, IL: University of Chicago Press.

Massey, Douglas S., and Nancy A. Denton. 1993. *American apartheid: Segregation and the making of the underclass.* Cambridge, MA: Harvard University Press.

National Advisory Commission on Civil Disorders. 1968. *Report of the National Advisory Commission on Civil Disorders.* Washington, DC: U.S. Department of Justice. Available from https://www.ncjrs.gov/pdffiles1/Digitization/8073NCJRS.pdf.

Park, Julie, and John Iceland. 2011. Residential segregation in metropolitan established immigrant gateways and new destinations, 1990–2000. *Social Science Research* 40 (3): 811–21.

Peterson, Ruth D., and Lauren J. Krivo. 2000. National neighborhood crime study (NNCS). [ICPSR27501-v1.] Ann Arbor, MI: Interuniversity Consortium for Political and Social Research.

Portes, Alejandro, and Julia Sensenbrenner. 1993. Embeddedness and immigration: Notes on the social determinants of economic action. *American Journal of Sociology* 98 (6): 1320–50.

Portes, Alejandro, and Steven Shafer. 2007. Revisiting the enclave hypothesis: Miami twenty-five years later. *Research in the Sociology of Organizations* 25:157–90.

Rugh, Jacob S., and Douglas S. Massey. 2014. Segregation in post-civil rights America: Stalled integration or end of the segregated century? *Du Bois Review: Social Science Research on Race* 11 (2): 205–32.

Sampson, Robert J. 2012. *Great American city: Chicago and the enduring neighborhood effect.* Chicago, IL: University of Chicago Press.

Sanders, Jimmy M., and Victor Nee. 1992. Problems in resolving the enclave economy debate. *American Sociological Review* 57 (3): 415–18.

Sharkey, Patrick, Amy E. Schwartz, Ingrid Gould Ellen, and Johanna Lacoe. 2014. High stakes in the classroom, high stakes on the street: The effects of community violence on student's standardized test performance. *Sociological Science* 1:199–220.

Small, Mario Luis. 2008. Four reasons to abandon the idea of "the ghetto." *City & Community* 7 (4): 389–98.

Small, Mario Luis. 2009. *Unanticipated gains: Origins of network inequality in everyday life.* New York, NY: Oxford University Press.

Small, Mario Luis, and Monica McDermott. 2006. The presence of organizational resources in poor urban neighborhoods: An analysis of average and contextual effects. *Social Forces* 84 (3): 1697–1724.

Telles, Edward E., and Vilma Ortiz. 2008. *Generations of exclusion: Mexican Americans, assimilation, and race.* New York, NY: Russell Sage Foundation.

Tienda, Marta, and Norma Fuentes. 2014. Hispanics in metropolitan America: New realities and old debates. *Annual Review of Sociology* 40:499–520.

Wilson, Kenneth, and Alejandro Portes. 1980. Immigrant enclaves: An analysis of the labor market experiences of Cubans in Miami. *American Journal of Sociology* 86 (2): 295–319.

Wilson, William Julius. 1991. Another look at The Truly Disadvantaged. *Political Science Quarterly* 106 (4): 639–56.

Wilson, William Julius. 1997. *When work disappears: The world of the new urban poor.* New York, NY: Vintage Press.

The Income Divide

Neighborhood Income Composition by Household Race and Income, 1990–2009

SEAN F. REARDON,
LINDSAY FOX,
and
JOSEPH TOWNSEND

Residential segregation, by definition, leads to racial and socioeconomic disparities in neighborhood conditions. These disparities may in turn produce inequality in social and economic opportunities and outcomes. Because racial and socioeconomic segregation are not independent of each other, however, any analysis of their causes, patterns, and effects must rest on an understanding of the joint distribution of race/ethnicity and income among neighborhoods. In this article, we use a new technique to describe the average racial composition and income distributions in the neighborhoods of households with different income levels and race/ethnicity. Using data from the decennial censuses and the American Community Survey, we investigate how patterns of neighborhood context in the United States over the past two decades vary by household race/ethnicity, income, and metropolitan area. We find large and persistent racial differences in neighborhood context, even among households with the same annual income.

Keywords: neighborhood economic conditions; racial segregation; socioeconomic segregation; segregation measurement

For the past four decades, residential racial segregation in the United States has been slowly declining, yet it remains very high. At the same time, residential segregation by income, which was very low in 1970, has risen sharply (Logan 2011; Reardon and Bischoff 2011a; Watson 2009; Jargowsky 1996). Both of

Sean F. Reardon is the endowed Professor of Poverty and Inequality in Education and professor (by courtesy) of sociology at Stanford University. His research focuses on the causes, patterns, trends, and consequences of social and educational inequality.

Lindsay Fox is a doctoral candidate in the Economics of Education program and an Institute of Education Sciences fellow at Stanford University's Center for Education Policy Analysis. Her research interests include teacher contributions to student learning, teacher labor markets, income inequality, and methods for causal inference.

DOI: 10.1177/0002716215576104

these trends are well-documented. Less well understood is how the two types of segregation interact. For example, how different are the neighborhoods of different race/ethnic groups with the same incomes? Does the decline in racial segregation coupled with the rise in income segregation lead to low-income black and Hispanic families living in higher or lower income neighborhoods than in the past?

Understanding the joint patterns of racial and socioeconomic segregation is important for two reasons. First, socioeconomic conditions may influence both neighborhood social processes and opportunities for social mobility. Income and racial segregation result in individuals of different socioeconomic backgrounds or different races/ethnicities living in neighborhoods that differ in their socioeconomic characteristics. To the extent that (1) segregation patterns lead to racial or socioeconomic disparities in neighborhood conditions and (2) neighborhood conditions affect opportunities and outcomes, it follows that segregation patterns may lead to racial or socioeconomic disparities in social mobility and well-being. Understanding racial disparities in neighborhood socioeconomic conditions is therefore essential to understanding how context shapes racial disparities in other dimensions.

Second, the policies and social forces that shape segregation do not shape racial and socioeconomic segregation independently. Indeed, racial and socioeconomic segregation patterns emerge from a complex interplay of many factors: racial disparities in income and wealth; racial differences in residential preferences, conditional on income; socioeconomic differences in residential preferences, conditional on race; the structure of the housing market; and patterns of racial prejudice and discrimination (Lareau and Goyette 2014; Krysan, Crowder and Bader 2014). Therefore, to fully understand the forces shaping racial and socioeconomic segregation patterns, it is necessary to consider them together. Conventional descriptions of segregation, however, typically consider income and racial segregation separately.

Both of these concerns suggest the need for a detailed description of the joint patterns of racial and socioeconomic context. This article is a step toward that aim. In particular, our goal here is to describe trends and patterns in racial and socioeconomic differences in neighborhood context over the last two decades. We use a set of newly developed methods to do so.

Prior Research on Neighborhood Socioeconomic Composition

Neighborhoods in the United States vary widely in both racial and socioeconomic composition, among many other dimensions. Sociological theory posits that

Joseph Townsend is a doctoral candidate in the Educational Policy program at Stanford University's Center for Education Policy Analysis. His research interests include schools' efforts at evidence-informed decision-making, and the use and development of technology-based research tools.

neighborhood socioeconomic composition (often operationalized as median income, poverty rates, or a composite measure called "concentrated disadvantage"), in particular, affects a number of educational, social, health, and political processes and outcomes (Sampson, Morenoff, and Gannon-Rowley 2002; Leventhal and Brooks-Gunn 2000). Moreover, economic context may affect individuals both directly and through a variety of secondary contextual factors that are shaped in part by economic conditions, including social norms, collective efficacy and social control, and exposure to violence (Sampson, Raudenbush, and Earls 1997; Sampson, Morenoff, and Gannon-Rowley 2002; Harding 2010; Sharkey 2010; Gorman-Smith and Tolan 1998). Empirical research on the effects of neighborhood socioeconomic conditions is somewhat mixed. Studies of the Moving to Opportunity program found little effect of neighborhood poverty levels on many child and family outcomes (Ludwig et al. 2013). A growing body of evidence, however, suggests that long-term exposure to neighborhood poverty has strong effects on cognitive and educational outcomes and teen pregnancy (Chetty, Hendren, and Katz 2015; Harding 2010; Sampson, Sharkey, and Raudenbush 2008).

Several studies have examined the joint patterns of neighborhood racial and socioeconomic conditions. Research on how economic segregation differs by race or ethnicity (see, for example, Jargowsky 1996; Watson 2009; Reardon and Bischoff 2011a; Wodtke 2013; Wodtke, Harding, and Elwert 2011) shows that income segregation among blacks and Hispanics (e.g., the extent to which middle- and low-income blacks and Hispanics live near one another) is higher than among whites and has increased more rapidly than among whites (Reardon and Bischoff 2011a; Bischoff and Reardon 2014). This research, however, does not describe the extent to which members of different racial groups are exposed to high- or low-income neighbors, regardless of race.

More relevant to our purposes here is research that explicitly measures racial differences in the exposure of households of different racial/ethnic groups to neighbors of various income levels. Black and Hispanic households are located, on average, in neighborhoods where the poverty rate is significantly higher than that of non-Hispanic whites (Firebaugh and Farrell 2012; Logan 2011). In particular, predominantly black neighborhoods, regardless of socioeconomic composition, continue to be spatially isolated in areas of severe disadvantage (Sharkey 2014). These racial disparities in neighborhood socioeconomic conditions persist even when comparing households of the same income. Although low-income households of all races are located disproportionately in low-income neighborhoods, the patterns are more pronounced for black and Hispanic households (Fry and Taylor 2012; Lichter, Parisi, and Taquino 2012; Logan 2011). This pattern of racial neighborhood disadvantage extends into the upper income categories for black and Hispanic minority households (Sharkey 2014). Logan (2011), for example, shows that the average affluent (earning more than $75,000 year) black or Hispanic household is located in a poorer neighborhood than the average lower-income (earning less than $40,000) white household. In part, these patterns are a result of the fact that U.S. metropolitan areas are substantially segregated by race, even when controlling for family income (Massey and Fischer 1999; Iceland and Wilkes 2006).

This body of research clearly shows that black and Hispanic households are located in more disadvantaged neighborhoods than white households with roughly similar levels of income. Nonetheless, most of this research relies on relatively broad categories of income ("poor," "middle-class," "affluent") that are not exactly comparable over time. This imprecision in the categorization of income limits the possibility of detailed descriptions of trends and patterns in racial differences in neighborhood socioeconomic context. We use newly developed methods to provide much more detailed and comparable measures of neighborhood income exposure.

Measuring Segregation and Neighborhood Context

There are many ways of describing differences in socioeconomic conditions across neighborhoods. A number of studies measure segregation in terms of the extent to which households of different incomes are evenly distributed among neighborhoods (Jargowsky 1996; Reardon and Bischoff 2011b; Watson 2009; also see Owens 2015, this volume). The advantage of measuring segregation this way is that it characterizes the degree of segregation along a spectrum ranging from complete evenness (every neighborhood has the same income distribution as the population as a whole) to complete unevenness (no one lives in a neighborhood with anyone of a different income level). One disadvantage of this approach, however, is that it does not provide any concrete characterization of the typical neighborhood context of a given type of household. Summary measures of segregation, such as the Jargowsky's Neighborhood Sorting Index, Reardon and Bischoff's rank-order information theory index, and Watson's Centile Gap Index provide no disaggregated information about the neighborhoods in which households of different income levels are located. Another disadvantage of the evenness measures is that it is not clear that they are useful for simultaneously describing joint racial and socioeconomic segregation patterns; they typically are used to describe either income or racial segregation of the total population or in each of several (racial/ethnic or income) groups.

An alternative is to characterize segregation in terms of the extent to which households of a given income level share neighborhoods with households of some other specific income level. The advantage of this approach is that it allows one to characterize the income distribution in the neighborhood of a typical household of a specific type. For example, one might say that "the typical white, non-Hispanic household earning $28,000/year is located in a neighborhood where the median annual income is $48,000 and where the 25th and 75th percentiles of the income distribution are $25,000 and $83,000 per year." Such "exposure"-based approaches to measuring segregation are therefore both more concrete (because they describe the typical composition of neighborhoods) and more disaggregated or fine-grained (because they describe the typical neighborhoods of different types of households) than are summary evenness measures. Their drawback is that they do not provide a single summary statistic for describing segregation.[1]

Three features of publicly available census data hamper the measurement of income segregation. First, household income is reported categorically (in sixteen categories in the most recent census and the American Community Survey). Second, the number and location of the income categories have changed over time. And third, the income distribution itself changes over time (because of inflation or changing income inequality, for example), so that even stable income category definitions do not correspond to the same part of the income distribution at different times. These features pose a challenge for the consistent measurement of income segregation patterns. Existing research (e.g., Logan 2011; Massey and Fischer 2003) deals with these issues by trying to combine income categories into a small number of roughly comparable categories. We improve on this prior work by using smoothed interpolation methods and by measuring income in percentile ranks relative to the national income distribution.

Data

We use census tract household population counts from the 1990 and 2000 decennial censuses and the 2007–2011 American Community Survey (for convenience we refer to the American Community Survey data as "2009"). The data provide information on household characteristics, including income (measured categorically), race, and ethnicity.[2] We operationalize neighborhoods as tracts. Because census data typically do not provide full cross-tabulations of race/ethnicity by income, we use an iterative proportional fitting algorithm to estimate tract-specific race-by-Hispanic-by-income category cross-tabulations (Beckman, Baggerly, McKay 1996; for details, see online appendix).

Estimation of neighborhood income exposure measures

For each geographical area of interest (metropolitan areas, or the United States as a whole), our goal is to estimate a set of average cumulative distribution functions, each of which describes the average income distribution in the neighborhoods of those of a given income level and race/ethnicity. Because census data do not provide information on individuals' exact income or the exact income of their neighbors, we cannot observe these functions directly from the data. Instead, we estimate them from the parameters of a constrained multidimensional polynomial regression model (for details, see appendix; Reardon, Townsend, Fox 2014).

National patterns of neighborhood income composition

We begin by examining how average neighborhood income distributions vary as a function of one's own household income. Figure 1 provides a simple

FIGURE 1

Neighborhood Median Income, by Household Income, All Households in United States, 2009

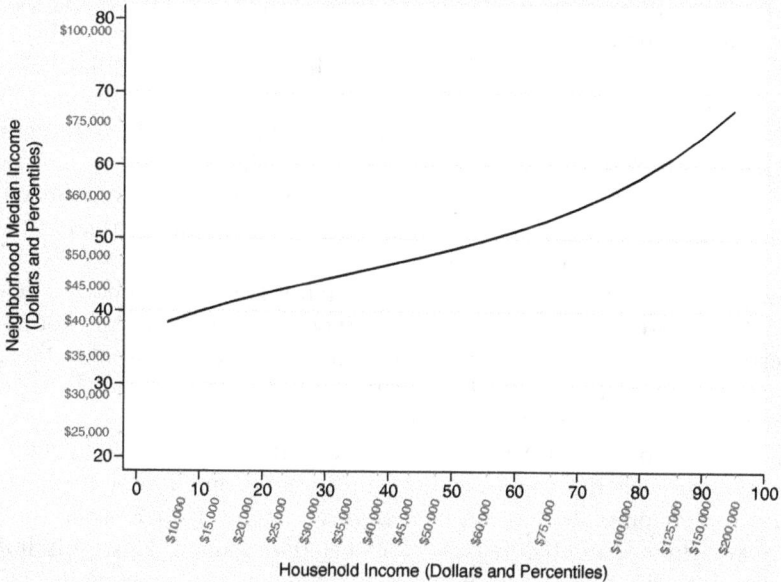

representation of this. Along the horizontal axis is a household's own income, expressed in terms of percentiles of the national household income distribution. On the vertical axis is median neighborhood household income, also expressed in terms of percentiles of the national income distribution. Both axes also show selected corresponding dollar figures (in 2008 dollars) for reference. The line indicates the median household income in the neighborhood of the average U.S. household at a given income level in 2009. For example, the average household with an income at the 25th percentile of the national income distribution (roughly $27,000) is located in a neighborhood where the median household income is at the 43rd percentile of the national income distribution (roughly $43,000). Similarly, the average household with an income at the 75th percentile is located in a neighborhood where the median income is at the 56th percentile.

The steepness of the line in Figure 1 can be thought of as an intuitive measure of segregation: a flat line would mean there is no association between one's own income and the median income of one's neighborhood (i.e., all households are located, on average, in neighborhoods with the same median income); a steep line would imply a strong association. Note also that the slope of the line (aver-aged over the income range) has a theoretical maximum value of one. The aver-age slope of the line in Figure 1 is roughly 0.3, which gives some sense of the magnitude of household income segregation in the United States relative to its theoretical maximum.

With this in mind, it is apparent from Figure 1 that segregation in the upper half of the income distribution is more pronounced than at the lower end: the neighborhoods where middle-class families live are more economically similar to those where the poor live than to those where the rich live. The difference in neighborhood median income between households at the 10th and 50th percentiles of the income distribution is 8.6 percentile points, compared with 15.6 percentile points between households at the 50th and 90th percentiles.[3] Thus, the segregation of the affluent is greater than the segregation of the poor, a finding consistent with prior research (Reardon and Bischoff 2011b; Bischoff and Reardon 2014). Note that this finding is not an artifact of using income percentiles; in fact, the difference in steepness would be even more pronounced if the Y-axis were scaled in terms of dollars or logged dollars, rather than in terms of percentiles of the income distribution.[4]

The patterns in 1990 and 2000 (not shown in Figure 1 but reported in appendix Table A1)[5] are very similar to those of 2009. Segregation of the poor declined modestly in the 1990s, by about 9 percent, and changed little in the 2000s. Segregation of the affluent declined as well in the 1990s, but only by 6 percent, before rebounding to its 1990 level in 2009.

The absence of substantial change in these patterns from 1990 to 2009 would seem to contradict the trend reported by Bischoff and Reardon (2014), who found that economic segregation increased by roughly 10 percent in the 2000s. There are three potential reasons for this discrepancy. First, Bischoff and Reardon describe average within–metropolitan area trends among the 117 largest metropolitan areas in the United States; our findings here, in contrast, describe trends in the nation as a whole. When we examine average within–metropolitan area trends (see Table 2), we find trends similar to Bischoff and Reardon's, at least with respect to the segregation of the affluent from the middle class. Second, Bischoff and Reardon report trends in income segregation among families; we report segregation among all households (families and nonfamily households combined). Owens (2014) finds that income segregation grew much more sharply from 1990 to 2009 among families with school-age children than among childless families and households; this suggests that the difference between our results and those of prior research may in part be due to differences in the trends among family and nonfamily households. Third, our trends are based on measures of exposure as opposed to the evenness measures that Bischoff and Reardon use, though this is unlikely to produce a substantial difference in trends.[6] The first two reasons likely account for the observed differences in trends.

National patterns of neighborhood racial composition

We next examine how the patterns evident in Figure 1 differ by race. First, however, it is informative to describe the typical racial composition of the neighborhoods of households of different races and incomes.[7] Figure 2 shows the average racial composition of the neighborhoods where households of different races and incomes reside. Each panel of the figure shows, for households of a given race, the average racial composition (summing to 100 percent on the vertical axis) of the neighborhoods of households of different income levels (on the horizontal axis).

Figure 2 makes evident that the racial composition of one's neighborhood depends much more on one's race than on one's income. Indeed, for all four racial/ethnic groups shown, the racial composition of neighborhoods depends remarkably little on one's household income. For example, white households—whether poor or affluent—are typically located in neighborhoods that are roughly 80 percent white. Black and Hispanic households, in contrast, are typically located in neighborhoods that are 40–50 percent white and 30–50 percent black or Hispanic. Even affluent black and Hispanic households typically are located in neighborhoods that are less than 50 percent white and that are 30–40 percent black or Hispanic. The patterns are similar for Asian households, which tend to locate in neighborhoods that are roughly 50–55 percent white and 20–25 percent Asian, regardless of income. In sum, Figure 2 illustrates the severity of racial residential segregation in the United States, even controlling for household income. These disparities in neighborhood racial composition foreshadow the economic disparities in neighborhood context discussed below.

Racial differences in average neighborhood income composition

Next, consider neighborhood socioeconomic composition by race and household income. The top panel of Figure 3 has the same axes as Figure 1 but shows one line for each race/ethnic group: Asian, white, Hispanic, and black. The panel below the figure indicates the proportion of the population made up of each group across the income distribution. The most notable feature of Figure 3 is that, conditional on having the same income, Asian and white households are typically located in neighborhoods with much higher median incomes than Hispanic and black households. The differences are substantial and relatively constant across the income distribution. This does not imply that all white and Asian households are located in neighborhoods with higher median household incomes than all black and Hispanic households of the same income. On average, however, they are.

One way to compare the neighborhood conditions of households of different racial/ethnic groups is to examine the vertical distance between the lines in Figure 3. Table 1 reports trends from 1990 to 2009 in specific values associated with the lines in Figure 3 (columns 1–4), as well as the vertical differences between the lines for each group and that of whites (columns 5–7). For Asians and whites at the 10th percentile of the national income distribution (i.e., those earning about $13,000/year), the median household income in their neighborhoods is above the 40th percentile of the national income distribution in all three time periods (roughly $45,000–48,000/year in 2009), while it is around the 30th percentile (roughly $32,000) for blacks and 35th percentile ($36,000) for Hispanics. More directly: neighborhood median income for poor black and Hispanic households is roughly two-thirds that of equally poor white and Asian households.

FIGURE 2

Average Neighborhood Racial Composition, by Household Income and Race, 2009

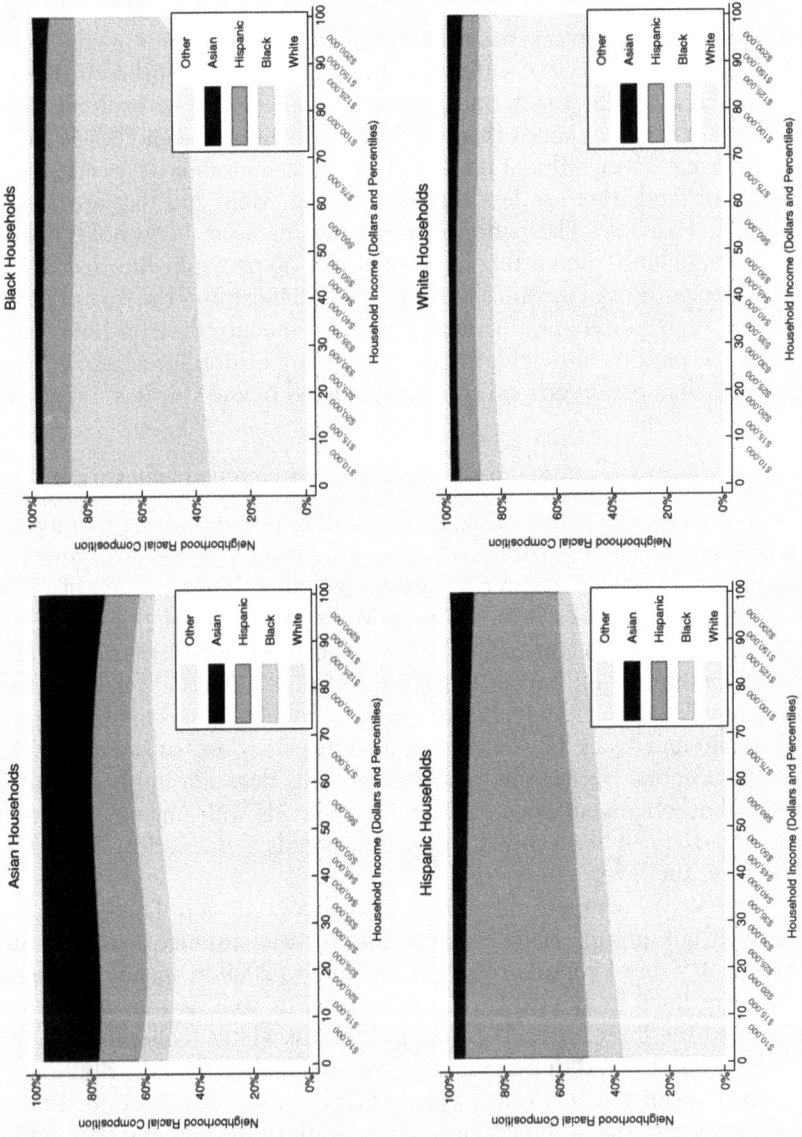

TABLE 1

Neighborhood Median Income, by Household Income and Race, 1990–2009

Households at 10th	Neighborhood Median Income				Difference from White		
Percentile Income	White	Black	Hispanic	Asian	Black	Hispanic	Asian
1990	42.2	28.4	34.5	42.5	−13.8	−7.7	0.3
2000	43.3	31.0	35.2	43.4	−12.3	−8.1	0.1
2009	43.4	31.3	36.2	45.3	−12.1	−7.2	1.9
Change, 1990–2009	1.2	2.9	1.6	2.8	1.7	0.5	1.6
Households at 50th							
Percentile Income	White	Black	Hispanic	Asian	Black	Hispanic	Asian
1990	50.0	41.7	45.1	55.2	−8.3	−4.9	5.3
2000	50.2	41.7	44.2	54.4	−8.6	−6.1	4.2
2009	50.1	42.2	45.2	55.2	−7.9	−4.9	5.1
Change, 1990–2009	0.1	0.5	0.1	0.0	0.4	0.0	−0.1
Households at 90th							
Percentile Income	White	Black	Hispanic	Asian	Black	Hispanic	Asian
1990	64.8	53.8	59.1	70.2	−10.9	−5.7	5.5
2000	64.2	53.7	56.7	69.1	−10.5	−7.5	4.9
2009	64.3	57.1	59.5	69.8	−7.2	−4.8	5.6
Change, 1990–2009	−0.5	3.2	0.4	−0.4	3.7	0.9	0.1

NOTE. Table 1 reads, for example, "white households at the 10th percentile of the national income distribution in 1990 lived in neighborhoods where the median income was at the 42.2 percentile of the national income distribution. In 1990, black households at the 10th percentile of the national income distribution lived in neighborhoods where the median income was 13.8 percentile points lower than that of white households with incomes at the 10th percentile of the national income distribution."

Similar patterns hold for households at the 50th and 90th percentiles of the national income distribution. The largest absolute changes over time occurred for black households. Black households at the 10th percentile in 2009 are located in neighborhoods with median incomes almost 3 percentile points higher than in 1990. Similarly, for black households at the 50th percentile, neighborhood median income increased half of a percentile point, and for blacks at the 90th percentile, neighborhood median income increased over 3 percentile points since 1990. At the 10th percentile, all groups experienced positive change between 1990 and 2009.[8] At the 90th percentile, however, only blacks and Hispanics experienced an increase in neighborhood median income.

The final three columns of Table 1 quantify the differences in the neighborhood median incomes of blacks, Hispanics, and Asians with whites at various income levels. In general, the patterns evident in Figure 3 are stable across years:

FIGURE 3

Neighborhood Median Income, by Household Income and Race, All Households in United States, 2009

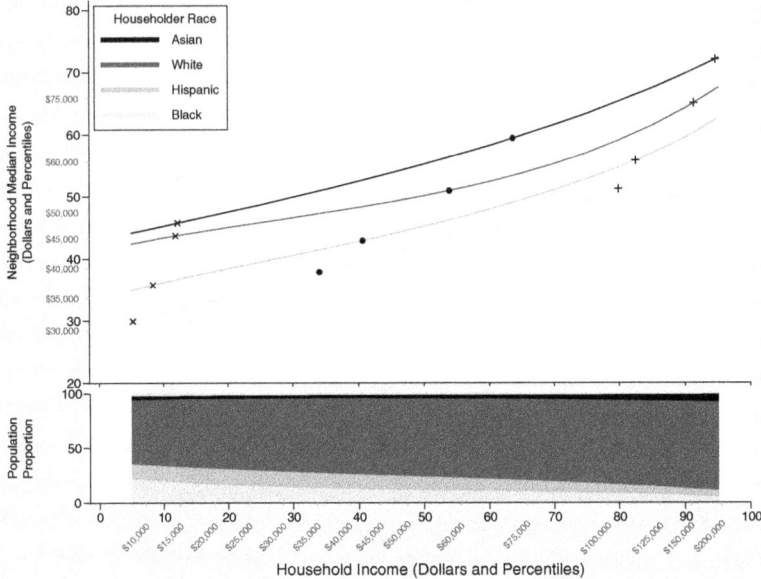

conditional on household income, black and Hispanic households are in neighborhoods with median incomes substantially lower than white households; Asian households are in higher-income neighborhoods. These patterns have changed relatively little over time, save for a moderate reduction in the white-black gap in neighborhood median incomes. For affluent black and white households, for example, the difference in neighborhood median income declined by a third (from 11 to 7 percentage points) between 1990 and 2009.

The steepness of the lines in Figure 3 indicates the degree of income segregation within each group. In the upper half of the income distribution, the degree of segregation is higher for all groups; the difference in neighborhood median income between the 90th and 50th percentile income households is at least 12 percentile points for all groups. The trends over time are consistent with those reported by Bischoff and Reardon (2014): we find that segregation in the upper half of the income distribution increased sharply among black households and modestly among Hispanic households from 2000–2009 (see Table A2 in the appendix for detail).[9]

The level and steepness of the lines shown in Figure 3 give a sense of group differences in neighborhood conditions and segregation, conditional on household income. Another way to describe these differences is to examine the horizontal distance between the lines. Read this way, Figure 3 illustrates that blacks and Hispanics must have household incomes that are substantially higher than

TABLE 2

Metropolitan Variation in Neighborhood Median Income, by Household Income, 250 Largest Metropolitan Areas by Population, 1990–2009

Year		Neighborhood Median Income			Difference in Neighborhood Median Income	
		Households at 10th Percentile Income	Households at 50th Percentile Income	Households at 90th Percentile Income	Between 10th and 50th Percentiles	Between 50th and 90th Percentiles
1990	Mean	41.7	49.4	58.8	7.7	9.3
	(Standard Deviation)	(8.2)	(7.5)	(8.9)	(3.2)	(3.5)
2000	Mean	42.2	49.7	58.8	7.5	9.1
	(Standard Deviation)	(7.5)	(7.1)	(8.5)	(2.9)	(3.5)
2009	Mean	41.5	49.3	59.7	7.9	10.3
	(Standard Deviation)	(7.4)	(6.6)	(7.9)	(2.8)	(3.7)
	Change in Mean 1990–2009	−0.2	−0.1	0.9	0.1	1.0°°
	Change in SD 1990–2009	−0.7	−0.9	−1.1	−0.4	0.2

NOTE: Each cell in Table 2 is computed by first estimating, within each of the largest 250 metropolitan areas, the neighborhood median income for households at a given percentile of the national income distribution. The cells show the (unweighted) mean and standard deviation of these metropolitan area–specific neighborhood median incomes. The upper left cells of the table, for example, are read as follows: "In the average metropolitan area in 1990, households at the 10th percentile of the national income distribution live, on average, in neighborhoods where the median income is at the 41.7th percentile of the national income distribution. The standard deviation (across metropolitan areas) of neighborhood median income for 10th percentile households is 8.2 percentile points." Similarly, the cells in the top of the fourth column read "In the average metropolitan area in 1990, households at the 50th percentile of the national income distribution lived in neighborhoods where the median income was 7.7 percentile points higher than that of households at the 10th percentile of the national income distribution. The standard deviation of this difference is 3.2 percentile points." Stars on the estimated changes in means indicate the p-value associated with the t-test of the null hypothesis that the average change in means from 1990–2009 was zero

°°°$p<0.001$. °°$p<0.01$. °$p<0.05$.

FIGURE 4

Metropolitan Variation in Neighborhood Median Income, by Household Income, Ten Largest Metropolitan Areas by Population, 2009

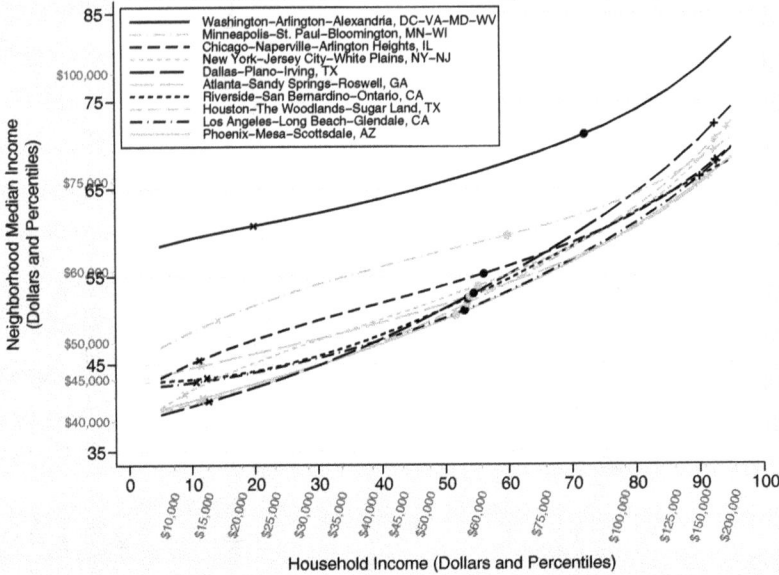

those of white or Asian households to live in neighborhoods with the same median income. For example, the income of a household at the 10th percentile of the national income distribution in 2009 is $11,800. Figure 3 shows that white households at this income level lived, on average, in neighborhoods where the median income was roughly $45,000. The income of black households that corresponds to this same average neighborhood median income level is roughly $60,000, five times the income of whites living in comparable neighborhoods. For Hispanic households, the corresponding income is roughly $45,000, 3.7 times that of whites. In other words, the average white household, earning $11,800, lives in a neighborhood with a similar income distribution to the average Hispanic household earning $45,000 and the average black household earning $60,000. Table A3 in the appendix shows these differences in more detail; in particular, it shows that these disparities narrowed slightly in the 1990s, but grew again to their 1990 levels by 2009.

Metropolitan variation in average neighborhood income composition

The figures and tables thus far describe patterns of neighborhood socioeconomic composition in the United States as a whole. However, these patterns may differ substantially across the country because of differences in local income distributions and patterns of residential segregation. Figure 4 shows average neighborhood median income, by household income, for the ten largest U.S. metropolitan areas for 2009.[10] The lines in this figure are analogous to those in Figure 1, but are shown for each metropolitan area separately. Among these ten metropolitan areas, the lines vary considerably in both their levels and their slopes.

For example, note that households in the Washington-Arlington-Alexandria, DC-VA-MD-WV metropolitan area (henceforth referred to as Washington, DC) are located in neighborhoods with very high average median incomes, relative to similar income families in other large U.S. metropolitan areas. In fact, even the poorest households in Washington, DC, are typically located in neighborhoods where the average median income is above the 55th percentile of the national income distribution. In contrast, poor households in the Dallas, TX, metropolitan area are typically located in neighborhoods with lower median incomes than their similar income counterparts in other large metros. In part, this variation is a result of the fact that the income distributions vary considerably among metropolitan areas; there are comparatively few poor households in the Washington, DC metropolitan area; as a result, many of the poor residents there live in relatively middle-class neighborhoods. But metropolitan areas also vary considerably in the degree of income segregation. Note, for example, the steepness of the line for the Dallas metropolitan area in comparison to the flatness of the line for the Minneapolis–St. Paul metropolitan area: low-income households in Dallas are located in poorer neighborhoods than in any other of the largest ten metros, but high-income households in Dallas are located in more affluent neighborhoods than are their counterparts in any other metropolitan area except Washington, DC.

Table 2 reports summary statistics for the 250 U.S. metropolitan areas with the largest household populations. In 2009, these metropolitan areas contained 78 percent of all households in the United States and 93 percent of all households in metropolitan areas. Table 2 shows the mean and standard deviation, across metropolitan areas, of neighborhood median income for the average 10th, 50th, and 90th percentile income households. The means are, on average, similar to the national means from appendix Table A1, but there is considerable variation among metropolitan areas. The standard deviation of the means ranges from 6.6 to 8.9 percentile points. In 2009, for example, the neighborhood median income of households with incomes at the 10th percentile of the national income distribution ranged from the 25th percentile (for metropolitan areas two standard deviations below the mean metropolitan area) to the 58th percentile (for those two standard deviations above the mean).

Table 2 also reports the average slope of the association between household and neighborhood income, using the 10th-to-50th and 50th-to-90th percentile differences as above. On average, the within metropolitan area 10th-to-50th percentile slopes are lower than the 50th-to-90th percentile slopes, but not by nearly so much as in the national patterns (compare to appendix Table A1). The variation across metropolitan areas is substantial in comparison to the average slope: in 2009 the 95 percent intervals of the 10th-to-50th and 50th-to-90th slopes are (2.4, 13.4) and (3.0, 17.6), respectively. The association between household and neighborhood income is as much as six times greater in the most segregated metropolitan areas than in the least segregated areas. Average within-metropolitan area upper-tail income segregation appears to have increased significantly from 1990 to 2009, with most of this change happening since 2000, a trend that is consistent with the findings of Bischoff and Reardon (2014).

Table 3 disaggregates the information in Table 2 by race/ethnic group. Similar to Table 1, the first four columns report the average neighborhood median income, averaged across metropolitan areas, by race/ethnic group, year, and household income percentile. The means here are similar to those in Table 1 and are relatively stable across time, with the exception of significant increases of 1.6 and 4.0 percentile points in the neighborhood median incomes of low- and high-income black households, respectively, from 1990–2009. Note also that there is substantial variation among metropolitan areas in the average neighborhood median incomes, particularly for high-income households and nonwhite households. In other words, for high-income nonwhite households, one's exposure to high-income neighbors is very dependent on the metropolitan area in which one lives.

The last three columns of Table 3 report the average black-white, Hispanic-white, and Asian-white differences in neighborhood median income. Across metropolitan areas, black households are typically located in neighborhoods where the median income is consistently 7 to 12 percentile points below that of similar income white households. For Hispanic households, the difference is generally 5 to 8 percentile points. These within–metropolitan area racial differences vary considerably among places. Indeed, there are some metropolitan areas where black and Hispanic households are typically located in neighborhoods with median incomes 20 to 30 percentile points lower than their similar income white counterparts. In other metropolitan areas, there are essentially no racial differences in neighborhood median income.

The pattern of white-Asian differences is particularly notable here. Recall that Figure 3 and Table 1 show that, nationally, the average Asian household is in a neighborhood with a significantly higher median income than a similar-income white household. Within metropolitan areas, however, this is not true, suggesting that much of the pattern evident in Figure 3 is due to the fact that Asian households, in general, are concentrated in metropolitan areas with high median incomes. Within the average metropolitan area, however, the typical low- or middle-income Asian household is in a neighborhood with slightly lower median income than the typical white household of the same income. For high-income households, there is little or no difference within metropolitan areas between white and Asian households in neighborhood median incomes.

Discussion

The findings described here are far from a complete description of how neighborhood income is associated with household income and race/ethnicity, and how these associations vary across place and time. Nonetheless, several key patterns are evident.

TABLE 3

Metropolitan Variation in Neighborhood Median Income, by Household Income and Race, 250 Largest Metropolitan Areas by Population, 1990–2009

Households at 10th Percentile Income		Neighborhood Median Income				Difference from White		
		White	Black	Hispanic	Asian	Black	Hispanic	Asian
1990	Mean	45.0	32.7	38.3	41.4	−12.3	−6.6	−3.5
	(SD)	(8.3)	(9.0)	(8.5)	(11.2)	(7.0)	(6.9)	(7.7)
2000	Mean	45.7	34.3	38.5	41.2	−11.4	−7.2	−4.5
	(SD)	(7.6)	(8.2)	(7.7)	(9.8)	(6.3)	(5.9)	(6.0)
2009	Mean	45.5	34.3	37.7	41.3	−11.3	−7.9	−4.2
	(SD)	(7.8)	(8.1)	(7.3)	(9.8)	(6.3)	(5.7)	(6.2)
Change in Mean, 1990–2009		0.6	1.6°	−0.7	−0.1	1.0	−1.2°	−0.7
Change in SD, 1990–2009		−0.5	−0.9	−1.2	−1.4	−0.7	−1.2	−1.5

Households at 50th Percentile Income		White	Black	Hispanic	Asian	Black	Hispanic	Asian
1990	Mean	51.0	41.5	45.8	49.2	−9.6	−5.2	−1.9
	(SD)	(7.5)	(8.2)	(7.0)	(8.8)	(5.9)	(5.6)	(5.7)
2000	Mean	51.5	42.1	44.8	49.2	−9.4	−6.7	−2.3
	(SD)	(7.3)	(7.5)	(6.5)	(8.0)	(5.2)	(4.8)	(4.1)
2009	Mean	51.6	42.3	44.7	50.3	−9.3	−6.9	−1.3
	(SD)	(7.0)	(7.8)	(6.2)	(7.8)	(5.4)	(4.7)	(4.8)
Change in Mean, 1990–2009		0.6	0.8	−1.1	1.1	0.3	−1.7°°°	0.5
Change in SD, 1990–2009		−0.5	−0.5	−0.9	−0.9	−0.5	−0.9	−0.9

Households at 90th Percentile Income		White	Black	Hispanic	Asian	Black	Hispanic	Asian
1990	Mean	59.1	49.0	54.4	59.9	−10.1	−4.8	0.8
	(SD)	(9.2)	(12.5)	(11.1)	(11.9)	(9.9)	(8.7)	(7.7)
2000	Mean	59.7	50.5	53.9	59.1	−9.2	−5.8	−0.6
	(SD)	(8.7)	(10.0)	(8.8)	(9.4)	(9.3)	(6.9)	(4.8)
2009	Mean	60.2	53.0	55.2	60.3	−7.2	−5.0	0.1
	(SD)	(8.1)	(11.2)	(10.9)	(9.6)	(8.7)	(7.9)	(6.1)
Change in Mean, 1990–2009		1.1	4.0°°°	0.8	0.4	2.9°°°	−0.3	−0.7
Change in SD, 1990–2009		−1.0	−1.2	−0.2	−2.2	−1.2	−0.8	−1.6

NOTE: Each cell in Table 3 is computed by first estimating, within each of the largest 250 metropolitan areas, the neighborhood median income for households of a given race/ethnicity at a given percentile of the national income distribution. The cells show the (unweighted) mean and standard deviation (SD) of these metropolitan area–specific neighborhood median incomes. See note below Table 2 for example of how to read the table. Asterisks on the estimated changes in means indicate the p value associated with the t-test of the null hypothesis that the average change in means from 1990–2009 was zero.
°°°$p < .001$. °°$p < .01$. °$p < .05$.

First, middle-class households are typically located in neighborhoods that are more similar to those of low-income households than to those of high-income households. That is, high-income households are more segregated from middle-class and poor households than low-income households are from the middle class and the rich. This pattern is consistent with the findings in Reardon and Bischoff (2011b) and Bischoff and Reardon (2014).

Second, income segregation at the national level—at least as measured by the strength of the association between household and neighborhood median income—has changed little over the past two decades, even as income segregation within metropolitan areas grew by almost 10 percent during the 2000s (see Tables A1 [online] and 2). This increase was driven entirely by the increase in the segregation of affluence. Recall that Bischoff and Reardon's (2014) finding that both segregation of affluence and segregation of poverty grew by roughly 10 percent in the 2000s is based on measures of economic segregation among families. Because income segregation has increased much more among families with children than among households without children (Owens 2014), our household income segregation measures may not capture the trends in family segregation of poverty that Bischoff and Reardon (2014) described.

Third, there is substantial variation among metropolitan areas in these patterns of neighborhood economic composition. Our findings demonstrate that the income distribution in one's neighborhood is not only a function of one's own income, but also of the metropolitan area where one lives. Low-income households in the Washington, DC, or Minneapolis, MN, metropolitan areas, for example, are typically located in neighborhoods similar to those of middle- or higher-income households in Atlanta, GA, Los Angeles, CA, and other metropolitan areas. As a result, children growing up in poor households in metropolitan areas such as Washington and Minneapolis may have, on average, more access to high-quality schools and other forms of opportunity than equally poor (or middle-class) children in metropolitan areas such as Atlanta or Los Angeles. If neighborhood context affects opportunities for social mobility, this variation might help to explain some of the geographic variation in economic mobility rates that Chetty et al (2014) have reported.

Fourth, even among households with the same annual income, there are sizable racial/ethnic differences in neighborhood income composition. Black middle-class households (with incomes of roughly $55,000–$60,000), for example, are typically located in neighborhoods with median incomes similar to those of very poor white households (those with incomes of roughly $12,000). For Hispanic households the disparity is only slightly smaller. Moreover, even high-income black and Hispanic households do not achieve neighborhood income parity with similar-income white households.

These large racial disparities in neighborhood income composition are at least partly due to patterns of racial segregation. As is evident in Figure 2, black and Hispanic middle-class households tend to be located in neighborhoods that contain much larger proportions of black and Hispanic residents, respectively, than the neighborhoods of similar-income white households. Because average black and Hispanic households' incomes are substantially lower than white households' incomes, racial residential segregation will tend to lead to disparities in neighborhood economic context. These patterns of racial and economic segregation are also partly due to racial differences in wealth. White households have, on average, greater wealth than black households (Oliver and Shapiro 2006), enabling them to afford housing in higher-income neighborhoods than similar-income black households. However, as Sharkey (2008) shows, wealth differences alone do not explain the disproportionate concentration of black households in high-poverty neighborhoods. Other factors, such as differences in household structure, lingering racial discrimination in the housing market, the location of affordable and subsidized housing, and residential preferences, likely also play a role (for a thorough discussion of the factors that lead to segregation, see Krysan, Crowder, and Bader 2014).

Fifth, some racial disparities in neighborhood income distributions, particularly the black-white disparity, appear to have narrowed modestly in the past two decades. Among low-income households, the black-white difference in neighborhood median income declined by more than 10 percent from 1990 to 2009; among high-income families it declined by one-third. Nationally, Hispanic-white differences in neighborhood median income widened in the 1990s and narrowed in the 2000s, resulting in only modest declines over the whole time period. Within metropolitan areas, however, Hispanic-white disparities increased, on average, by roughly 20 percent from 1990 to 2009, meaning that in many metropolitan areas, particularly those with smaller Hispanic populations, the gaps in neighborhood context grew substantially. These changes, however, are small relative to the magnitude of persistent racial inequality in neighborhood income distributions.

The racial disparities in neighborhood income distributions are particularly troubling because these are differences that are present even among households with the same incomes. If long-term exposure to neighborhood poverty negatively affects child development, educational success, mental health, and adult earnings (and a growing body of research suggests it does, as noted above), then these large racial disparities in exposure to poverty may have long-term consequences. They mean that black and Hispanic children and families are doubly disadvantaged—both economically and contextually—relative to white and Asian families. Not only do black and Hispanic households have lower average incomes than do white and Asian households, but their lower incomes do not—for reasons beyond the scope of this article—result in access to the same neighborhoods as those of equally low-income white households.

Notes

1. For more on the distinction between evenness and exposure-based approaches to measuring segregation, see Massey and Denton (1988).

2. See online supplemental tables: http://ann.sagepub.com/supplemental.

3. These numbers can be found in the online appendix, Table A1.

4. To see this, note that the typical family at the 90th percentile of the income distribution is in a neighborhood with a median income of roughly $75,000, one-and-half times larger than the neighborhood median income (roughly $50,000) of a typical family at the 50th percentile. The difference in neighbor-

hood median incomes between families at the 10th and 50th percentiles of the income distribution is much smaller (median income is roughly $42,000 in poor families' neighborhoods, compared with $50,000 in middle-class families' neighborhoods).

5. See http://ann.sagepub.com/supplemental.

6. Trends in evenness and exposure measures of segregation tend to differ when the population composition changes over time (Reardon and Owens 2014). However, because we define income in percentile ranks, the population composition remains unchanged (a uniform distribution) across time, so evenness and exposure trends are unlikely to differ substantially.

7. Patterns of neighborhood racial composition for all households are shown in appendix Figure A1.

8. It may seem logically impossible that all groups could live, on average, in higher-income neighborhoods in 2009 than in 1990, given that income is measured in percentile ranks. Nonetheless the patterns in Table 1 are real; they result from the facts that the Hispanic and (to a lesser extent) black shares of the population have grown, and these groups' incomes have risen modestly relative to whites. Given these trends, it is logically possible for all group median incomes to rise even while the national median income stays—as it must—exactly at the 50th percentile of the income distribution.

9. See http://ann.sagepub.com/supplemental.

10. In our data, metropolitan areas are defined using metropolitan division codes, and these areas are ranked according to their total populations in 2010. For statistics on the largest fifty metropolitan areas, see online appendix Table A4.

References

Beckman, Richard J., Keith A. Baggerly, and Michael D. McKay. 1996. Creating synthetic baseline populations. *Transportation Research Part A: Policy and Practice* 30 (6): 415–29.

Bischoff, Kendra, and Sean F. Reardon. 2014. Residential segregation by income, 1970–2009. In *Diversity and disparities: America enters a new century*, ed. John. R. Logan, 208–33. New York, NY: Russell Sage Foundation.

Chetty, Raj, Nathaniel Hendren, and Lawrence F. Katz. 2015. *The effects of exposure to better neighborhoods on children: New evidence from the moving to opportunity experiment.* Cambridge, MA: Harvard University.

Chetty, Raj, Nathaniel Hendren, Patrick Kline, and Emmanuel Saez. 2014. Where is the land of opportunity? The geography of intergenerational mobility in the United States. National Bureau of Economic Research Working Paper 19843, Cambridge, MA.

Firebaugh, Glenn, and Chad R. Farrell. 2012. Still separate, but less unequal: The decline in racial neighborhood inequality in America. Unpublished manuscript.

Fry, Richard, and Paul Taylor. 2012. *The rise of residential segregation by income.* Washington, DC: Pew Research Center. Available from http://www.pewsocialtrends.org/files/2012/08/Rise-of-Residential-Income-Segregation-2012.2.pdf.

Gorman-Smith, Deborah, and Patrick Tolan. 1998. The role of exposure to community violence and developmental problems among inner-city youth. *Development and Psychopathology* 10 (1): 101–16.

Harding, David. 2010. *Living the drama.* Chicago, IL: University of Chicago.

Iceland, John, and Rima Wilkes. 2006. Does socioeconomic status matter? Race, class, and residential segregation? *Social Problems* 53 (2): 248–73.

Jargowsky, Paul A. 1996. Take the money and run: Economic segregation in U.S. metropolitan areas. *American Sociological Review* 61 (6): 984–98.

Krysan, Maria, Kyle Crowder, and Michael D. M. Bader. 2014. Pathways to residential segregation. In *Choosing homes, choosing schools*, eds. Annette Lareau and Kimberly Goyette, 27–63. New York, NY: Russell Sage Foundation.

Lareau, Annette, and Kimberly Goyette, eds. 2014. *Choosing homes, choosing schools.* New York, NY: Russell Sage Foundation.

Leventhal, Tama, and Jeanne Brooks-Gunn. 2000. The neighborhoods they live in: The effects of neighborhood residence on child and adolescent outcomes. *Psychological Bulletin* 126 (2): 309–37.

Lichter, Daniel T., Domenico Parisi, and Michael C. Taquino. 2012. The geography of exclusion: Race, segregation, and concentrated poverty. *Social Problems* 59 (3): 364–88.

Logan, John R. 2011. *Separate and unequal: The neighborhood gap for blacks, Hispanics and Asians in metropolitan America*. Project US2010 Report. New York, NY and Providence, RI: Russell Sage Foundation and Brown University.

Ludwig, Jens, Greg J. Duncan, Lisa A. Gennetian, Lawrence F. Katz, Ronald C. Kessler, Jeffrey R. Kling, and Lisa Sanbonmatsu. 2013. Long-term neighborhood effects on low-income families: Evidence from Moving to Opportunity. National Bureau of Economic Research Working Paper 18722, Cambridge, MA.

Massey, Douglas S., and Nancy A. Denton. 1988. The dimensions of segregation. *Social Forces* 67 (2): 281–315.

Massey, Douglas S., and Mary J. Fischer. 1999. Does rising income bring integration? New results for blacks, Hispanics, and Asians in 1990. *Social Science Research* 28:316–26.

Massey, Douglas S., and Mary J. Fischer. 2003. The geography of inequality in the United States, 1950–2000. *Brookings-Wharton Papers on Urban Affairs*, 1–40, Washington, DC: Brookings Institution Press.

Oliver, Melvin L., and Thomas M. Shapiro. 2006. *Black wealth, white wealth: A new perspective on racial inequality*. New York, NY: Taylor & Francis.

Owens, Ann. 2014. Inequality in children's contexts: Economic segregation between school districts, 1990 to 2010. Paper presented at the American Sociological Association annual meeting. San Francisco, CA, August 2014.

Owens, Ann. 2015. Assisted housing and income segregation between neighborhoods in U.S. metropolitan areas. *The ANNALS of the American Academy of Political and Social Science* (this volume).

Reardon, Sean F., and Kendra Bischoff. 2011a. *Growth in the residential segregation of families by income, 1970–2009*. Project US2010 Report. New York, NY, and Providence, RI: Russell Sage Foundation and Brown University.

Reardon, Sean F., and Kendra Bischoff. 2011b. Income inequality and income segregation. *American Journal of Sociology* 116 (4): 1092–153.

Reardon, Sean F., and Ann Owens. 2014. 60 Years after Brown: Trends and consequences of school segregation. *Annual Review of Sociology* 40:199–218.

Reardon, Sean F., Joseph B. Townsend, and Lindsay Fox. 2014. Characteristics of the joint distribution of race and income among neighborhoods. Unpublished manuscript.

Sampson, Robert J., Jeffrey D. Morenoff, and Thomas Gannon-Rowley. 2002. Assessing "neighborhood effects": Social processes and new directions in research. *Annual Review of Sociology* 28:443–78.

Sampson, Robert J., Stephen W. Raudenbush, and Felton Earls. 1997. Neighborhoods and violent crime: A multilevel study of collective efficacy. *Science (New York, N.Y.)* 277 (5328): 918–24.

Sampson, Robert J., Patrick Sharkey, and Stephen W. Raudenbush. 2008. Durable effects of concentrated disadvantage on verbal ability among African-American children. *Proceedings of the National Academy of Sciences of the United States of America* 105 (3): 845–52.

Sharkey, Patrick. 2008. The intergenerational transmission of context. *American Journal of Sociology* 113 (4): 931–69.

Sharkey, Patrick. 2010. The acute effect of local homicides on children's cognitive performance. *Proceedings of the National Academy of Sciences* 107:11733–38.

Sharkey, Patrick. 2014. Spatial segmentation and the black middle class. *American Journal of Sociology* 119 (4): 903–54.

Watson, Tara. 2009. Inequality and the measurement of residential segregation by income in American neighborhoods. *Review of Income and Wealth* 55 (3): 820–44.

Wodtke, Geoffrey T. 2013. Duration and timing of exposure to neighborhood poverty and the risk of adolescent parenthood. *Demography* 50 (5): 1765–88.

Wodtke, Geoffrey T., David J. Harding, and Felix Elwert. 2011. Neighborhood effects in temporal perspective: The impact of long-term exposure to concentrated disadvantage on high school graduation. *American Sociological Review* 76 (5): 713–36.

Assisted Housing and Income Segregation among Neighborhoods in U.S. Metropolitan Areas

By
ANN OWENS

Over the past 40 years, assisted housing in the United States has undergone a dramatic geographic deconcentration, with at least one unit of assisted housing now located in most metropolitan neighborhoods. The location of assisted housing shapes where low-income assisted renters live, and it may also affect the residential choices of nonassisted residents. This article examines whether the deconcentration of assisted housing has reduced the segregation of families by income among neighborhoods in metropolitan areas from 1980 to 2005–9. I find that the deconcentration of assisted housing resulted in modest economic residential integration for very low-income families. However, high-income families became even more segregated, as assisted housing was deconcentrated, potentially offsetting the economic integration gains and ensuring that very low-income families are living in neighborhoods with only slightly higher-income neighbors. I conclude by discussing features of housing policies that might promote greater income integration among neighborhoods.

Keywords: assisted housing; income segregation; housing policy; deconcentration; residential segregation

In the past four decades, assisted housing in the United States has undergone dramatic changes and become geographically deconcentrated across neighborhoods.[1] Historically, housing for low-income renters was subsidized in public housing developments, predominantly located in racially and economically segregated neighborhoods (Schwartz 2010). With the introduction of housing vouchers in 1974,

Ann Owens is an assistant professor of sociology at the University of Southern California. Her research examines the causes and consequences of inequalities among neighborhoods and in educational opportunities and outcomes.

NOTE: The author gratefully acknowledges the helpful comments of a reviewer and the editors and thanks the organizers and participants of the Penn State Stratification Conference on Residential Inequality in American Neighborhoods and Communities.

DOI: 10.1177/0002716215576106

low-income renters could rent private market apartments in nearly any neighborhood. Over the next 40 years, the voucher program grew, construction of large public housing developments ceased, many public housing units were demolished, and new housing programs created units for low-income renters in privately owned apartment buildings. As a result of these policy changes, assisted housing became geographically deconcentrated, located in many more and lower-poverty neighborhoods than in the 1970s.

Changes in assisted housing policy affect where some low-income residents live, providing assisted renters with access to higher-income neighborhoods than they could otherwise have afforded. Assisted housing may also influence the residential choices of residents who do not live in assisted housing. High-income residents may not want to live in a neighborhood where new assisted housing units are built or where voucher users are moving in, while low-income residents who do not live in assisted housing may move into neighborhoods with new assisted housing, attracted by the new investments in the neighborhood (Freeman 2003). The changing landscape of assisted housing, then, may shape where both high- and low-income residents live, affecting the income composition of neighborhoods and the degree of income segregation among neighborhoods in metropolitan areas. On one hand, income segregation could decline as assisted housing creates opportunities for low-income residents to live among higher-income neighbors. On the other hand, if higher-income residents avoid neighborhoods with assisted housing, assisted housing may not lead to much income integration.

In this article, I examine whether the geographic deconcentration of assisted housing units has reduced income segregation—the degree to which families are segregated by income among neighborhoods—within U.S. metropolitan areas from 1980 to the late 2000s. I find that the deconcentration of assisted housing reduced the segregation of very low-income residents from higher-income residents among neighborhoods but is not associated with overall income segregation among neighborhoods. I conclude by discussing policy features that might encourage greater income integration.

Income Segregation among Neighborhoods

Trends in segregation by income among neighborhoods have been well documented by social scientists. Economic segregation captures the degree to which families with different incomes live in the same neighborhood or are sorted by income among neighborhoods within a city or metropolitan area. Research shows economic segregation among neighborhoods increased during the 1970s and 1980s, stabilized or declined slightly during the 1990s, and increased again during the 2000s (Jargowsky 1996; Watson 2009; Reardon and Bischoff 2011; Bischoff and Reardon 2014). The rise in economic segregation among neighborhoods from 1970 to 2010 corresponds to a declining number of mixed- or middle-income neighborhoods and a growing number of either very poor or very affluent neighborhoods (Reardon and Bischoff 2011; Booza, Cutsinger, and Galster 2006).

Several economic and political changes have spurred the growth in income segregation. First, rising levels of income inequality since 1970 have led to increased income segregation among neighborhoods (Watson 2009; Reardon and Bischoff 2011). Second, land use regulations on building size and density have contributed to income segregation, especially between city and suburban neighborhoods (Pendall 2000a; Rothwell and Massey 2010). Third, economic restructuring has led to a lack of jobs for low-skilled workers and to economic growth outside of central cities, concentrating low-income residents in some urban areas (Wilson 1987). Finally, persistent racial discrimination and segregation among neighborhoods has contributed to economic segregation (Massey and Denton 1993). Black-white racial segregation declined during the period of rising economic segregation (Logan and Stults 2011), highlighting the rise of economic segregation within racial groups (Reardon and Bischoff 2011). Racial segregation does persist and remain high, however, and the association between income and race means that barriers to racial integration in the housing market contribute to economic segregation.

Assisted housing policy has long been implicated as a source of racial and economic segregation, as some have argued that the U.S. Department of Housing and Urban Development (HUD) perpetuated segregation by building public housing in predominantly low-income and minority communities (Hirsch 1983). Since the 1960s, HUD has attempted to address segregation concerns by integrating public housing by race, stopping the construction of large high-rise family public housing developments, and creating programs to geographically disperse residents to more integrated neighborhoods (Goetz 2003; Khadduri 2001). These new programs were seen as having the potential to encourage greater income mix both inside assisted housing developments and in neighborhoods where assisted housing was located, fostering income integration in metropolitan areas.

The Deconcentration of Assisted Housing

Historically, assisted units were in public housing developments located in a relatively small number of U.S. neighborhoods. Several assisted housing programs introduced since 1970 have resulted in the presence of assisted units in many more and lower-poverty neighborhoods. Housing Choice Vouchers (formerly Section 8 vouchers, established in 1974) subsidize rents on the private market, and about two million were in use by 2008 (compared with 1.1 million public housing units) (Schwartz 2010). In 1974, HUD also began to subsidize rents in privately owned buildings through Section 8 projects (where the subsidy is project-based, linked to a particular building, rather than tenant-based and portable, like vouchers), and nearly 850,000 Section 8 project units were subsidized. In 1986, the Low Income Housing Tax Credit (LIHTC) program began, with developers receiving tax credits for building affordable units, 1.6 million of which existed in 2008. Since the mid-1990s, there has been a 20 percent net loss in the number of public housing units as many public housing developments were demolished or sold, and former residents were often given vouchers to rent units on the private market (Schwartz 2010). Other public housing was redeveloped, often with an aim toward creating income diversity within the developments. Over 250 grants through the HOPE VI program funded the demolition and redevelopment of public housing into mixed-income communities between 1992 and 2010. In addition to new programs, HUD's policies began to focus on the geographic dispersal of both voucher and project-based units and income mix within developments, reflected in the Quality Housing and Work Responsibility Act of 1998 (QHWRA), which includes tenant allocation policies for public housing developments to maximize income diversity and prioritizes very low-income renters for voucher receipt.

This patchwork of programs alongside traditional public housing served to dramatically deconcentrate assisted housing: at least one unit of assisted housing is now located in nearly every tract in the metropolitan United States (Owens 2012). This dramatic geographic deconcentration provides opportunities for low-income, assisted renters to live in neighborhoods they may not otherwise be able to afford and to live alongside higher-income neighbors. Assisted housing is not an entitlement; today only about a quarter of those who are eligible receive it. However, the number of assisted housing units more than tripled from 1977 to 2008, so assisted housing has a bigger impact on where low-income households are located today than in the past. Therefore, assisted housing may increasingly play a role in income integration.

With the introduction of these new programs, assisted housing is now located in slightly lower-poverty neighborhoods compared with public housing in the 1970s. While voucher users live, on average, in lower-poverty neighborhoods than where public housing was historically located, they still do not live in very low-poverty areas (Devine et al. 2003; Pendall 2000b; Newman and Schnare 1997; Galster 2013). The Section 8 project–based program has mixed outcomes, with about a third of units located in neighborhoods with poverty rates below 10 percent but another third located in neighborhoods with poverty rates above 40 percent in 2007, about the same proportion as in the public housing program (Schwartz 2010). LIHTC units are most likely to be in low-poverty neighborhoods, with 40 percent located in neighborhoods with poverty rates below 20 percent and only 6 percent in neighborhoods with poverty rates over 40 percent (Schwartz 2010).

Demolition of the most troubled public housing projects and a focus on creating income mix within public housing resulted in a decline in the proportion of public housing units located in neighborhoods with poverty rates over 20 percent (Schwartz 2010). Residents displaced from public housing due to demolition or redevelopment typically moved to lower-poverty neighborhoods, but these neighborhoods are still typically racially segregated with higher poverty rates than the citywide average (Oakley and Burchfield 2009; Kingsley, Johnson, and Pettit 2003). Poverty rates declined in neighborhoods surrounding several HOPE VI sites, providing lower-poverty contexts for the public housing residents that live there (Holin et al. 2003). Overall, the changes to assisted housing since the 1970s have resulted in low-income assisted renters living in neighborhoods with fewer poor residents. Has this been enough to reduce income segregation?

Assisted Housing and Income Mixing

The deconcentration of assisted housing may affect income segregation among neighborhoods because it shapes where some residents, especially low-income residents, live. The deconcentration of assisted housing may have direct effects on income segregation because low-income families can now use vouchers or find project-based units in neighborhoods with higher-income neighbors than they could otherwise have afforded. This could reduce segregation by income among neighborhoods and create a greater income mix within a neighborhood. The demolition of some large public housing developments may have broken up clusters of low-income residents, creating more income-diverse neighborhoods and leading to more income integration among neighborhoods. However, the transformation of assisted housing's impact on income segregation depends on low-income assisted residents leaving lower-income neighborhoods and moving to higher-income neighborhoods, and this does not always occur, as noted above. While assisted housing policy is a state intervention in the housing market, the location of assisted housing is largely left up to the private market, potentially limiting assisted housing's integrating effects. Further, assisted housing accounts for only about 3 percent of housing units in the United States, and only about 25 percent of the poor live in assisted units, so assisted housing's impact on income segregation may be modest since it does not account for the location of the majority of low-income renters.[2]

The ability of assisted housing to reduce income segregation also depends on the residential location of nonassisted residents. Past research suggests that assisted housing may have indirect effects on the residential choices of nonassisted households that may undermine income integration. Higher-income, non-assisted residents may move out of or avoid neighborhoods with assisted housing because of concerns about property values, safety, or stereotypes linked to assisted renters (Freeman 2003; Owens, forthcoming). These negative perceptions of assisted housing may linger in regard to the neighborhood even after assisted housing has been removed (Owens, forthcoming). Low-income, nonassisted residents are more likely to move into neighborhoods with newly developed assisted housing than other similar neighborhoods, attracted by the investment in the neighborhood (Freeman 2003). Therefore, the deconcentration of assisted housing may simply move pockets of low-income residents from one neighborhood to another, rather than integrating low- and high-income residents within neighborhoods.

Little research has examined the impact of assisted housing on metropolitan-area income segregation. Three studies have examined how the changing location of assisted housing affects poverty concentration, focusing on the degree to which poor residents are segregated from all other households. Two studies used simulations. Quillian (2005) found that under various relocation criteria, "vouchering out" public housing residents did not substantially reduce poverty concentration because the number of public housing residents was not large enough to have an impact on the distribution of poor residents. Moving beyond public housing demolition, Kucheva (2011) found that the deconcentration of assisted housing units from 1977 to 2008 may have increased segregation of poor and nonpoor residents in the eight counties she studied. In past work (Owens 2012), I have used measures of assisted housing concentration similar to those used here and found that, overall, the deconcentration of assisted housing from 1977 to 2008 only modestly reduced poverty concentration in the 100 largest metropolitan areas.

No study has looked at the relationship between assisted housing and income segregation aside from looking at poverty concentration, which only measures segregation between two groups: those with incomes above and those with incomes below the poverty threshold. The effect of assisted housing on segregation at other points in the income distribution is thus not captured. For example, past research shows that assisted housing is now located in lower-poverty neighborhoods than in the past, but these neighborhoods still often have poverty rates above the national average, so the deconcentration of assisted housing may lead to income integration of the very poor and working poor, both with incomes below the poverty threshold. Alternatively, the deconcentration of assisted housing may lead very affluent residents to become even more segregated if the presence of assisted housing causes them to move out of or avoid neighborhoods, since assisted housing may make low-income neighbors more visible. The following analyses explore these possibilities.

Data and Methods

This article explores how the deconcentration of assisted housing has shaped income segregation among neighborhoods in metropolitan areas from 1980 to 2005–9, using longitudinal regression models.

Income segregation

I estimated income segregation in U.S. metropolitan areas in 1980, 2000, and 2005–9. The 1980 and 2000 U.S. Censuses and the 2005–9 American Community Survey (ACS) provide counts of families in income categories (seventeen in 1980, and sixteen in 2000 and 2005–9).[3] I use the 1999 Office of Management and Budget's definition of Primary Metropolitan Statistical Areas (MSAs) and analyze all 331 MSAs in the United States.[4]

To capture income segregation across the full income distribution, rather than examining poor versus nonpoor or some other measure of segregation between just two groups, I use the rank-order information theory index (H). H compares the entropy (variation) in family incomes within census tracts to the entropy in family incomes in the metropolitan area. Entropy is calculated at each income threshold, comparing the number of families with incomes above and below each category cutoff point in the tract to the number in the metro area. Then, the rank order information theory index H is estimated by taking a weighted average of the binary H computed at every threshold between the income categories (Reardon 2009). The online methodological appendix provides more details on estimating H. In theory, H can range from zero (no segregation) to one (total segregation). A value of zero indicates that the family income distributions are identical in all tracts (and therefore identical to the metropolitan area distribution). In contrast, a value of one indicates that every family in each tract has incomes in the same category as all other families in the tract—there is no income diversity.

Table 1 presents the mean and standard deviation of H in 1980, 2000, and 2005–9. Reardon and Bischoff (2011) report mean values of H between 0.13 and 0.16 among the 100 largest metropolitan areas from 1980 to 2000. The estimates here are slightly lower because they include all 331 MSAs in the United States, but the magnitude and trend are similar.

Because H is a weighted average of binary H at each income category, H can be estimated at each percentile in the income distribution by fitting a polynomial regression through the estimates at each category (see Reardon and Bischoff [2011] for details). Therefore, in addition to H averaged across all income categories, I also estimate H at the 10th, 50th, and 90th percentiles of the income distribution to capture the segregation of the poor, between the bottom and top halves of the income distribution, and of the affluent from all others. Table 1 presents the mean and standard deviation of H at these points in the income distribution. Consistent with past research (Reardon and Bischoff 2011), income segregation is highest for the affluent, between those with incomes above and below the 90th percentile. It is also high among the poor, between those with incomes above and below the 10th percentile of the income distribution.

Assisted housing concentration

I measured the segregation of assisted housing in 1977, 2000, and 2008 using data from HUD's *Picture of Subsidized Households*, which provides the number of assisted housing units located in each census tract.[5] The 1977 file provided the number of units in public housing projects. HUD's "Subsidized Housing Projects' Geographic Codes, Form HUD-951" database provided addresses for the projects, which I geocoded to assign projects to census tracts. About 30 percent of assisted units were missing address data. I reduced missing data among eleven MSAs that represent about one-fifth of all assisted units (Chicago, Cleveland, Dallas, Los Angeles, Louisville, New York, Phoenix, Pittsburgh, Providence, Seattle, and Washington, DC) by finding project addresses in HUD documentation and through communication with housing authorities. This reduced the missing data to about 25 percent of assisted units, and in total I assigned approximately 700,000 public housing units to tracts.[6] Programs other than public

TABLE 1
Descriptive Statistics for Income Segregation and Control Variables
at the MSA Level, 1980 to 2005–9

Dependent variables	1980	2000	2005–9
MSA income segregation: Rank-order information theory Index H	0.095	0.107	0.116
	(0.031)	(0.033)	(0.034)
MSA H, 10th percentile	0.109	0.124	0.138
	(0.033)	(0.035)	(0.036)
MSA H, 50th percentile	0.088	0.098	0.106
	(0.030)	(0.033)	(0.034)
MSA H, 90th percentile	0.128	0.144	0.154
	(0.040)	(0.042)	(0.045)
Control variables			
Poverty rate	0.121	0.124	0.140
	(0.046)	(0.044)	(0.043)
Income inequality (Gini)	35.926	39.259	39.616
	(2.451)	(2.629)	(2.457)
Proportion non-Hispanic white	0.813	0.754	0.719
	(0.146)	(0.168)	(0.172)
Proportion non-Hispanic black	0.105	0.109	0.107
	(0.098)	(0.106)	(0.106)
B-W dissimilarity	0.609	0.514	0.541
	(0.148)	(0.130)	(0.107)
Population density	394.021	437.668	459.983
	(910.996)	(927.571)	(923.890)
Proportion rental units below fair market rate	0.536	0.447	0.475
	(0.127)	(0.081)	(0.068)
Vacancy rate	0.081	0.079	0.081
	(0.029)	(0.029)	(0.030)
Number of affordable housing units	2542	9153	10734
	(6764)	(20,840)	(25,501)
N	265	331	331

NOTE: Cells present means with standard deviations in parentheses below. Statistics for 1980 are for MSAs with assisted housing units in 1977, as those without are excluded from regression analyses.

housing were not included in the 1977 data even though about 150,000 Section 8 project–based units and more than 100,000 voucher units were in use by 1977 nationwide (Schwartz 2010). Therefore, the 1977 affordable housing units (AHUs) concentration measures include only public housing, so the change from 1977 to 2000 can be interpreted as deconcentration of assisted housing due to introducing new programs other than public housing as well as changes to the public housing program.

The *Picture of Subsidized Households* provided the number of units in census tracts in 2000 and 2008 for all federally assisted housing programs: public housing, vouchers, Section 8 project–based, LIHTC, and other small programs.[7] Some vouchers and project-based units were not assigned to tracts in these years. I could not assign missing tracts for vouchers, as street addresses for individual units were not available, but I again reduced missing data for project-based units among eleven MSAs, after which only about 5 percent of assisted units could not be geocoded. I identified tracts for about 1.9 million project-based units and 1.7 million vouchers in the metropolitan United States in 2000 and 2008. The 2000 and 2008 AHU concentration measures included the federally funded assisted housing programs listed above, and the change from 2000 to 2008 captures dispersal efforts among all these programs.

I measured the degree to which assisted housing is evenly spread throughout the MSA using the binary information theory index (H). This is analogous to the rank-order information theory index described above, except that it compares the distribution of two nominal categories (assisted and nonassisted housing units) rather than many ordered categories. An H of zero suggests the AHU rate in all tracts is equal to the overall MSA AHU rate, while H equals one when some tracts had an AHU rate of 100 percent and other tracts had AHU rates of 0 percent, so no tract had both assisted and nonassisted housing units. Declining levels of segregation of AHUs indicates their deconcentration. In the case of segregation between two groups, the value of H is partially conditioned by the relative number of minority members (AHUs in this case; Massey and Denton 1988; Reardon and Firebaugh 2002). However, it has advantages over other evenness measures like the dissimilarity index, which is affected only by the redistribution of AHUs from areas where they were overrepresented to where they are underrepresented and which is particularly sensitive when the number of minority group members is small compared with the number of units, which is the case in some metros here.

Table 2 presents descriptive statistics for AHU H in 1977, 2000, and 2008. AHU H declined dramatically from 1977 to 2000 due largely to the addition of units in new AHU programs that tended to be located in many more neighborhoods than public housing, like voucher, Section 8 project–based, and LIHTC units. The AHU H in 1977 was very high—0.59 on a zero to one scale (in comparison to segregation levels under 0.12 for income)—and declined to 0.25 by 2000. The decline of nearly 60 percent suggests that assisted housing has been dispersed into more neighborhoods—but residential income desegregation will likely follow only if AHUs have moved not just to more but to higher-income neighborhoods. AHU segregation continued to decline, by an additional 20 percent, from 2000 to 2008, as dispersal became an even bigger focal point of HUD policies after the QHWRA of 1998 and as the LIHTC program—which tends to be located in lower-income neighborhoods—took off, doubling in size after 2000 (Schwartz 2010).

Table 2 also explores how changes in AHU concentration vary across metropolitan areas. Assisted housing was most concentrated in the Midwest in 1977, and the largest declines by 2008 were in western MSAs, where housing stock is

TABLE 2
Descriptive Statistics for AHU Segregation, 1977 to 2008

Independent variable	1977	2000	2008
MSA AHU segregation: Binary information	0.589	0.249	0.197
theory index H	(0.161)	(0.076)	(0.058)
By region			
Northeast	0.571	0.277	0.235
Midwest	0.610	0.259	0.195
South	0.590	0.257	0.205
West	0.583	0.194	0.148
By population (2009)			
0–250,000	0.578	0.226	0.176
250,001–500,000	0.571	0.244	0.207
500,001–1,000,000	0.567	0.267	0.211
1,000,001+	0.643	0.293	0.221
By AHU type (2008)			
Majority voucher	0.566	0.207	0.160
Majority public housing	0.491	0.288	0.227
Majority LIHTC	0.631	0.259	0.193
Mix	0.580	0.265	0.219
By AHU prevalence (2008)			
< 2%	0.638	0.247	0.199
2–4%	0.587	0.247	0.194
4–6%	0.580	0.249	0.200
> 6%	0.565	0.274	0.210

NOTE: Cells present means with standard deviations in parentheses below. Statistics for 1977 exclude MSAs without assisted housing.

lower density, perhaps facilitating the distribution of assisted housing. Assisted housing segregation was highest in all years in large metropolitan areas, with more than 1,000,000 residents, but the size of the decline in AHU segregation among these MSAs was second largest, behind small metros with under 150,000 residents. AHU concentration also varies by the type of AHU programs used and by the proportion of housing units that are AHUs (AHU rate). I classified metropolitan areas by whether the majority of their AHUs were vouchers, public housing, LIHTC, or no majority (a mix of programs) in 2008. Metropolitan areas that had more vouchers and LIHTC units had the least segregated AHUs in 2008 and the largest declines since 1977. Metropolitan areas with the highest AHU rates in 2008 had slightly higher levels of AHU segregation in 2008 compared with metropolitan areas with a lower AHU rate in 2008, perhaps because high AHU rates correspond to large developments that cluster AHUs in one neighborhood (and because binary H reflects, in part, the AHU rate). Overall, AHU concentration has declined dramatically across metropolitan areas, with the largest declines occurring in western MSAs, very large and very small MSAs, those that rely on vouchers and LIHTC units, and those with lower AHU rates.

Analyses

I conducted metropolitan-level analyses and explored whether AHU segregation predicts metropolitan-area income segregation using regression with MSA and time-period fixed effects. Income segregation and the sociodemographic control variables are measured in 1980, 2000, and 2005–9 while AHU concentration is measured in 1977, 2000, and 2008. I treated 1977 and 1980 data as the first time point, 2000 data as the second, and the 2008 and 2005–9 data as the third, and pooled observations from all three time points. I also examined the two time periods separately. The online appendix provides more detail on the models.

I controlled for MSA characteristics associated with assisted unit deconcentration and poverty deconcentration: (1) MSA economic characteristics (MSA poverty rate and income inequality; measured with the Gini coefficient); (2) racial composition (proportion non-Hispanic black and non-Hispanic white) and racial segregation (the black-white dissimilarity index, or the proportion of black residents that would have to move to achieve an even distribution across tracts); (3) population density; and (4) housing market characteristics (the proportion of vacant rental units, the proportion of units renting below fair market rate, and the rental limit for voucher use, which is typically set at the 40th percentile of the area rent distribution). Data on the sociodemographic characteristics come from the censuses and ACS. The MSA and time fixed effects account for time invariant characteristics of metros, such as region, and secular time trends across metros. Other time-variant observed or unobserved variables that I do not account for bias causal validity. Descriptive statistics for the control variables are presented in Table 1.[8]

Results

Table 3 presents results for longitudinal regressions predicting MSA-level income segregation from assisted housing segregation with MSA and time fixed effects. The first column presents results from the first time period, the second column presents results from the second, and the third column presents the pooled model with all three waves. A positive coefficient indicates that metropolitan areas with declining geographic concentration of AHUs (where the AHU segregation index declines) have declining income segregation (a positive coefficient could also indicate rising levels of income segregation in MSAs with rising levels of AHU segregation, but since AHU segregation declined in nearly all MSAs, I interpret the results in terms of geographic deconcentration of AHUs).

The first row presents results for average metropolitan-area income segregation across the entire income distribution. Looking at the pooled model with all

TABLE 3
Longitudinal Regression Analyses Predicting Metropolitan-Area Income Segregation
from Assisted Housing Segregation with Metropolitan Area and Time Fixed Effects

	1980 to 2000	2000 to 2005–9	Pooled
Average income segregation (H)			
AHU *H*	−0.0004	0.008	0.006
	(0.005)	(0.009)	(0.004)
Income segregation (H), 10th percentile			
AHU *H*	0.006	0.037*	0.014*
	(0.007)	(0.017)	(0.006)
Income segregation (H), 50th percentile			
AHU *H*	−0.003	0.010	0.003
	(0.006)	(0.010)	(0.005)
Income segregation (H), 90th percentile			
AHU *H*	0.000	−0.032*	0.003
	(0.007)	(0.015)	(0.007)
MSA × Time controls	Y	Y	Y
MSA FE	Y	Y	Y
Time FE	Y	Y	Y
N Obs (MSAs × Year)	596	662	927
N MSAs	331	331	331

NOTE: All models include MSA × time control variables shown in Table 1 as well as MSA and time fixed effects. In 1977, some MSAs had no AHUs and are missing observations at that time point, so the number of observations is less than 331×3 (993). All significance tests are two-tailed.
*$p \leq .05$.

three time periods, the segregation of assisted housing does not significantly predict income segregation among neighborhoods within MSAs, though the direction of the coefficient is consistent with the hypothesis that MSAs with lower concentration of AHUs have lower levels of income segregation. There is not a significant effect in either time period.

Rows 2–4 in Table 3 show that the concentration of assisted housing has a significant and positive relationship with income segregation among neighborhoods at specific points in the income distribution. The change in AHU segregation from 1977 to 2008 significantly and positively predicts the change in income segregation at the 10th percentile. Very low-income families with incomes below the 10th percentile became more likely to share neighborhoods with higher-income families (those with incomes above the 10th percentile) in MSAs where assisted housing was geographically deconcentrated. The magnitude of the coefficient indicates that a one-unit decline in AHU segregation (moving from complete segregation to no segregation) would reduce income segregation by 1.4 percent (0.014 points on a zero to one scale), a small effect. The change in average AHU concentration from 1977 to 2008 was –0.392 (Table 2), so the corresponding decline in income segregation would be only about 0.5 percent. During this time, income segregation at the 10th percentile increased by about 0.03, or 3 percent (Table 1). Therefore, AHU deconcentration created an integrating effect that offset about one-sixth of the increase in income segregation experienced during this period.

The relationship is not significant in the earlier time period, and the pooled result is driven by the post-2000 time period. AHU concentration declined, on average, by 0.048 points after 2000, suggesting an average impact on income segregation of 0.002 (0.048×0.037), about one-seventh of the rise in income segregation at the 10th percentile during this time. Income segregation increased, on average, in nearly every MSA, suggesting that other segregating forces (for example, income inequality, racial segregation, or the Great Recession) counteracted the integrating effect of deconcentrating assisted housing. However, income segregation increased less in MSAs where greater AHU deconcentration occurred.

AHU segregation may more strongly shape income segregation at the bottom of the income distribution because low-income families are the most likely to be living in assisted housing and thus are most directly affected by its concentration as it determines where some live. (In 2009, the median income of both voucher and public housing residents was just over $10,000 [Schwartz 2010].) The deconcentration of assisted housing units appears to integrate very low-income families into at least slightly higher-income neighborhoods. This finding suggests "direct" effects of deconcentrating assisted housing—that it contributes to income integration through its effect on those who may live in assisted housing—but I also found evidence of a "spillover" or indirect effect on families not living in assisted housing, in the opposite direction. Table 3 presents a negative and significant relationship between AHU segregation and neighborhood income segregation of the affluent after 2000. In MSAs with declining AHU concentration, families in the top decile of the income distribution increasingly segregated themselves from all others. As AHUs entered into more neighborhoods, affluent residents responded by isolating themselves into high-income neighborhoods. There is no relationship between AHU concentration and income segregation at the top of the income distribution from the bottom (H at the 50th percentile).

AHUs make up a minority of housing units in all metropolitan areas—no MSA has an AHU rate above 15 percent—and AHUs compose a very small amount—less than 3 percent—of housing units in half of all metropolitan areas. That AHUs are only a small proportion of all households may limit their impact—both direct and indirect—on income segregation. Therefore, I examined metropolitan areas by AHU rate, comparing those with AHU rates above and below the median of 3.4 percent in 2008, to see whether the relationship was more robust in MSAs with more AHUs in proportion to total housing units. Table 4 presents the results of the same models estimated in Table 3 but limited to MSAs with AHU rates above the median.

Row 1 of Table 4 shows that, when examining only the 165 MSAs with high AHU rates (above the median), there is a borderline significant and positive association between AHU segregation and income segregation across the income

TABLE 4

Longitudinal Regression Analyses Predicting Metropolitan Area Income Segregation from Assisted Housing Segregation among MSAs with High AHU Rates

	1980 to 2000	2000 to 2005–9	Pooled
Average income segregation (H)			
AHU *H*	−0.003	0.004	0.011^
	(0.008)	(0.013)	(0.006)
Income segregation (H), 10th percentile			
AHU *H*	0.002	0.054°	0.023°
	(0.011)	(0.027)	(0.010)
Income segregation (H), 50th percentile			
AHU *H*	−0.006	0.002	0.008
	(0.009)	(0.014)	(0.007)
Income segregation (H), 90th percentile			
AHU *H*	0.001	−0.041^	0.008
	(0.011)	(0.022)	(0.010)
MSA × time controls	Y	Y	Y
MSA FE	Y	Y	Y
Time FE	Y	Y	Y
N Obs (MSAs × Year)	305	330	470
N MSAs	165	165	165

NOTE: All models include MSA × time control variables shown in Table 1 as well as MSA and time fixed effects. In 1977, some MSAs had no AHUs and are missing observations at that time point. All significance tests are two-tailed.
°$p \leq .05$. ^$p \leq .10$.

distribution. This suggests that for the deconcentration of AHUs to have an impact on overall income segregation, there must be a larger presence of AHUs—there were no significant relationships between AHU segregation and income segregation among MSAs with AHU rates below the median. Like Table 3, Row 2 shows the effect of AHU concentration is most robust for segregation of the poor and significant only after 2000. The coefficients are about 50–60 percent larger among high-AHU MSAs compared with all MSAs. Segregation of the affluent is again negative but only borderline significant after 2000 (perhaps due to small sample size). A higher presence of AHUs may have more robust effects on income segregation both because the location of more low-income, assisted renters is affected (direct effects) and because nonassisted residents may move in response to assisted units (indirect effects). If there are proportionally more AHUs in an MSA, their presence may be even more salient and visible to nonassisted residents.

In both analyses, the relationship between AHU concentration and income segregation is driven by results after 2000 and is not significant from 1980 to 2000. Several factors may account for the stronger relationship after 2000.

Policies aimed at deconcentrating assisted housing into lower-poverty neighborhoods (not just more neighborhoods) may have had most of their effect, or reached a tipping point at which an impact on income segregation is observed, after 2000. First, the QHWRA of 1998 instituted explicit provisions for income mix in HUD programs and the focus on dispersal became stronger (Goetz 2003). Second, redevelopment of public housing takes time, and many HOPE VI–funded projects did not begin housing residents until after 2000. Demolition of public housing also continued into the 2000s. Third, the LITHC program nearly doubled in size from 2000 to 2008. More LIHTC units are located in low-poverty neighborhoods than other programs, and LIHTC developments house residents with a greater income mix, so the increase in the number of LIHTC units during this time may have tempered income segregation. In addition, the post-2000 period includes the onset of the Great Recession, and deconcentrated AHUs may have helped poor residents to stay in their homes and neighborhoods compared with others without rental subsidies. Housing assistance may have been critically important in creating or maintaining income segregation during this precarious time.

Altogether, the results tell a mixed story. On one hand, income segregation among very low-income families is lower in metropolitan areas where assisted housing has been spread more evenly across neighborhoods. On the other hand, AHU deconcentration has a negative association with the segregation of very high-income residents and does not affect segregation between those with incomes in the top and those in the bottom halves of the income distribution. This suggests that very low-income residents are not being integrated with middle- or upper-income families but likely with other poor or near-poor families. Overall, AHU deconcentration has not substantially reduced income segregation across the income distribution among neighborhoods in metropolitan areas, having only small integrating effects for the very poorest residents and having small segregating effects for the most affluent residents after 2000, and preventing low-income families from accessing very high-income neighborhoods.

Discussion

Over the past several decades, income segregation between neighborhoods has increased, and the number of mixed income neighborhoods has declined. While HUD's main responsibility is to provide decent, safe, and affordable housing for low-income households, assisted housing policy was seen as an opportunity for policy-makers to combat income segregation, particularly the concentration of poverty, and to facilitate income mix and economic integration of low-income families. HUD policies and programs address this goal through a patchwork of programs that focuses on the dispersal of assisted renters to many more neighborhoods than traditional public housing allows for. These programs have succeeded in geographically deconcentrating assisted housing units, and a greater number of assisted units exist today than in the mid-1970s. That said, assisted housing still accounts for the location of only about 3 percent of households, tempering expectations for a large impact on income integration. However, assisted housing is the main policy promoting the mobility of lower-income families to higher-income neighborhoods, so assessing whether it has made an impact on income segregation is critical.

This article examined whether the geographic deconcentration of assisted housing units has been enough to foster economic integration among neighborhoods in the metropolitan United States. The results indicate that the declining segregation of assisted housing fostered a small amount of economic integration among neighborhoods, mainly for very low-income families and in MSAs where there is a greater prevalence of assisted housing. I also found indirect effects: high-income families in metropolitan areas with lower concentrations of assisted housing have become even more segregated, potentially offsetting the economic integration gains and ensuring that very low-income families are living in neighborhoods with only slightly higher-income neighbors.

Overall, then, assisted housing has made some progress in slowing the growth in income segregation among neighborhoods in an era when income segregation has increased. How can the effect be greater? Given the constellation of housing programs in place today, the location of assisted housing largely depends on the decisions of private market landlords and developers. On one hand, the private market has not prevented assisted housing from becoming geographically deconcentrated over the past 40 years, with at least one unit of assisted housing now located in nearly every neighborhood. On the other hand, assisted housing is still not predominantly located in higher-income neighborhoods, which may be necessary to achieve more income integration. Changes to the voucher program, such as landlord outreach, higher voucher values in high-cost areas, property tax rebates for landlords who accept vouchers, and more information about high-income areas for voucher users, may facilitate voucher holders' residence in higher-income areas (Khadduri and Wilkins 2008). When siting project-based assisted housing, local policy-makers should work closely with developers to target stably mixed or higher-income neighborhoods. However, investment in very low-income neighborhoods is also important, and LIHTC developments, which tend to house higher-income residents than other types of assisted housing, may reduce poverty rates in the poorest neighborhoods (Ellen, O'Regan, and Voicu 2009). To avoid reactive moves by high-income residents when assisted housing is built in their neighborhood, more research must be done on the architectural features, development size, income mix, and place-based investments that can retain high-income residents in mixed-income neighborhoods.

The magnitude of the effect of assisted housing policy on income integration may be limited by the fact that assisted housing serves only a minority of residents. My results show that assisted housing has larger effects on income segregation when more residents live in assisted housing, but it is unlikely that assisted housing will ever be expanded to a majority of households, given the political and economic climate in the United States. Therefore, other national and local policies in the urban, antipoverty, and education arenas must also be aimed at promoting income integration. Rent control, zoning policy, and property taxes are all tools that can be adjusted to limit income segregation and maintain mixed-income neighborhoods. School quality also shapes where families choose to live, and economic integration may be difficult to achieve if school resources and attendance are linked to real estate prices. The poverty deconcentration and mixed-income paradigms have been part of assisted housing policy for several decades, but without other economic, urban, and educational policies, assisted housing's effect on income segregation among neighborhoods will be limited.

Notes

1. I use the term *assisted housing* to refer to means-tested federally funded programs supported by both project- and tenant-based subsidies.

2. Nationally, about 1.1 percent of housing units were assisted units in 1977, 3.1 percent in 2000, and 3.3 percent in 2008 (author's calculations based on HUD and census/ACS data).

3. The Neighborhood Change Database provides 1980 data on family income in 2000 tract boundaries (GeoLytics 2003). I followed past research and use family income to estimate income segregation between tracts (Reardon and Bischoff 2011; Watson 2009). I also predicted household income segregation in 2000 and 2005–9 from AHU segregation from 2000 to 2008, and results are substantively identical to those presented here (data on 1980 household income are not available in 2000 boundaries).

4. Other research on income segregation focuses only on the 100 largest metropolitan areas. Results limited to this sample are presented in the online appendix, Table A1. See http://ann.sagepub.com/supplemental.

5. The *Picture of Subsidized Housing* includes 1993 data on public housing but excludes other housing programs that include millions of units (see http://www.huduser.org/portal/datasets/picture/yearlydata .html). Therefore, I cannot estimate a relationship between assisted housing and income segregation in the 1990s.

6. I replicated analyses on the 100 largest MSAs for which missing data have been reduced to less than 10 percent (see Table A1 in online appendix).

7. Data included mortgage subsidy programs (Section 236 and 221(d)3) and housing for disabled (Section 202) and elderly residents (Section 811). Data excluded the U.S. Department of Agriculture's rural housing program (Section 515) as well as Indian Housing, and the HOME and Community Development Block Grant programs.

8. I found a higher average dissimilarity index in 2005–9 than 2000, in contrast to research showing declining racial segregation in the 2000s (Logan and Stults 2011). This may be because Logan and Stults use census rather than ACS data, document trends through 2010 rather than 2005–9, use 2010 tract and metropolitan area boundaries rather than 2000 or 1999, count multiracial black individuals as blacks, and present averages weighted by group members rather than unweighted averages.

References

Bischoff, Kendra, and Sean F. Reardon. 2014. Residential segregation by income, 1970–2009. In *Diversity and disparities*, ed. John R. Logan, 208–33. New York, NY: Russell Sage Foundation.

Booza, Jason C., Jackie Cutsinger, and George Galster. 2006. *Where did they go? The decline of middle-income neighborhoods in metropolitan America.* Washington, DC: The Brookings Institution. Available from www.brookings.edu.

Devine, Deborah J., Robert W. Gray, Lester Rubin, and Lydia B. Taghavi. 2003. *Housing choice voucher location patterns: Implications for participants and neighborhood welfare.* Washington, DC: U.S. Department of Housing and Urban Development. Available from www.huduser.org.

Ellen, Ingrid Gould, Katherine M. O'Regan, and Ioan Voicu. 2009. Siting, spillovers, and segregation: A reexamination of the low income housing tax credit program. In *Housing markets and the economy: Risk, regulation, and policy*, eds. Edward L. Glaeser and John M Quigley, 233–67. Cambridge, MA: Lincoln Institute of Land Policy.

Freeman, Lance. 2003. The impact of assisted housing developments on concentrated poverty. *Housing Policy Debate* 14 (1–2): 103–41.

Galster, George. 2013. U.S. assisted housing programs and poverty deconcentration: A critical geographic review. In *Neighbourhood effects or neighbourhood based problems? A policy context*, eds. David Manley, Maarten van Ham, Nick Bailey, Ludi Simpson, and Duncan Maclennan, 215–49. Dordrecht: Springer Netherlands.

GeoLytics. 2003. CensusCD neighborhood change database. East Brunswick, NJ: GeoLytics, Inc.

Goetz, Edward G. 2003. *Clearing the way: Deconcentrating the poor in urban America*. Washington, DC: The Urban Institute Press.

Hirsch, Arnold R. 1983. *Making the second ghetto: Race and housing in Chicago, 1940–1960*. Chicago, IL: University of Chicago Press.

Holin, Mary Joel, Larry Buron, Gretchen Locke, and Alvaro Cortes. 2003. *Interim assessment of the HOPE VI program cross-site report*. Washington, DC: U.S. Department of Housing and Urban Development. Available from www.huduser.org.

Jargowsky, Paul A. 1996. Take the money and run: Economic segregation in U.S. metropolitan areas. *American Sociological Review* 61 (6): 984–98.

Khadduri, Jill. 2001. Deconcentration: What do we mean? What do we want? *Cityscape* 5 (2): 69–84.

Khadduri, Jill, and Charles Wilkins. 2008. Designing subsidized rental housing programs: What have we learned. In *Revisiting rental housing: Policies, programs, and priorities*, eds. Nicholas P. Retsinas and Edward S. Belsky, 161–91. Washington, DC: Brookings Institution Press.

Kingsley, G. Thomas, Jennifer Johnson, and Kathryn L. S. Pettit. 2003. Patterns of Section 8 relocation in the HOPE VI program. *Journal of Urban Affairs* 25 (4): 427–47.

Kucheva, Yana Andreeva. 2011. Subsidized housing and the concentration of poverty: 1977–2008: A comparison between eight U.S. cities. CCPR Online Working Paper Series PWP-CCCR-2011-008. Los Angeles, CA: California Center for Population Research.

Logan, John R., and Brian J. Stults. 2011. *The persistence of segregation in the metropolis: New findings from the 2010 Census*. US2010 Project. New York, NY, and Providence, RI: Russell Sage Foundation and Brown University. Available from www.s4.brown.edu/us2010.

Massey, Douglas S., and Nancy A. Denton. 1988. The dimensions of residential segregation. *Social Forces* 67 (2): 281–315.

Massey, Douglas S., and Nancy A. Denton. 1993. *American apartheid: Segregation and the making of the underclass*. Cambridge, MA: Harvard University Press.

Newman, Sandra J., and Ann B. Schnare. 1997. "... And a suitable living environment": The failure of housing programs to deliver on neighborhood quality. *Housing Policy Debate* 8 (4): 703–41.

Oakley, Deirdre, and Keri Burchfield. 2009. Out of the projects, still in the hood: The spatial constraints on public housing residents' relocation in Chicago. *Journal of Urban Affairs* 31 (5): 589–614.

Owens, Ann. 2012. The new geography of subsidized housing: Implications for urban poverty. PhD diss., Harvard University.

Owens, Ann. Forthcoming. Assisted housing and neighborhood poverty dynamics, 1977 to 2008. *Urban Affairs Review*.

Pendall, Rolf. 2000a. Local land-use regulation and the chain of exclusion. *Journal of the American Planning Association* 66 (2): 125–42.

Pendall, Rolf. 2000b. Why voucher and certificate users live in distressed neighborhoods. *Housing Policy Debate* 11 (4): 881–910.

Quillian, Lincoln. 2005. Public housing and the spatial concentration of poverty. Paper presented at the annual meeting of the Population Association of America, 31 March–2 April 2005. Philadelphia, PA.

Reardon, Sean F. 2009. Measures of ordinal segregation. *Research on Economic Inequality* 17:129–55.

Reardon, Sean F., and Kendra Bischoff. 2011. Income inequality and income segregation. *American Journal of Sociology* 116 (4): 1092–153.

Reardon, Sean F., and Glenn Firebaugh. 2002. Measures of multigroup segregation. *Sociological Methodology* 32 (1): 33–67.

Rothwell, Jonathan T, and Douglas S. Massey. 2010. Density zoning and class segregation in U.S. metro-
        politan areas. *Social Science Quarterly* 91 (5): 1123–43.
Schwartz, Alex F. 2010. *Housing policy in the United States.* 2nd ed. New York, NY: Routledge.
Watson, Tara. 2009. Inequality and the measurement of residential segregation by income in American
        neighborhoods. *Review of Income and Wealth* 3:820–44.
Wilson, William Julius. 1987. *The truly disadvantaged: The inner city, the underclass, and public policy.*
        Chicago, IL: University of Chicago Press.

Housing Unit Turnover and the Socioeconomic Mix of Low-Income Neighborhoods

BRETT THEODOS,
CLAUDIA J. COULTON,
and
ROB PITINGOLO

A number of place-based policies attempt to deconcentrate poverty, yet not enough is known about how the socioeconomic mix of low-income neighborhoods evolves nor the role of residential mobility in this evolution. This study focuses on changes in low-income neighborhoods as they transpire at the micro level of housing unit turnover. Using a unique panel survey of low-income neighborhoods, the study examines how characteristics of previous occupants, housing units, and neighborhoods affect the chances that units will transition into or out of poverty. Results show that although turnover rates are high, the poverty status of occupants changes infrequently. Occupant, unit, and neighborhood factors help to explain the changes that do occur. Improving low-income neighborhoods is challenging because the poverty status of occupants tends not to change, but there are aspects of the built and social environment that can affect occupancy transitions in ways that reduce poverty concentration.

Keywords: housing unit turnover; residential mobility; neighborhood stability; neighborhood change; concentrated poverty

The negative effects of geographically concentrated poverty on people and places are well documented. Living in high poverty neighborhoods is associated with educational and health deficits and diminished life chances for residents, especially for children and youth (Ellen and Turner 1997; Leventhal and Brooks-Gunn 2000; Sampson, Sharkey, and Raudenbush 2008). Metropolitan areas with highly concentrated poverty fare poorly as well. Economically divided cities experience a drag on their

Brett Theodos is a senior research associate at the Urban Institute.

Claudia J. Coulton is the Lillian F. Harris Professor of Urban Research and Social Change in the Mandel School of Applied Social Sciences at Case Western Reserve University.

Rob Pitingolo is a research associate at the Urban Institute.

DOI: 10.1177/0002716215576112

economic growth and bear significant costs of crime, blight, and other problems that emerge in high poverty areas (Downs 1994; Orfield 2002).

Over the past several decades a number of policies and programs have tackled the challenging conditions in high poverty neighborhoods. Collectively referred to as place-based initiatives, these efforts work to improve the built environment and quality of local services, create opportunities for residents, and provide amenities that can attract more affluent residents. Although these efforts have made strides on various fronts, they have had little discernible effect on concentrated poverty (Cytron 2010; Turner et al. 2014).

Underappreciated in the implementation of these place-based policies, though, are processes of residential mobility and housing unit turnover. Residential mobility is often dismissed as a nuisance because it is easier to design programs for stable populations, and community engagement relies on an experienced cadre of resident leaders and established social networks. Yet residential mobility is an element of households' adaptation to changing needs and aspirations and can be a driving force of neighborhood vitality and dynamic housing markets. Practically speaking, residential mobility can contribute to the reduction of concentrated poverty if newcomers are of higher income than the previous residents. Given that the effect of residential mobility on place-based initiatives is seldom simple, further information is needed about how the processes of housing units being vacated and reoccupied contribute to the socioeconomic profile of the population and to the problem of poverty concentration.

With these purposes in mind, this study takes an in-depth look at housing unit turnover in some seventy neighborhoods in ten cities that were part of one place-based initiative, the Annie E. Casey Foundation's Making Connections (MC) program. The study draws on a unique panel of housing units and their occupants that was representative of these neighborhoods over a seven-year period. The article examines housing unit turnover to pinpoint occupant, housing unit, and neighborhood characteristics that are predictive of whether the poverty status of out-mover and newcomer households differs. A better understanding of these microlevel processes of socioeconomic transitions at the housing-unit level is an essential backdrop for planning and evaluating place-based policies and programs.

Background

The policy context

Concentrated poverty in U.S. metropolitan areas has increased over the past four decades, with a precipitous rise in recent years (Bishaw 2014; Jargowsky

NOTE: Funding for this analysis was provided by the Annie E. Casey Foundation. The authors are especially grateful for the support of the foundation's associate director for policy research, Cindy Guy. Also, this article would not have been possible without the partnership of the Making Connections site teams and NORC's survey collection efforts. We also thank Doug Wissoker for analytic advice.

2013). In response, government agencies and philanthropic organizations have launched a variety of place-based programs that invest in the physical, economic, and institutional infrastructure of these poor neighborhoods or the needs of the poor people in their midst (O'Connor 1999). Examples include federal neighborhood revitalization programs such as Empowerment and Enterprise Zones and HOPE VI public housing transformation and foundation directed comprehensive community initiatives. More recently, as exemplified by the federal Promise and Choice Neighborhood programs, place-based initiatives have broadened their vision to incorporate health and human development goals alongside traditional investments in housing and neighborhood development.

Although recognition of the harmful effects of concentrated poverty is their raison d'être, place-based policies have seldom been explicit about the role of residential mobility in shaping neighborhood poverty trends (Turner et al. 2014). However, improvements in housing and services will have limited impact on concentrated poverty unless higher income households can be attracted as a result. Moreover, neighborhood poverty rates will not fall if residents who benefit from the initiative use their newfound resources to leave the neighborhood.

Housing unit transitions and neighborhood change

Much of the research on change in neighborhood poverty has relied on stock data from the decennial census (Gramlich, Laren, and Sealand 1992; Kingsley and Pettit 2007; Ellen and O'Regan 2008), but these snapshots do not show precisely how residential turnover shapes the socioeconomic mix. A few studies have attempted to pinpoint the contribution of residential mobility to socioeconomic change in neighborhoods. High rates of residential mobility in poor areas often do not register much of an impact on poverty rates because residents who move up and out are replaced by equally poor in-movers, while distressed households simply relocate nearby (Coulton, Theodos, and Turner 2012). However, selective residential mobility—in which newcomers differ from out-movers—along with incumbent processes whereby stayers change their economic status, can produce a gradual shift in a neighborhood's socioeconomic mix (Coulton, Theodos, and Turner 2012; Teernstra 2014). For example, neighborhoods that upgraded economically in the decade of the 1990s were a combined result of richer households moving in, the disproportionate exit of poorer households, and income increases for low-income households that stayed in place (Ellen and O'Regan 2008). Once begun, the direction of neighborhood change can be reinforcing over time as individual and institutional investment decisions affect social and political processes (Temkin and Rohe 1996).

It is the transition of housing units from poor to nonpoor occupants that most directly contributes to the reduction of concentrated poverty, yet only a few studies have focused directly on housing unit transitions and compared demographic or socioeconomic characteristics of the departing and incoming occupants.[1] Most of the studies of housing unit transitions have focused on racial and ethnic change using panel data from the American Housing Survey. Spain (1980), Marullo (1985), and Ellen (2000) assess the extent to which housing units are likely to transition to new occupants of the same racial group, finding that there is a significant level of racial stickiness in unit transitions. Hipp (2012) extends this analysis further by controlling for amenities and the racial/ethnic composition in both the microneighborhood (residents' ten closest neighbors) and the census tract. He finds that the race and ethnicity of the previous occupants of the housing unit remain strongly related to the race and ethnicity of the new occupants. Ellen (2000) and Hipp (2012) offer several possible explanations for the stability of race and ethnicity in housing units that transition, including the role that social networks play in the housing search and the possibility that prospective residents pick up on the race and ethnicity of previous occupants as a signal of whether the unit is appropriate for them.

Housing unit transitions have also been examined in relation to neighborhoods that are upgrading economically. Ellen, Horn, and O'Regan (2011) undertook a study of in-movers whose household incomes were high relative to the neighborhoods that they moved into. They found that renters, childless couples, first-time homebuyers, and minority households were more likely to make these types of pioneering moves (Ellen, Horn, and O'Regan 2011). At the same time, the movement of more affluent households into low-income neighborhoods has raised concerns about the dangers of gentrification. Studies of housing unit transitions show little evidence of wholesale displacement of the poor and also suggest some signs of their benefiting from slower residential turnover in upgrading areas (Freeman and Braconi 2004; Vigdor 2002), However, as pressure on property values mounts, a growing number of low-income renters have been found to face prohibitive rent burdens and need to resort to housing assistance programs, which are in short supply (Newman and Wyly 2006). These mixed results suggest that place-based initiatives need to guard against inadvertently contributing to spiraling housing costs or the displacement of poor households in their efforts to reduce concentrated poverty.

Conceptual approach

The current study examines housing unit turnover to understand the micro-level processes underlying change in concentrated neighborhood poverty. We adopt a multilevel conceptual framework to explain whether the poverty status of newcomers differs from the poverty status of the out-movers that they replace. At the occupant level, we hypothesize that the likelihood of a newcomer household being poor is related to characteristics of the previous occupants. Occupant concordance on socioeconomic characteristics may be due to the role that social networks play in the housing search (Krysan 2008), or through signs that prospective occupants pick up about the race/ethnicity of prior occupants to determine whether the unit is right for them (Hipp 2012). We expect that poverty status of housing units' occupants is sticky, with a high frequency of poor households replacing poor households that leave. Concordance on race and ethnicity is also expected to be high, and given economic inequality between groups, it will be an additional contributor to the stickiness of poverty. Elderly occupants, possibly due to a lack of updating in their units (Aurand and Reynolds 2013), are also expected to have an increased chance of being replaced by lower income in-movers. Housing units that are owner occupied are expected to have a lower chance of transitioning to poor occupants, possibly due to the greater invest-ments in upkeep (Galster 1987). Finally, if the previous occupant is a Housing Choice Voucher holder, the likelihood that the newcomer household will be poor is raised, possibly because the landlord has already participated in the voucher program. Conversely, landlords who initially refused to accept vouchers as a form of payment are likely to continue to do so.

Housing unit characteristics are also hypothesized to influence the chances that units will upgrade economically. Single-family homes are expected to be more likely than those located in multifamily buildings to transition from or remain out of poverty, in part because they are generally considered more desir-able than attached housing (Clark, Deurloo, and Dieleman 1984; Pendall, Theodos, and Hildner, forthcoming). Housing units showing signs of external disrepair or damage are more likely to be reoccupied by poor households. Controlling for other factors, larger units in low-income neighborhoods are likely to attract large families, and family size is a risk factor for poverty. Moreover, large units are unsuitable for single individuals and childless couples, a group that research suggests is frequently in the vanguard of gentrification (Ellen, Horn, and O'Regan 2011). Finally, units with relatively low estimated values are more likely to be affordable for poor occupants.

The neighborhood context is also expected to affect the likelihood that hous-ing units will transition out of poverty or remain nonpoor. Neighborhoods with high concentrations of African American and Latino residents have been found to have fewer signs of gentrification than mixed race/ethnicity neighborhoods (Hwang and Sampson 2014). Neighborhood homeownership levels are inversely linked with rising poverty levels (Pendall, Theodos, and Hildner, forthcoming). A reputation that a neighborhood is headed in a positive direction may increase the chances of attracting newcomers with incomes above the poverty line (Feijten and van Ham 2009). Finally, neighborhoods with lower poverty levels are expected to be more attractive to nonpoor residents (see Bader and Krysan, this volume, for further discussion of the importance of a community's reputation).

Methods

Study sites and sample

The housing units in this study come from three waves of a household survey that was carried out as part of the MC initiative, a program of the Annie E. Casey Foundation. Through MC, the Annie E. Casey Foundation sought to better understand the neighborhoods that disadvantaged children inhabit. To do so, the foundation made an unparalleled investment in data collection in these commu-nities. This research takes advantage of those data.

Data collection for the MC initiative took place in selected target areas of ten cities: Denver, Des Moines, Hartford, Indianapolis, Louisville, Milwaukee,

Oakland, Providence, San Antonio, and Seattle/White Center. The MC target areas were chosen through a deliberative process involving the foundation and local representatives and included distressed neighborhoods with concentrations of poverty and disproportionately representative of immigrants or racial and ethnic minority groups. Although each target area consisted of numerous census tracts, we employed a methodology described below to craft resident-defined neighborhood units.

Data sources

Data for this analysis come from three waves of a household survey conducted in the MC target areas in the ten cities.[2] The interviewers returned to the same housing units at each wave and attempted to interview the household that was living there at the time. They established in each instance whether the household was the same one as interviewed previously. Wave 1 was completed in 2002–3, wave 2 in 2005–6, and wave 3 in 2008–9.

All data were collected jointly by NORC and the Urban Institute. In designing and selecting the samples, NORC performed list-assisted probability sampling of households using as a basis the United States Postal Service master list of delivery addresses. The sample design was directed at obtaining a representative sample of households and housing units in each target area. An adult respondent provided information about the housing unit, household, and neighborhood. Surveyors asked about household composition, income, neighborhood dynamics, and other topics. Respondents were also asked to draw maps of their neighborhoods and to report the names that were used for their neighborhoods.

Housing unit turnover

To best take advantage of the available data, we created a housing unit period dataset. Period 1 (P1) is the unit at the beginning of the wave and period 2 (P2) is the unit at the following wave. Excepting public housing, we included all housing units in the study that were in the study for at least two waves. There are 2,302 housing units in the dataset.

The principal dynamic of interest in this article is whether the housing unit turned over between waves. A housing unit was defined as turning over if the respondent and all other members of the household at the previous wave were not present in the next wave. Otherwise, the unit was defined as a no-turnover unit. We use the terms turnover and residential mobility interchangeably, but we do acknowledge that turnover of some units may instead involve mortality. Forty-three percent of the units turned over between periods according to this definition.

Neighborhood context

The neighborhood units were carefully crafted to overcome a major limitation of most research on neighborhood change; that is, the use of census geography as proxies for neighborhoods without confirmation that the units are consistent with local perceptions. Instead, we drew on previous analysis of resident perceived neighborhoods in MC to craft neighborhood units (Coulton, Chan, and Mikelbank 2011). That study used resident-drawn neighborhood maps and neighborhood names to identify resident-informed neighborhood units in each of the MC sites. Beginning with these maps, we identified the portions of census block groups and census tracts that intersected with the resident-informed neighborhoods. There was not perfect contiguity with census geography, and we therefore apportioned census geography into neighborhoods where necessary. Although the resident neighborhoods defined in the previous study were somewhat overlapping, for this research we constrained all blocks to fall into only one neighborhood unit. This process resulted in the identification of seventy "resident-informed" neighborhoods for analysis in this article.

Variables and measures

The dependent variable of interest is poverty status of the P2 occupant, the newcomer. Poverty is defined by comparing reported household income against the federal poverty level as defined by the U.S. Department of Health and Human Services. Poverty status of the P1 occupant, the out-mover, is used as an independent predictor in the model and is calculated in the same manner.

Household-level variables are measured for the P1 occupant. They include poverty status (1 = yes, 0 = no), housing tenure (dichotomous; 1 = owner, 0 = renter), race and ethnicity of the householder (three dichotomous variables included: non-Hispanic black ["black"], Hispanic, and other; 1 = yes, 0 = no; non-Hispanic white ["white"] omitted), and age of householder (two dichotomous variables included: under 30 years old, 65 years and older; 1 = yes, 0 = no; 30–64 years old omitted).

Housing unit characteristics included structure type (dichotomous: 1 = single family, 0 = multifamily). The size of the occupant's unit was not directly collected in the field. Instead, we used a proxy variable based on the number of people living in the unit and calculated an average for that unit across all waves. Housing unit damage comes from interviewer ratings during the wave 1 survey.[3] The unit is marked as damaged (dichotomous; 1 = yes, 0 = no) if it had any of the following: a crumbling or cracked foundation, a hole in the roof, material missing from the roof, a sagging roof, sloping walls, bricks missing from an outside wall, materials missing from an outside wall, broken or loose steps in the inside or outside hallway, or a broken window. Although independently verified market values were not available, the relative value of the housing unit is calculated as a ratio of the occupants' total rent over the average rent in the neighborhood for all renters[4] and self-reported housing value over the average self-reported housing value in the neighborhood for all owners.

Neighborhood characteristics included calculations based on variables collected from the 2000 U.S. Census: dichotomous measures of whether a neighborhood had poverty rates above 40 percent (1 = yes, 0 = no), was majority black (1 = yes, 0 = no), was majority Hispanic (1 = yes, 0 = no), and whether the housing stock was majority single family (1 = yes, 0 = no). We also included a measure of the share of housing that was owner occupied. Additionally, we relied on one neighborhood variable from MC survey data, neighborhood expectations, which has three values: –1 (neighborhood getting worse), 0 (neighborhood staying the same), and 1 (neighborhood getting better), and is calculated as the average values from all respondents in each neighborhood in P1.[5]

Analytic approach

We constructed a multilevel random coefficient logistic model, where the dependent variable of interest is whether the P2 occupant was poor (1 = yes/0 = no). Independent variables of interest were drawn from three levels of observation: households occupying the units, the housing unit, and the neighborhood.

The types of phenomena we are interested in understanding are multilayered and complex. The benefit of this approach is that it allows for an analysis that teases apart the independent influence of individual and neighborhood factors in predicting turnover. Random coefficient models allow both the intercept and slope of the parameters of interest to vary across higher levels. This approach also takes into account clustering and the resulting violation of the independence assumption. Without this correction, clustering by housing unit and neighborhood would result in biased parameter estimates.

We run the models in Stata with the *xtmelogit* command using adaptive quadrature with twelve quadrature points. Below is a reduced form equation for all three levels,

$$
\log\left[\frac{p_{ijk}}{1 - p_{ijk}}\right] = \beta_0 + \beta_1 x_{1,ijk} + \ldots + \beta_{21} x_{21,k} + \zeta_{jk} + \zeta_k
$$

where i, j, and k respectively index levels 1, 2, and 3, and where ζ_{jk} is a random intercept varying over housing units and ζ_k is a random intercept varying over neighborhoods.

Missing data were generated in the MC sample for questions where the respondent did not know the answer or elected not to respond. To make use of cases with data missing from one or more covariates, we used the *mi impute* procedure in Stata, creating five estimates for each missing value and incorporating these into the regression models. Continuous variables were limited to their theoretical minimum and maximum values, while dichotomous variables were assigned to either 0 or 1.

Results

Prevalence and direction of turnover

Housing unit turnover was defined as the complete departure of the P1 household, replaced by new residents at P2. Over the roughly three years between

TABLE 1
Continuity and Change in Poverty Status for Turned-Over Units

Poverty status of period 1 and period 2 occupants	Share of turned over units with P1 status	Share of all turned over housing units	Number of units
P1 = poor			
Poor → poor	63%	28%	639
Poor → nonpoor	37%	16%	381
Total	100%	44%	1,020
P1 = nonpoor			
Nonpoor → poor	39%	22%	497
Nonpoor → nonpoor	61%	34%	785
Total	100%	56%	1,282
Total		100%	2,302

NOTE: Turned-over units exclude apartments in public housing.

periods, 43 percent of units turned over according to this definition, roughly the rate that is anticipated by extrapolating one-year mobility rates for households in this income range. There is evidence that mobility rates varied by local context, however. Looking across the seventy study neighborhoods, 14 percent (ten neighborhoods) saw fewer than 30 percent of their units turn over, and another 29 percent (twenty neighborhoods) saw between 30 and 40 percent turn over. Some neighborhoods were highly unstable: 32 percent (twenty-two neighborhoods) experienced half of their residents moving between study periods.

Given the importance of residential mobility in generating shifts in neighborhood poverty rates, it is important to examine the continuity and change in poverty status for housing units that turned over. Of the 44 percent of units occupied by poor households in P1, 63 percent remained occupied by poor households by P2 (Table 1, column 2). A similar proportion of nonpoor units at P1 were occupied by nonpoor households by P2.

Poverty persistence and transition will be explored in greater detail through the multivariate model below, but the descriptive statistics provide evidence that poverty is likely to be an enduring aspect of a housing unit's characteristic. Whether the drivers relate to the previous occupants, unit characteristics, or neighborhoods, it is clear that poor and nonpoor units are likely to remain in their respective status even with turnover. It is, however, important to acknowledge that a unit's previous poverty status is not the sole determinate of its trajectory. Further, poverty status is more apt to change than is race/ethnicity of the occupants.

Descriptive statistics on study variables

Focusing solely on the turned-over units, we see in Table 2 that a considerable share of residents at P1 was poor: 44 percent. This poverty rate, more than three

TABLE 2
Description and Summary Statistics of Study Variables

Variable	Description	Mean (SD)	Range
Level 1: Household level			
Poverty status of out-mover	1 = poor, 0 = nonpoor	.44	0/1
Black householder	1 = yes, 0 = no	.33	0/1
Hispanic householder	1 = yes, 0 = no	.28	0/1
Other race householder	1 = yes, 0 = no	.11	0/1
White householder (omitted)	1 = yes, 0 = no	.29	0/1
Householder under 30 years	1 = yes, 0 = no	.34	0/1
Householder 30–64 years (omitted)	1 = yes, 0 = no	.59	0/1
Householder 65 years or older	1 = yes, 0 = no	.07	0/1
Housing Choice Voucher use	1 = voucher user, 0 = market rate	.13	0/1
Owned home	1 = owner occupied, 0 = renter occupied	.20	0/1
Level 2: Housing unit level			
Home value	Unit value/Making Connections site value	1.15 (0.30)	0.49–1.96
Housing unit size	Housing unit size, proxy = number of people living in household averaged across waves	3.04 (1.5)	1–10
Single-family home	1 = single family, 0 = multi-family	.59	0/1
Damage to housing unit	1 = yes, 0 = no	.20	0/1
Level 3: Neighborhood level			
Poverty rate > 40%	1 = yes, 0 = no	.24	0/1
Majority black	1 = yes, 0 = no	.21	0/1
Majority Hispanic	1 = yes, 0 = no	.35	0/1
% Owner occupancy	Continuous, min = 0, max = 1	.41 (0.17)	.03–.81
Majority single-family homes	1 = yes, 0 = no	.59	0/1
Neighborhood expectations	Neighborhood getting better (= 1), staying the same (= 0) or getting worse (= −1)	.41 (.17)	(.22)–.84

NOTE: There were 2,302 level 1 observations; 1,947 level 2 observations; and 70 level 3 observations.

times the national average, conveys the serious and concentrated economic disadvantage for these residents. (For context, it is worth noting that poverty rates were somewhat lower among those who stayed in place, which is consistent with other research finding higher mobility rates among lower- than upper-income households [Ihrke and Faber 2012].)

While this study population's poverty rate is well above the national average, owner occupancy, at 20 percent, is well below the national average. A total of 13 percent of out-movers rented their unit with the support of a housing voucher. Occupants in turned-over units were racially mixed: one-third were black, 28 percent were Hispanic, 28 percent were white, and 11 percent of another or multiple races. Elderly householders represented a small share of turned-over units, while younger householders, those under 30, were well represented, consistent with previous research showing that residential mobility rates are higher among younger than older adults (Ihrke and Faber 2012).

Turning to housing unit variables, 59 percent of residents in units that turned over lived in single-family homes, with the remainder in multifamily buildings. The average number of people living in a housing unit over all available waves (a proxy for unit size) was three, ranging from one to ten. Survey administrators rated 20 percent of units as damaged, a large share relative to national levels (Eggers and Moumen 2013). The average MC housing unit was valued at 115 percent of its surrounding area and ranged from 49 to 196 percent.

At the neighborhood level, poverty rates (i.e., not just of the turned-over units) for the average out-mover were 31 percent, with considerable range from just 8 percent to as high as 72 percent. Owner occupancy rates for the average out-mover were 41 percent and also ranged widely from 3 percent to 81 percent. Single-family homes constitute, on average, 40 percent of all units in the neighborhoods, ranging from neighborhoods that are almost entirely single family to neighborhoods that are almost entirely multifamily. The average out-mover's neighborhood was 51 percent black and 29 percent Hispanic, though, again, there was a considerable spread across neighborhoods. Neighborhood expectations ranged from –0.22 to 0.84, averaging 0.41.

Models of housing unit turnover

We examine housing unit turnover incorporating three levels of explanatory factors, those pertaining to occupants, housing units, and neighborhoods. Table 3 presents results from this analysis, beginning with a model displaying just the level 1 factors, then a second model incorporating the level 1 and 2 factors, and finally, a model including all three levels. The authors conducted tests of significance and established that the addition of each level provides additional explanatory power. As such, we concentrate our discussion on the third model, which includes all three levels. The results are presented as odd ratios. Holding constant other variables, an odds ratio of greater than 1.0 signals a positive effect on P2 poverty, while an odds ratio of less than 1.0 indicates a negative effect on P2 poverty.

In all three models we find that the poverty status of the out-mover is strongly associated with the poverty status of the newcomer. Looking at the third model, we see that units occupied by poor households at P1 have 53 percent higher odds of being occupied by a poor household at P2 than units occupied by nonpoor households at P1, controlling for other factors. This finding demonstrates the

TABLE 3
Multilevel Logistic Model Predicting P2 Poverty Status

Variables	Level 1			Level 1 + 2			Level 1 + 2 + 3		
	Odds Ratio	SE	P > |t|	Odds Ratio	SE	P > |t|	Odds Ratio	SE	P > |t|
Level 1: Occupant (P1) Level									
Poverty status of out-mover	1.90	0.17	0.00	1.65	0.15	0.00	1.53	0.14	0.00
Black householder	1.73	0.18	0.00	1.61	0.17	0.00	1.29	0.16	0.04
Hispanic householder	1.92	0.21	0.00	1.55	0.18	0.00	1.60	0.20	0.00
Other race householder	1.43	0.22	0.02	1.16	0.18	0.35	1.21	0.19	0.23
Householder under 30 years	0.95	0.09	0.54	0.93	0.09	0.52	0.95	0.09	0.57
Householder 65 years or older	1.04	0.19	0.83	1.40	0.27	0.10	1.44	0.28	0.06
Housing Choice Voucher use	1.65	0.23	0.00	1.81	0.26	0.00	1.74	0.28	0.00
Owned home	0.62	0.07	0.00	0.57	0.07	0.00	0.58	0.07	0.00
Level 2: Housing Unit Level									
Home value				0.35	0.07	0.00	0.37	0.07	0.00
Housing unit size				1.40	0.06	0.00	1.38	0.05	0.00
Single-family home				0.93	0.09	0.52	0.82	0.09	0.06
Damage to the housing unit				1.19	0.12	0.17	1.18	0.12	0.10
Level 3: Neighborhood Level									
Poverty rate > 40%							1.49	0.29	0.04
Majority black							1.37	0.27	0.10
Majority Hispanic							1.07	0.18	0.70
% Owner occupancy							0.48	0.39	0.25
Majority single-family homes							1.39	0.34	0.16
Neighborhood expectations							0.09	0.11	0.00
Constant	0.48	0.05	0.00	0.64	0.14	0.02	3.84	3.14	0.04

NOTE: Omitted reference category for level 1 black and Hispanic householder is white house-holder. Omitted reference category for level 1 householder under 30 years and householder 65 years or older is householder age 30 to 64.

relative stickiness of household poverty, apart from housing unit and neighborhood factors.

Other P1 occupant characteristics are also important predictors of P2 poverty status—and some as much or more than P1 poverty. Hispanic occupancy is strongly predictive of P2 poverty—these units have 60 percent higher odds of being occupied by a poor household than white-occupied units. Black-occupied units have 29 percent higher odds of being occupied by a poor newcomer than white-occupied units. (As a reminder, we are examining whites living in the low-income neighborhoods targeted for the MC intervention, and these findings may hold differently for middle- and upper-income neighborhoods.)

Consistent with our expectations, older householders had 44 percent higher odds of having their units occupied by a poor household at P2 than middle-age households (significant at the 10 percent, but not 5 percent threshold). There was no difference between younger and middle-age householders. Tenure matters greatly. Units occupied by homeowners at P1 have 42 percent lower odds of being occupied by a poor household at P2 than rented units. And units rented with a housing voucher had 74 percent higher odds of being occupied by a poor household at P2.

Characteristics of the housing unit matter in predicting whether a poor or nonpoor resident occupies the home at P2. Most notably, home value is an important predictor, where homes that are higher in price relative to their surrounding area have higher odds of being occupied by poor newcomers. We expect that home value captures (i.e., capitalizes) much of the physical attributes of a home. But, controlling for other factors, larger housing units showed higher odds of P2 poverty (38 percent for a 1-increment change). Damaged units had modestly higher odds of being occupied by a poor newcomer than those that are not damaged, and single-family homes had modestly lower odds of transitioning into poverty than multifamily units (both significant at the 10 percent threshold).

Turning to the neighborhood-level variables it is evident that neighborhood poverty rate is an important driver of selecting poor newcomers into units. This finding verifies other work establishing the stickiness of neighborhood-level poverty (Pendall, Theodos, and Hildner, forthcoming). Living in a majority black neighborhood was associated with 37 percent higher odds of living in poverty (significant at the 10 percent threshold). There was not a similar effect for majority Hispanic neighborhoods. The share of units that were occupied by owners or units in neighborhoods that were majority single-family units were no more or less likely to be occupied by a poor newcomer. A shared sense of a neighborhood's trajectory is also important in predicting whether units are occupied by poor newcomers, even after controlling for other neighborhood attributes. Units located in neighborhoods where residents perceived that their local context was on an upward trajectory had lower odds of being occupied by a poor newcomer.

Discussion

Summary of findings

To explain newcomer poverty, we created a conceptual framework that relies on attributes of the neighborhood and also the housing unit. The influence of these components is likely intuitive to many readers. But we have also included a range of factors that have escaped notice in most research—the influence of the previous occupant.

We hypothesized that the likelihood of a newcomer household being poor was related to characteristics of the previous occupants and find substantial evidence to support this hypothesis. We find that poor households replace poor households with high frequency. Previous occupant characteristics matter in predicting poverty of the next occupant as well: tenure, voucher status, race/ethnicity, and older-age householders. Housing unit characteristics were demonstrated to matter as well. We examined a fairly robust, though not exhaustive, set of housing unit variables, and found home value and average housing unit population (a proxy for housing unit size) to be meaningful predictors of newcomer poverty status. Building structure (single versus multifamily) and damage appear to matter modestly as well. We examined several neighborhood factors for their influence on newcomer poverty. High poverty neighborhoods, highly black neighborhoods, and neighborhood expectations mattered in predicting P2 poverty.

Study limitations and caveats

The MC dataset provides the rare ability to track changes for households, housing units, and neighborhoods longitudinally, in many ways combining the strengths of the American Community Survey and American Housing Survey. Yet there are limitations to the dataset and study worth noting. The MC data inform the experience of a small sample of low-income neighborhoods in ten midsize and large cities in the United States; they are not generalizable to all communities. For example, the factors predicting which housing units receive poor in-movers in middle- and upper-income communities may be different than the ones identified here.

Additionally, there may be other factors, especially at the housing unit or neighborhood level, that we were not able to capture and include in this analysis. Largely absent were physical attributes of the housing unit (e.g., fireplaces, square feet, acreage, central air conditioning, condition of the appliances, etc.). Theory suggests that the economic value of these attributes should be captured in the rent or value of the home, a measure we include, but perhaps there are other features of a home that matter in predicting poverty that were not adequately quantified here. And neighborhood attributes such as walkability, access to transit, and access to shopping may be important in predicting whether a unit is inhabited by a poor or nonpoor household, but were not available for inclusion in this analysis.

Areas for further study

Three conclusions emerge from this analysis that are worthy of further investigation. First, and most important, we need a better understanding of the mechanisms through which the out-mover households influence, or are linked to, newcomers; in particular, how poverty status is so sticky even after controlling for housing unit and neighborhood attributes. This finding extends previous work, which finds stickiness for race/ethnicity (Ellen 2000; Hipp 2012; Marullo 1985; Spain 1980).

Concordance between out-movers and newcomers on socioeconomic characteristics may be due, in part, to the role that social networks play in a housing search. Social networks have been shown to play a role in housing search and seem to matter more for African American renters than for whites (Krysan 2008).

Another potential explanation for why poverty is sticky is the role of landlords, leasing agents, and realtors. The behavior of these actors has been demonstrated as important in determining which types of potential residents are made aware of, shown, invited to apply for, and accepted to reside in a home (Fischer and Massey 2004). This behavior can sometimes be unlawful discrimination: recent research has shown that discrimination, while diminished somewhat from previous decades, is still alive in the U.S. housing market (Turner et al. 2013). But other types of selective discrimination (e.g., source or amount of income) are not illegal in many jurisdictions.

Finally, Hipp (2012) theorizes that newcomers assess the appropriateness of a place to live based on the characteristics of the previous occupant. Prospective buyers typically interact with sellers and thereby learn information about and from them. While race can be directly observed, poverty status would need to be inferred rather than directly observed. Further, contact with previous occupants for newcomer renters is less likely.

Making sense of these various mechanisms, then, and their individual and overlapping influence, is an important area for further research. The real world is more complicated than we sometimes give it credit for. Observational and qualitative research may prove especially helpful in this regard.

A second conclusion from this research is that the receipt of housing choice vouchers for out-movers is linked with poverty for newcomers. From a policy design standpoint, this represents a challenge. Our findings are important in confirming that something is different (i.e., worse) about the units these residents occupy. But there is little current research on the location decisions of Housing Choice Voucher recipients, and further study is needed. Additionally, it is likely that policy changes are required to better help voucher recipients access high-quality housing.

Third is the role of neighborhoods. While it has proven surprisingly difficult to establish empirically, neighborhood attributes are believed to influence residential mobility decisions (Bayoh, Irwin, and Haab 2006; Nechyba and Strauss 1997; Quigley 1985; Rapaport 1997). But while neighborhood attributes may be important in selecting which neighborhood to live in, are they as important in selecting a specific unit? Concentrated neighborhood poverty and majority black (though not majority Hispanic) measures were meaningful in predicting the poverty status of newcomer households. Other measures, such as the composition of the housing stock and tenure, appeared to matter less. We were also able to incorporate a measure of residents' shared sense of their neighborhoods' trajectory. While this perception is interesting and important, it has been understudied, likely because a similar metric is not available through census data. Building from the work of qualitative researchers (DeLuca, Wood, and Rosenblatt 2011), further research is needed to explore what and how neighborhoods matter to making residential location decisions, especially for the poor, as their choice set is much more constrained compared with higher-income individuals and families.

Implications for policy and practice

There is a growing federal and philanthropic emphasis on place-based efforts to upgrade the social and physical capital of low-income neighborhoods, likely both because of increased awareness that local context can have important implications for individual's well-being, and also because concentrated poverty is again on the rise. (Nearly 30 million people live in high-poverty neighborhoods, and, of these, over nine million lived in tracts with "extreme poverty" rates of 40 percent or greater [Pendall et al. 2011].) While a primary driver of variations in concentrated poverty is shifts in the national poverty rate, mobility patterns play an important role in strengthening or lessening economic segregation.

A general weakness of place-based initiatives is their lack of attention to the realities of mobility. Roughly 13 percent of those earning below $15,000 move annually, while just 7 percent of those earning above $100,000 move each year.[6] These figures mean that low-income neighborhoods are constantly sending out and receiving residents. Yet despite the high levels of mobility, neighborhoods generally change infrequently and slowly in their aggregate composition; poverty is sticky.

Place-based programs can be better designed and implemented in a way that prepares for a high rate of turnover, and even to capitalize on mobility when it occurs. Some residents will need help from initiatives to stay in place. Other residents will need help to leave the neighborhood, making opportunity moves to new areas—especially if they are to help their children reach better schools (Theodos, Coulton, and Budde 2014). However, where units become available for rent or purchase in economically distressed neighborhoods, practitioners and policy-makers should give thought to the household, housing unit, and neighborhood factors that are important in encouraging nonpoor households to join the neighborhood, while not seeking the rapid upscaling of neighborhoods that results in displacement of the poor, nor suggesting that poor households should be excluded from even the low-income neighborhoods they can afford. It is nevertheless likely that concentrated poverty will fall both by encouraging the movement of the poor into affluent neighborhoods and by encouraging the affluent to move to areas of concentrated poverty. Place-based initiatives can take a more proactive stand in managing mobility.

Our research also supports the conclusion that the state of a neighborhood's physical housing stock matters greatly for its development. Few long to return to the days where neighborhood redevelopment was purely a strategy of physical demolition and rebuilding. But people-based solutions, while important, are unlikely to be sufficient in deconcentrating poverty in neighborhoods consisting of low-quality, undesirable housing. Taking liberties with Tolstoy's adage, it is likely that every distressed neighborhood is distressed in its own way. Some lack ready transit access to business districts, some consist of degraded public housing or private housing stock, some are located in cities with limited abilities to provide basic services. Yet, despite these differences, the physical housing stock will likely need to be attended to if poverty is to be deconcentrated. While policymakers and practitioners have too often focused either on people-based interventions or physical redevelopment (Kubisch et al. 2010), attention to both is necessary if concentrated poverty is to be sustainably addressed.

Notes

1. Of course, neighborhoods can change in socioeconomic mix because of demolition of an existing stock or eviction or displacement of residents, or also through new construction. The effects of displacement or eviction/foreclosure on neighborhoods have received wider treatment; for example, describing the effects on neighborhoods of the demolition of distressed public housing through the HOPE VI program or through foreclosures.

2. Three of the MC sites were surveyed only in wave 1 and wave 2: Hartford, Milwaukee, and Oakland.

3. It is possible for housing unit damage to be repaired or previously undamaged units to become damaged; however, there is no way to know from the MC data.

4. Public housing rents are excluded from neighborhood rent calculation.

5. We also investigated the effects of perceptions of neighborhood safety and perceptions of neighborhood services and amenities (including schools and policing). Neither of these factors was statistically related to the outcome of interest, and they were not included in the final model specifications due to a limited number of level 3 observations (70).

6. Authors' calculations of mobility rates from 2012 to 2013 using the American Community Survey.

References

Aurand, Andrew, and Angela Reynolds. 2013. Elderly mobility and the occupancy status of single-family homes. *Housing Studies* 28 (5): 661–81.

Bader, Michael D. M., and Maria Krysan. 2015. Community attraction and avoidance in Chicago: What's race got to do with it? *The ANNALS of the American Academy of Political and Social Science* (this volume).

Bayoh, Isaac, Elena G. Irwin, and Timothy Haab. 2006. Determinants of residential location choice: How important are local public goods in attracting homeowners to central city locations? *Journal of Regional Science* 46 (1): 97–120.

Bishaw, Alemayehu. 2014. *Changes in areas with concentrated poverty: 2000 to 2010.* American Community Survey Reports. Washington, DC: U.S. Census Bureau. Available from http://www.census.gov/content/dam/Census/library/publications/2014/acs/acs-27.pdf.

Clark, William A., Marinus C. Deurloo, and Frans M. Dieleman 1984. Housing consumption and residential mobility. *Annals of the Association of American Geographers* 74 (1): 29–43.

Coulton, Claudia J., Tsui Chan, and Kristen Mikelbank. 2011. Finding place in community initiatives: Using GIS to uncover resident perceptions of their neighborhoods. *Journal of Community Practice* 19 (1): 10–28.

Coulton, Claudia J., Brett Theodos, and Margery Austin Turner. 2012. Residential mobility and neighborhood change: Real neighborhoods under the microscope. *Cityscape* 14 (3): 55–89.

Cytron, Naomi. 2010. *Improving the outcomes of place-based initiatives.* San Francisco, CA: Federal Reserve Bank of San Francisco.

DeLuca, Stephanie, Holly Wood, and Peter Rosenblatt. 2011. How do poor people move? Paper presented at the annual meeting of the American Sociological Association, August, Las Vegas.

Downs, Anthony. 1994. *New visions for metropolitan America*. Washington, DC: Brookings Institution Press.

Eggers, Frederick J., and Fouad Moumen. 2013. *Housing adequacy and quality as measured by the AHS*. Washington, DC: U.S. Department of Housing and Urban Development.

Ellen, Ingrid Gould. 2000. *Sharing America's neighborhoods: The prospects for stable racial integration*. Cambridge, MA: Harvard University Press.

Ellen, Ingrid Gould, Karen Horn, and Katherine O'Regan. 2011. Urban pioneers: Why do higher income households choose lower income neighborhoods? *Urban Studies* 50 (12): 2478–95.

Ellen, Ingrid Gould, and Katherine O'Regan. 2008. Reversal of fortunes? Lower-income urban neighbourhoods in the U.S. in the 1990s. *Urban Studies* 45 (4): 845–69.

Ellen, Ingrid Gould, and Margery Austin Turner. 1997. Does neighborhood matter? Assessing recent evidence. *Housing Policy Debate* 8 (4): 833–66.

Feijten, Peteke, and Maarten van Ham. 2009. Neighbourhood change . . . reason to leave? *Urban Studies* 46 (10): 2103–22.

Fischer, Mary J., and Douglas S. Massey. 2004. The ecology of racial discrimination. *City and Community* 3 (3): 221–41.

Freeman, Lance, and Franck Braconi. 2004. Gentrification and displacement: New York City in the 1990s. *Journal of the American Planning Association* 70 (1): 39–52.

Galster, George C. 1987. *Homeowners and neighborhood reinvestment*. Durham, NC: Duke University Press.

Gramlich, Edward, Deborah Laren, and Naomi Sealand. 1992. Moving into and out of poor urban areas. *Journal of Policy Analysis & Management* 11 (2): 273–87.

Hipp, John R. 2012. Segregation through the lens of housing unit transition: What roles do the prior residents, the local micro-neighborhood, and the broader neighborhood play? *Demography* 49 (4): 1285–1306.

Hwang, Jackelyn, and Robert J. Sampson. 2014. Divergent pathways of gentrification racial inequality and the social order of renewal in Chicago neighborhoods. *American Sociological Review* 79 (4): 726–51.

Ihrke, David K., and Carol S. Faber. 2012. *Geographical mobility: 2005 to 2010*. Washington, DC: U.S. Census Bureau.

Jargowsky, Paul A. 2013. *Concentration of poverty in the new millennium: Changes in prevalence, composition, and location of high-poverty neighborhoods*. New York, NY and Camden, NJ: The Century Foundation and the Center for Urban Research and Education, Rutgers-Camden.

Kingsley, G. Thomas, and Kathryn L. S. Pettit. 2007. *Concentrated poverty: Dynamics of change*. Washington, DC: The Urban Institute.

Krysan, Maria. 2008. Does race matter in the search for housing? An exploratory study of search strategies, experiences, and locations. *Social Science Research* 37 (2): 581–603.

Kubisch, Anne C., Patricia Auspos, Prudence Brown, and Tom Dewar. 2010. *Voices from the field III*. Washington, DC: Aspen Institute.

Leventhal, Tama, and Jeanne Brooks-Gunn. 2000. The neighborhoods they live in: The effects of neighborhood residence on child and adolescent outcomes. *Psychological Bulletin* 126 (2): 309–37.

Marullo, Sam. 1985. Targets for racial invasion and reinvasion: Housing units where racial turnovers occurred, 1974–77. *Social Forces* 63 (3): 748–74.

Newman, Kathe, and Elvin Wyly. 2006. The right to stay put, revisited: Gentrification and resistance to displacement in New York City. *Urban Studies* 43 (1): 23–57.

Nechyba, Thomas J., and Robert P. Strauss. 1997. Community choice and local public services: A discrete choice approach. *Regional Science and Urban Economics* 28 (1): 51–73.

O'Connor, Alice. 1999. Swimming against the tide: A brief history of federal policy in poor communities. In *Urban problems and community development*, eds. Ronald F. Ferguson and William T. Dickens, 77–138. Washington DC: Brookings Institution Press.

Orfield, Myron. 2002. *American metropolitics: The new suburban reality*. Washington, DC: Brookings Institution Press.

Pendall, Rolf, Elizabeth Davies, Lesley Freiman, and Rob Pitingolo. 2011. *A lost decade: Neighborhood poverty and the urban crisis of the 2000s*. Washington, DC: Joint Center for Political and Economic Studies.

Pendall, Rolf, Brett Theodos, and Kaitlin Hildner. Forthcoming. Why high-poverty neighborhoods persist: The role of precarious housing. *Urban Affairs Review.*

Quigley, John M. 1985. Consumer choice of dwelling, neighborhood and public services. *Regional Science and Urban Economics* 15 (1): 41–63.

Rapaport, Carol. 1997. Housing demand and community choice: An empirical analysis. *Journal of Urban Economics* 42 (2): 243–60.

Sampson, Robert J., Patrick Sharkey, and Stephen W. Raudenbush. 2008. Durable effects of concentrated disadvantage on verbal ability among African-American children. *Proceedings of the National Academy of Sciences* 105 (3): 845–52.

Spain, Daphne. 1980. Black-to-white successions in central-city housing: Limited evidence of urban revitalization. *Urban Affairs Review* 15:381–95.

Teernstra, Annalies. 2014. Neighbourhood change, mobility and incumbent processes: Exploring income developments of in-migrants, out-migrants and non-migrants of neighbourhoods. *Urban Studies* 51 (5): 978–99.

Temkin, Kenneth, and William Rohe. 1996. Neighborhood change and urban policy. *Journal of Planning Education and Research* 15 (3): 159–70.

Theodos, Brett, Claudia Coulton, and Amos Budde. 2014. Getting to better performing schools: The role of residential mobility in school attainment in low-income neighborhoods. *Cityscape* 16 (1): 61–84.

Turner, Margery Austin, Peter Edelman, Erika Poethig, and Laudan Aron. 2014. *Tackling persistent poverty in distressed urban neighborhoods.* Washington, DC: Urban Institute.

Turner, Margery Austin, Robert Santos, Diane K. Levy, Douglas A. Wissoker, Claudia Aranda, and Rob Pitingolo. 2013. *Housing discrimination against racial and ethnic minorities 2012: Full report.* Washington, DC: U.S. Department of Housing and Urban Development.

Vigdor, Jacob. 2002. Does gentrification harm the poor? *Brookings-Wharton Papers on Urban Affairs* (2002):132–82.

Contested Space: Design Principles and Regulatory Regimes in Mixed-Income Communities in Chicago

ROBERT J. CHASKIN
and
MARK L. JOSEPH

At the center of Chicago's large-scale public housing transformation is a stated emphasis on economic integration. Based on interviews, field observations, and documentary research in three new, mixed-income communities that were built on the footprint of former public housing developments in Chicago, this article examines how design choices and regulatory regimes militate against the effective integration of public housing residents in these contexts. We find that the strategies used to maintain social order contribute to redirecting the integrationist aims of the development policy toward a kind of *incorporated exclusion*, in which physical integration reproduces marginalization and leads more to withdrawal and alienation than to the engagement and inclusion of relocated public housing residents and other low-income residents.

Keywords: public housing; integration; social exclusion; regulation

C hicago is currently implementing the largest and most ambitious effort in the United States to redevelop inner-city neighborhoods and address the problems of urban poverty through the "transformation" of public housing. Chicago's effort is part of a broader policy trend, nationally and internationally, focused on deconcentrating urban poverty and addressing the problems that have become endemic to many public housing communities over the past half century. At the center of this effort is a stated emphasis on *integration*—on remediating the negative effects of racial and economic segregation that were so starkly exacerbated

Robert J. Chaskin is associate professor and deputy dean for strategic initiatives at the University of Chicago School of Social Service Administration, and an affiliated scholar at Chapin Hall at the University of Chicago.

Mark L. Joseph is associate professor at the Mandel School of Applied Social Sciences at Case Western Reserve University and director of the National Initiative on Mixed-Income Communities.

DOI: 10.1177/0002716215576113

and reproduced by past public housing policy (see Owens, this volume). Using large-scale demolition, redevelopment, and the relocation of thousands of public housing residents, the effort seeks to reshape urban space, remake urban neighborhoods, and reverse the isolation of public housing residents through their integration into new neighborhoods and into the broader contexts, institutions, and opportunities provided by the city as a whole.

Emblematic of neoliberal urban policy, Chicago's Plan for Transformation relies to a large extent on market processes, operating through public-private partnerships to reclaim and rebuild neighborhoods while fundamentally remaking public housing's role in responding to the needs of the urban poor. It also relies on the design principles and theoretical orientations of new urbanism, which assumes that particular aspects of the built environment can support social objectives associated with diversity and community building, as well as maximize the use and informal surveillance of public spaces and promote care and defense of private space. This dual orientation toward community and control contains an inherent tension, however, and generates complex dynamics and significant contention in the new mixed-income communities that are at the center of the transformation.

This article examines the ways in which design choices and regulatory regimes militate against the effective integration of public housing residents in these contexts, contributing instead to new forms of residential inequality, exclusion, and alienation. We find that the strategies used to maintain social order have contributed to the implementation of a kind of *incorporated exclusion*, in which physical integration reproduces marginalization and leads more to withdrawal and alienation than to the engagement and inclusion of relocated public housing residents and other low-income residents.[1]

Theoretical Foundations and Expectations

Reshaping neighborhood context through redevelopment responds to concerns about how "neighborhood effects" operate in areas of concentrated disadvantage. A large body of research, for example, finds associations between living in high-poverty neighborhoods and a range of social problems, including high rates of child abuse, teenage and out-of-wedlock births, school dropout rates, crime and delinquency, and adult unemployment (for reviews, see Gephart 1997; Sampson, Morenoff, and Gannon-Rowley 2002). Mixed-income development approaches to public housing reform seek to address the problems created by concentrated disadvantage by changing the composition, structural circumstances, and social

NOTE: This research was supported with funding from the John D. and Catherine T. MacArthur Foundation, with additional support from the Annie E. Casey Foundation. We are grateful to the many individuals who have helped to facilitate this research project, including our research team and representatives of the Chicago Housing Authority (CHA), development team members at the study sites, community leaders, and most importantly, the residents of the mixed-income developments who discussed their experiences with us.

process dynamics in these communities through the wholesale redevelopment of the built environment; the screening out of problematic residents; the integration of higher-income renters and homeowners; the provision of some services and supports for low-income residents; and the establishment of organizations and processes to establish rules, monitor compliance, and respond to problems as they emerge.

Briefly, arguments for the potential benefits of mixed-income public housing reform fall into four broad categories (Joseph, Chaskin, and Webber 2007). Social capital arguments suggest that integrating public housing residents into economically diverse neighborhoods may connect them to the relational networks of their higher-income neighbors, promoting access to information and opportunity that is not available through their own relatively closed networks. Social control arguments suggest that, since crime is highly correlated with socioeconomic status, residential stability, and homeownership—and higher-income people may be more likely to exert pressure to maintain order and enforce rules—the presence of higher-income residents may promote a context of greater safety and a foundation for more harmonious community dynamics. Middle-class "role model" arguments suggest that the presence of higher-income people may contribute to the modification of aspirations and behavior among those who have been living in isolated poverty toward more prosocial engagement in community and society. And arguments about political and market influence suggest that the presence of higher-income residents can attract greater investment and the provision of higher quality and more responsive services from both public- and private-sector sources, leading to improvements in the physical, service, and organizational infrastructure of local communities.

In addition to these arguments, mixed-income public housing reform draws on new urbanist planning principles, which argue that urban planning and design can play a facilitative role in shaping a built environment—with a focus on scale, walkability, mixed-use, and civic and transitional space—that (among other things) supports diversity, promotes social interaction, and ensures safety and civic engagement (Leccesse and McCormick 2000). Although many of the principles espoused by new urbanism focus on civic space, demographic and functional diversity, and the promotion of pedestrian presence and public activity, in mixed-income public housing redevelopments, relatively more emphasis has been placed on design elements meant to create "defensible space" (Newman 1972) by providing the clear spatial delineation of private and public spheres and facilitating informal surveillance and individual and collective responsibility.

Increasing the proportion of higher-income residents, promoting homeownership and residential stability, and shaping defensible space may counteract the effects of social isolation and contribute to higher levels of social control and reductions in crime, but they may also generate conflict, particularly in light of how issues of race, class, and other dimensions of difference inform social interaction in the context of rapid neighborhood change (Chaskin and Joseph 2013; Freeman 2006; Hyra 2008; Pattillo 2007). Indeed, the social dynamics and organizational responses that have been generated as a consequence of redevelopment and demographic change have also produced a set of fundamental tensions that contribute to serious contestation about the nature of community in these contexts and the rights, privileges, and responsibilities that are shared or differentially enjoyed by community members within them.

Contexts and Methods

The analysis presented here is based on in-depth interviews, field observation, and a review of documentary data over the course of six years of field work focused on Oakwood Shores, Park Boulevard, and Westhaven Park, three mixed-income developments that have been built on the footprint of public housing complexes demolished as part of Chicago's Plan for Transformation (see Table 1 for a summary comparison of the three developments).

Multiple in-depth interviews were conducted with a panel of eighty-five residents living in the mixed-income developments, and focus groups were conducted with an additional sample of 102 residents. Resident interviewees were randomly selected from developer occupancy lists in each development, and respondents included residents across income levels and housing tenures. Interviews were also conducted with a panel of eighty-four development professionals (private developers, property managers, service providers, leaders of community organizations, and other civic stakeholders) over the course of three waves of data collection. In addition to interview data, field observations of approximately 500 community meetings, programs, and events were used to contextualize interview and focus group data within the specific dynamics of each site and provide both a check on and new insight into the dynamics described by the sample of respondents. Interviews and focus groups were recorded digitally, and transcripts and field notes were coded for analysis based on a set of deductively derived thematic codes and refined based on inductive interim analysis. Coding and analysis were done using NVivo qualitative analysis software.

Shaping Regulatory Regimes: Rationales and Motivating Considerations

In thinking about their desire for neighborhood order and the need for standards, rules, and mechanisms to ensure its maintenance, both residents and development professionals expressed a number of different concerns that motivated their support for regulation and enforcement. These fall principally into three broad categories.

The first was concern about crime, safety, and disorder. Concerns about violent crime and criminal activity that bring with them the threat of violence, such as drug trafficking and gang activity, were shared by development professionals and residents across sites regardless of income, race, or housing tenure. As salient as they were, however, these concerns were far from overriding. More prevalent were complaints about property crimes and, especially, about a broad range of

TABLE 1
Mixed-Income Developments

	Oakwood Shores	Park Boulevard	Westhaven Park
Former public housing site	Ida B. Wells, Madden Park	Stateway Gardens	Henry Horner Homes
Developers	National nonprofit (rental); local for-profit (for sale)	Four local for-profits	Two regional and national for-profits
Social service providers	Nonprofit, delivered by developer and later contracted out to local	Nonprofit, created by developer	Nonprofit, contracted out to local
Total projected units	3,000	1,316	1,317
Units built to date:	806	367	1,060
Relocated public housing units (%)	263 (33)	127[a] (34)	766[b] (72)
Affordable rental units (%)	289[c] (36)	106[d] (29)	80 (8)
Market-rate rental units (%)	188 (23)	29 (8)	75 (7)
Homeownership units (%)	66 (8)	105 (29)	139 (13)
Initial occupancy dates	Renters: 2005 Homeowners: 2006	Renters: 2007 Homeowners: 2007	Renters: 2003 Homeowners: 2006
Select site specific criteria	30 hours per week work requirement; 5-year criminal background check; credit screening; residential history check; annual drug test	30 hours per week work requirement; 5-year criminal background check; credit screening; residential history check	20 hours per week engagement requirement; criminal background check; credit screening; residential history check
Guiding legal authority for returning residents	Relocation rights contract	Relocation rights contract	Consent decree
Neighborhood	Bronzeville/North Kenwood Oakland, Southside Chicago	Bronzeville, Southside Chicago	Near West Side, Westside Chicago
Neighborhood amenities and institutions	Near Lake Michigan, public parks, Hyde Park, and University of Chicago	Near public transit corridor, Illinois Institute of Technology, White Sox stadium, major highway	Near downtown central business district, public transit stop, United Center stadium

a. Includes twenty-seven public housing replacement units in an off-site rental building, The Pershing.
b. Includes the Villages, a 200-unit "superblock" of 100 percent public housing units located in the middle of the mixed-income development; the Annex, a 90-unit rehabilitated public housing building nearby; and 261 scattered-site public housing units in the surrounding neighborhood.
c. Includes seventy-five units of affordable senior rental housing.
d. Includes fifty-three units of affordable rental housing in The Pershing.

"incivilities" that stop short of criminality but contribute to residents' assessment of neighborhood quality of life. (Indeed, total reported crime—especially violent crime—has declined significantly over the course of the transformation across sites.) Concerns along these lines, primarily expressed by homeowners and higher-income renters, focused on a broad range of what they often described as "ghetto" behaviors, such as hanging out in groups, playing loud music on the street and in cars, yelling and arguing in public, and littering. Although not in themselves criminal, these behaviors are often conflated with concerns about crime and seen as issues to be confronted in the name of safety and security.

The second motivating orientation that drove the development of regulatory regimes in these contexts, shared principally by development professionals and homeowners, concerned the market viability of the communities and the need to protect investment and exchange values. The concern here focused on maintaining a community that is well ordered, well maintained, and stable, where the resources a homeowner invests are likely to increase in value over time and where market-rate renters will feel they are getting comfort and value for their money. These concerns tie directly to calls for rules, surveillance, and enforcement that can contribute to the maintenance of order and the protection of investment. As a development professional noted:

> If there isn't some consistent enforcement, then the property values will go down, the appearance of the property will be degraded, the desirability of living there, the comfort of people living there will diminish.

Seen in this light, evidence of "disorder" provides negative cues for potential investors and higher-income renters, and these concerns were often mentioned in both development and homeowner association meetings and at a range of public forums in the broader neighborhood. Often, the issues raised focused on objections to "loitering" and the very presence of people, especially black men and unsupervised youth, on the street. The comments of a homeowner in the neighborhood surrounding Westhaven Park illustrate this point:

> Last night there were seventeen guys down at this place just hanging, seventeen people in front of two of the CHA homes, just hanging around talking and so forth. If anyone drives through this neighborhood and they see that, they're not going to buy a home next door to that. It's not gonna happen.

But while exchange-value orientations were clearly operative and contributed to homeowners' and development professionals' orientations toward social control expectations and responses, even more salient—particularly emphasized by homeowners and market-rate renters—were expectations regarding *use value.*

This leads to the third motivation, which concerned the need for clear community norms and standards of behavior. Like concerns about safety, there was wide embrace of the need for neighborly norms among development professionals and across residents of different backgrounds, but also some disagreement about what those norms should be, how they should be enforced, and what processes should be used to establish them. The range of (noncriminal) behaviors to be changed in this regard (and in response to which rules and sanctions are developed) ranged from generally agreed upon incivilities such as those noted above to activities that are far more innocuous, such as storing personal items or hanging laundry in plain view on balconies, washing or repairing cars in the street, and barbequing in public (cf. Freeman 2006; Hyra 2008; Pattillo 2007).

Market norms and broken windows

Acting on these considerations, development professionals at each site have embraced particular design principles and established specific rules and processes to establish and maintain regulatory regimes oriented toward safety, security, and the controlled use of public space. These are driven by the dominance of "market norms" and "broken windows" orientations that frame such responses and lead to a range of rules and sanctions and the establishment of specific mechanisms and processes to monitor behavior and compel compliance.

The focus on market norms was driven both by the need to attract and retain higher-income residents and by the effort to acclimate relocated public housing residents to the expectations of behavior and engagement required of them in the market and civil society, outside the institutional framework and exceptional, isolated circumstances provided by the "projects" from which they came. As a development professional explained:

> We can set certain rules that are basic management rules, but they have to be market-norm management rules. ... The point of this is you're making, for public housing families you're transitioning into the market.

This market orientation was coupled with an embrace of the "broken windows" theory of crime and disorder (Kelling and Wilson 1982), in which outward signs of disorder (litter, broken windows, graffiti) and expressions of incivility (loitering, panhandling, cursing, unruly behavior, public drinking) are seen to indicate more fundamental problems with safety and crime, leading residents to assume that they are at greater risk of victimization and providing "cues" to youth and others inclined to crime and antisocial behavior that such action will be tolerated. While visual cues certainly matter,[2] the empirical basis for the causal link between disorder and crime rates proposed by the broken windows thesis has been challenged (Sampson and Raudenbush 1999). Further, in the new mixed-income communities replacing public housing complexes, there is some clear disagreement about what "counts" as disorder and what should be viewed as normative enjoyment of community space. As a public housing resident at Westhaven Park argues:

> They're acting like we're the problem when our community has been like this. They have a problem with us standing on the corner. We're colored. That's what we do. We gather in groups. We don't have to be no drug activity or nothing like that for us to gather around.

Broken-windows orientations to preserving order in the name of safety and security are generally embraced by development professionals and higher-income residents in these contexts and receive renewed emphasis in times of heightened conflict or perceived threat. But beyond energized responses to particular spikes in concern over crime and antisocial behavior, much of the focus has been on routinely curtailing access to public space and proscribing daily activities that are deemed unsafe in some way, or that are seen as potentially generative of more serious problems down the road. Further, the conflation of a broad range of incivilities with more serious concerns about crime creates a kind of gray area in which the one is linked, seamlessly, to the other. The comments of a renter of a market-rate unit make this point unselfconsciously:

> The security is very, very important to [property management]. They take that very seriously. Just—you can't do this. You can't hang clothes outside. You can't barbeque in those common areas and then things like that.

These concerns about maintaining order and the orientations that guide strategies for doing so were reflected in specific rules and sanctions and codified in specific instruments—perhaps most instrumentally the rental lease agreement—that establish the foundation for organized responses to residents' infractions when they occur.

Rules, Regulations, and Mechanisms of Enforcement

Many of the rules developed within the context of these mixed-income public housing redevelopments were no different from those that govern any rental community or condominium: on-time payment of rent and fees, keeping noise down after a certain hour at night, and maintaining property upkeep, for example. For homeowners, condominium associations were responsible for establishing rules and regulations for their members. These extended as well to renters living in buildings that included homeownership and, in many cases, to common areas both in and around the building. Homeowner associations thus held wide discretion and responsibility for setting rules by which all residents must abide.

For renters, rules and regulations were codified in rental leases and associated "house rules" that elaborated expectations for residents' conduct. These were meant (at least formally) to extend to all renters, although there were some exceptions that applied only to relocated public housing residents. These included reporting requirements regarding changes to household composition, employment, or income; community service or "self-sufficiency" requirements for all adult residents in the household (at Westhaven Park and Oakwood Shores);[3] and "zero tolerance" responses to criminal activity—the definition of which includes both specific criminal activity (such as the use or distribution of illegal substances) and more general infractions (such as parole violations and the "abuse or pattern of abuse of alcohol")—by the leaseholder or, importantly, any member of the household or their guests.

The list of rules across sites was substantial—"they have a dictionary-sized Rules and Regulations," as one relocated public housing resident put it. These rules were cited by both development professionals and residents—particularly relocated public housing residents and other low-income renters—as tools for ensuring compliance and triggering penalties. While formally applicable to all renters (if not to homeowners), and despite development professionals' insistence that enforcement was equitable—"a complaint is a complaint, a violation is a violation, the rules and regulations are the same" as one put it—the lion's share of concern regarding rule adherence and responding to infractions focused on low-income renters, especially relocated public housing residents. Indeed, low-income renters were seen as likely to cause fundamentally different kinds of problems than were homeowners and other higher-income residents. As a development stakeholder put it:

> There's a huge distinction between owners and renters. So from in this building, what the owners do that is annoying to other owners are things that you'd expect in a condo building. ... The renters on the other hand ... listen, it happens all the time; you know that the tax, you can get a tax-credit renter that's selling drugs out of their unit.

Monitoring and enforcement of these rules was accomplished through a number of mechanisms, both formal and informal. "When you put all these rules in place," a homeowner noted, "you really needed to bring an army to enforce it."

Formal monitoring and enforcement was largely the domain of development team members, especially property management staff. Monitoring and enforcement occurred in part through collective engagement with renters at tenant meetings, almost exclusively attended by relocated public housing and other low-income renters. It also occurred through more individualized engagement, such as walking the streets during the day and driving around the site at night, talking to residents about how things were going, calling residents into the office for meetings, sending notices about expected behavior or reported violations, and documenting offenses. And it occurred, increasingly, through formal surrogates (such as private security personnel) and electronic surveillance. Indeed, beginning in 2010, all three sites deployed closed-circuit television cameras as both a deterrent and a tool to facilitate prosecution.[4] By far the most extensive system is at Oakwood Shores, where perceived increases in crime and youth antisocial behavior have led to particularly vigorous efforts to address resident concerns about security. A total of 150 cameras have been installed, focused on all rental buildings, parking lots, alleyways, open spaces (with the exception of the public parks, where the city has installed cameras), and the perimeter of the development site. In addition, cameras are designed to be able to respond to certain behaviors—such as groups of people standing around for certain periods of time—with recorded messages to warn people away.

Formal mechanisms also included the police, who were often sought out by development professionals at each site to increase their presence on the streets and in the parks and, in keeping with the broken windows orientation described above, to be more vigilant in responding to both serious crime and a broad range of incivilities. Discussion at CAPS[5] meetings also frequently turned into requests for more active and aggressive policing, additional patrols, asking for youth to be stopped and checked for identification (one resident suggested issuing armbands to identify resident youth versus outsiders), and adhering to a "zero-tolerance policy" toward loitering. Describing a police detail that would provide extra patrols in the neighborhood surrounding Westhaven Park,[6] for example, a police officer noted that its primary focus would be on these kinds of concerns about incivilities—drinking, hanging-out—"the kind of crap you see every day."

Beyond formal mechanisms, residents were also encouraged by both property management and the police to engage in informal surveillance and to report infractions. Police, for example, routinely emphasized the importance of calling 911, creating phone trees, and becoming involved in volunteer patrols. Property management communicated systematically with residents to enlist their help in monitoring infractions (such as leaving notes in mailboxes or posting notices to call management if residents notice anything problematic) and worked through relationships with specific residents around specific issues. As a development professional explained:

> According to my visual inspection, I can look at a building and look at it and say I need to go visit that person on the second floor, just by seeing—I mean for me personally, I can look at the outside sidewalk and see if there's been too much traffic inside of the building going on, and when I start noticing, hey, this property has a pickup on traffic, I know there's something wrong, and I need to find out.

For many relocated public housing residents, these processes had a kind of Panopticon quality, and they expressed discomfort at the level and pervasiveness of surveillance in place in these contexts and at the sense that their behavior was under constant scrutiny. As one relocated public housing resident put it: "it's like everything you do, they know about it."

Targets, Impacts, and Responses

It should be noted that most people in these communities—development professionals and residents both—recognized that the issues of concern, whether clearly criminal or more focused on incivilities, were likely generated by a relatively few "problem households" and their guests, or (particularly in the case of crime) by low-income residents in the surrounding neighborhood who may or may not be connected to current residents. That said, relocated public housing residents and other low-income renters (the distinction between which is virtually never made by higher-income residents)[7] were the principal focus of the regulatory regimes put in place in these contexts. The leader of a community organization explained:

> The target becomes people in public housing; it's just easier to lump them in as a group. … It's a clash unlike anything I've seen, and to get anywhere remotely close to that, you'd have to go back to when blacks were trying to integrate communities back in the

60s, to get that kind of venom and rabid anger that comes out when people are talking about the neighborhood.

Relocated public housing residents themselves expressed some ambivalence about rules and rule enforcement. On one hand, many recognized their importance generally (as noted above) and credited their enforcement with contributing to improvements in these communities (safety, sanitation, quality of the built environment) as compared to the circumstances in public housing, especially to the extent they reined in the behavior of disruptive youth or led to the removal of the tenants they recognized as problematic. For some, the more restrictive regime was a fair trade for the improvement in living standards provided by the new development. As one resident noted:

> The rules are what is expected. I mean what can you say? You come from the projects and you get blessed with a brand new apartment that's built from the ground. What more can you ask for? You come out of the projects where there's rats, roaches, floods, no heat half the time, no lights half the time. So I'm grateful. I have no complaints.

Most, however, found the nature and extent of surveillance and regulation highly invasive, often excessive, and a significant source of stress. Indeed, while the majority of homeowners and market-rate renters with whom we spoke advocated more stringent rules and more rigorous enforcement, virtually no relocated public housing residents shared this view. Beyond some basic disagreements about the appropriateness of some rules—restricted access to what they viewed as public space, injunctions against "congregating," prohibitions against barbequing—there was general concern about the extent to which certain rules singled them out as likely transgressors, were differentially enforced, and had more significant—and potentially detrimental—impacts on both their rights to community enjoyment and their ultimate housing stability. These concerns were noted, sometimes with resignation, sometimes with rancor, with regard to both private behavior and public space.

Policing private behavior

Rules governing aspects of private behavior and the mechanisms put in place to enforce them often overlapped with concerns about the use of public space, but were also oriented toward specific aspects of self-sufficiency, self-control, and lifestyle behaviors—from employment (or, in the absence of employment, training and community service) to drug use to personal hygiene to housekeeping. At one condominium association meeting, for example, contention around the smell of cigarette smoke emanating from people's apartments—the source of which was explicitly presumed to be relocated public housing residents—led to discussion about the possibility of prohibiting smoking in apartment units by declaring the building a no-smoking zone. At a tenants meeting at another site, development professionals shared with relocated public housing residents homeowner complaints about residents stepping outside their units "not looking acceptable for public presentation" (uncombed, barefoot, in pajamas), and encouraged them to "take away [homeowner's] ammunition" by thinking beyond the specific rules codified in their lease. Instead, they should remember that "people are watching" and avoid behaviors that would be perceived as negative by their higher-income neighbors.

One way in which these expectations for resident behavior were monitored was through periodic unit inspections. Formally, all renters regardless of income or nature of subsidy were required to allow inspection of their units at least annually. In practice, however, as interviews with both residents and development professionals made clear, unit inspections were disproportionally focused on low-income residents and often took place far more frequently than the annual requirement.[8]

The weight of surveillance and the rigor with which infractions of rules were sought out and enforced led many relocated public housing residents to feel both overly confined and under constant threat—"walking on eggshells" as one put it—as well as ultimately demeaned. As one relocated public housing resident put it, "believe me, you are being watched. The cameras, the cameras. And if anything goes wrong they pull you in the office, they're gonna tell you every detail."

The pressure of neighbor complaints and the responses of property management to them were often seen by relocated public housing residents in contrast to the more flexible, tolerant stance that they believed they took to living with neighbors in these contexts. "We're adaptable to noise, to people walking when we can hear it at 4:00 in the morning," noted a public housing resident leader. "We should tell each other. Maybe we should complain, but we figure it's their business."

These concerns about overzealous and unfair enforcement extended beyond the stigma of the scrutinized behavior of leaseholders to the feeling that their children were a focal concern, unwelcome in the community and unfairly targeted by efforts to enforce rules, especially by those who controlled access to public space. Even more problematic was the extent to which they might be held responsible for the actions—indeed, even the presence—of visitors and nonresident relatives. As the leader of a local community organization in one neighborhood noted, "We're programmed that when we see a group of young people, particularly teens just hanging out somewhere, it's a cause for concern." This led, as noted above, to blanket calls for a "zero tolerance" policy toward loitering, with a particular emphasis on eliminating the presence of young people in public spaces and, more recently, on efforts to enlist residents to police their presence. In some cases, this occurred through service-oriented efforts to build and engage local resident leadership. In Park Boulevard, for example, development professionals sought, as one explained, to organize "leadership teams" of resident young people to intervene, "get[ting] the teens to say [to other teenagers congregating in public], 'hey, it's not cool to hang out.'" In others, policing such activity took a more punitive turn, making clear the consequences of inaction. At one neighborhood meeting, for example, a CHA staff member shared with the participants that as he was walking up to the meeting location he noticed two young men sitting on the front steps of a house. On discovering they did not live there, he spoke with the leaseholder and told her that "you can't have young black guys hanging out in front of your house," and that she would be held responsible for their behavior and could be evicted if they did something wrong. This kind of message was increasingly sent to relocated public housing residents by property management staff and other development professionals across sites.

Privatizing the public

As this last example suggests, the focus on policing individual behavior was often connected with concerns about maintaining order in public spaces. These concerns informed a range of responses, most of which hinge on different approaches to privatization. On one hand, as noted with reference to the tenets of new urbanism, the importance of public and transitional space was generally recognized by development professionals and residents both. On the other hand, monetary considerations, concerns about safety, homeowner preferences for privacy, and disagreements regarding normative expectations for behavior in public space have led to both design and management choices that for the most part limit the amount of public space available, separate it from the main concentrations of residential living, and regulate access to these spaces and the kinds of use to which they can be put.

Regarding design, the principal focus was on the provision of housing for individuals and families, and the vast majority of space was set aside for the development of private rental or for-sale units. In part this was driven by financial considerations and desires for density, and by preferences, especially those of homeowners. But it was also, importantly, driven by concerns about safety. The design implications of this draw on the seminal arguments regarding "defensible space" put forth by Oscar Newman in the 1970s, which were adopted by new urbanists promoting "traditional neighborhood design" and, in turn, informed thinking about HOPE VI public housing redevelopment nationally. The emphasis here is on promoting a sense of territoriality, ownership and responsibility, the demarcation of "safe zones," and increasing the likelihood of informal surveillance (Leccesse and McCormick 2000; Newman 1972). Thus, design across these sites has, for the most part, privileged private (and privately controlled) space over common areas, including a preference for individual entrances and private balconies as well as the demarcation of common spaces that can be effectively monitored and managed.

Such privatization incorporates both design choices and management strategies. In some cases, privatization is explicit, by creating civic space (such as "community meeting" rooms) that are privately managed and staffed, or by designating particular common spaces as private that, to the general observer, might reasonably be seen as public space. Regarding the former, both Oakwood Shores and Westhaven Park have created such spaces, access to which is provided through formal request, approval, and scheduling and the use of which is regulated by development staff. The benefits of such spaces in terms of social control are noted by a development professional:

> The indoor public spaces are easy because the indoor public spaces we can monitor and we staff and we maintain. The outdoor public spaces are more challenging just because they require the police to do their job.

Regarding the second, Park Boulevard, provides a case in point. Here, designers created a kind of town square area, with green space and a playground, around which townhouses (along the long sides of the park) and multiunit dwellings (at the corners of these blocks) are located. Access to and use of this park, and particularly the playground within it, has been the site of some contention, including a dramatic event early in the development's history in which several young people (each 10 or 11 years old) were actually arrested by an "overzealous security guard," in the words of one development professional there, for being too old to play on the equipment and for playing too loudly. The incident raised issues about the nature of public space and who could use it and about appropriate responses to policing such space. It led to both the clear designation—through signage explicitly stating rules of access and use—of the park as for the enjoyment of Park Boulevard residents only ("the reality is that it is a private park," noted another development professional there, "it's a private park for 150 residents"), and to more measured responses to addressing concerns about youth presence and public activities.

As suggested by the foregoing example, the privatization of space in the name of security is as much a function of regulation as it is of formal privatization (Ruppert 2006). Across sites, regulations have sought to essentially redefine public space by limiting resident access to common areas not explicitly designated for social uses and prohibiting their use for such purposes. Thus, prohibitions have been put in place across sites that seek to keep people off the streets and sidewalks and away from the fronts of buildings and to control the use of parking lots, boulevards, and parks to limit disturbances and curtail visible "hanging out." Relocated public housing residents and other low-income renters frequently commented on the draconian nature of these prohibitions and their explicit focus on controlling the behavior of low-income renters. As one relocated public housing resident stated:

> They must have been sitting out on their porch or sitting outside on the crate or something but they put notices in all their mailboxes telling them that was very ghetto. You know: "You're not allowed to congregate in front of the property." Well, where do you want me to go? Where do you want me to go?

More fundamentally, these preferences reflect a profoundly different orientation, and different expectations of neighborhood space as places of sociability, than those shared by most relocated public housing residents and other low-income renters. As one explained:

> They want us to sit in the back because they thought it's unsightly to have us out here, but we don't see anybody in the back. In the front you can see people coming and going. … People drive by and they stop and they talk, but in the back you not going to get that.

They are also in stark contrast to the orientations toward public space that relocated public housing residents had experienced in public housing. There, public space was seen as providing essential sites for socialization and building community, contributing to the building of social networks and enduring relationships, despite the important concerns about safety that they often also discussed. "You knew the whole—everybody's body, mamas, cousins," as one relocated public housing resident noted, "their second generations, their third generations."

The privatization of space and the enforcement of rules to support it, while partially effective at curtailing some of the behaviors development professionals and higher-income residents wished to limit, also led to a countervailing process in which such privatized space was reappropriated for social interaction, recreation, and leisure. Often, this manifested as individuals or small-group gatherings, standing together in front of buildings or on street corners, sitting on front stoops, or pulling up chairs to socialize outside. In other cases, the appropriation of space was more active—kids running up and down the street and between cars in the parking lots or playing in the alleys; parties being held on the street to eat, drink, and listen to music. Particularly in the warmer months, homeowners across sites complained of these activities; one homeowner, for example, complained of "mobs of people" setting up a late-night party behind her building, "totally invad[ing] the parking area."

While some relocated public housing residents responded by pushing back against what they saw as unfair restrictions in this way, most responded to these regulatory regimes instead by withdrawing, seeking to minimize the possibility of punitive action and choosing instead to "keep their head down" and stay out of trouble. As one relocated public housing resident explained:

> I say it's best to just mind your own business and just speak to people hi and bye and not socialize or fraternize with them, then that way you won't be one of the ones that they calling into the office on.

Such a response was reflective of the differential impact that rule infractions have on relocated public housing residents and other low-income renters compared with their higher-income neighbors.

Differential impact

In addition to a general sense of being targeted by these kinds of regulations and complaints about the appropriateness of some of them, relocated public housing residents and other low-income renters complained as well about the ways in which they were unfairly enforced and about the lack of recourse available to act on complaints about their higher-income neighbors. "The only people that have to abide by the rules is us as the low-income people," as one relocated public housing resident complained, "[homeowners] don't have rules."

But perhaps more important is the extent to which enforcement had a fundamentally different impact on relocated public housing residents and other low-income renters. This was particularly the case in the context of the adoption of "three strike" rules across sites and the effort to step up evictions in response to complaints. Low-income people, after all, have significantly fewer housing options available to them in the market than those with greater financial resources, and since relocated public housing residents had taken up residence in these communities as their "permanent" housing choice, eviction from their current unit could mean losing their right to public housing subsidy entirely. Thus, in addition to the perceived leniency with which their higher-income neighbors were treated and the broader array of rights they were seen to enjoy, the dangers presented by enforcement were palpable.

Indeed, enforcement of three-strike provisions and efforts to move quickly to eviction have become more common across sites over time. This is a shift from earlier in the history of these developments when, on one hand, the desire to keep apartments rented contributed to development professionals' willingness to work through some challenges with residents not presenting serious problems or causing complaints from neighbors and, on the other, the barriers to moving forward with eviction proceedings for relocated public housing residents were particularly stringent in an effort to protect their rights under the Relocation Rights Contract. More recently, efforts have been made to enlist CHA's help in addressing the barriers to evicting problematic tenants, and more vigorous enforcement of three-strike provisions that could move to eviction has been adopted. "We have to rack up the lease violations," a development professional noted at a local CAPS meeting, "if you breathe hard, that's a lease violation."

Most evictions have been for nonpayment of rent and these, in the context of the Great Recession, have included a number of market-rate renters. But other eviction proceedings, especially when involving relocated public housing residents and tax-credit renters, were more likely to be in response to rule violations and problematic behavior. Some of this has included criminal convictions, an explicitly evictionable offense (though not without contention depending on the offender's relationship to the household, as noted above). Many, however, focus on more generally problematic behavior, from positive drug tests (not a crime, but a lease violation) to unauthorized guests to multiple complaints about noise or other incivilities. "I'm cracking down on residents who allow people that don't live here to come and mess up what you have," a development professional noted, "if your guest causes problems, you will reap the repercussions of it." Similarly, at a management meeting at another site, discussion about pending eviction proceedings focused on the importance of targeting households that were responsible for causing a range of "disturbances"—from substance abuse to loud music to a large number of visitors to dropping cigarette butts off the balcony. Thus, when seeking to respond to complaints about incivilities, a link was made between a broad range of behaviors and actionable lease violations, and residents were enlisted in the effort to make the case.

Conclusion

From the standpoint of improved safety, order, and the quality of the built environment, the communities that have emerged on the footprint of public housing complexes under the Plan for Transformation show considerable success compared with what they have replaced. While relocated public housing residents and other low-income renters acknowledge these improvements and appreciate, in particular, the relative safety of the neighborhood and quality of the housing in which they now live, homeowners and newly arrived market-rate renters are not entirely satisfied, and their concerns about safety, disorder, and neighborhood quality of life are largely grounded in complaints about their low-income neighbors, especially relocated public housing residents. This has led to significant contention about both place and space—both the kind of communities each of these neighborhoods will become and the ways in which the space each provides can be shared equally or is to be differentially enjoyed by different community members based on income or housing tenure. Responding to this contention, regulatory regimes have been put in place that disproportionately impact relocated public housing residents and other low-income renters in these contexts, and that are grounded in the logic of contemporary "poverty governance" (Soss, Fording, and Schram 2011). This orientation privileges market norms and paternalistic orientations to managing the poor, linking eligibility of benefit receipt with evidence of individual responsibility and adherence to societal expectations for particular kinds of (positive) behavior, as well as to a range of punitive responses for noncompliance.

Rather than promoting effective integration of the poor into well-functioning, mixed-income neighborhoods, the mechanisms put in place to implement these regimes—both formal and informal, individual and organizational—and the perceived differential targeting and inequitable enforcement of rules in their service have instead produced what we term *incorporated exclusion*, in which spatial integration is accompanied by new forms of inequality and marginalization and, in the case of many relocated public housing residents in particular, withdrawal.

Given the policy goals of inclusion and integration at the center of the transformation, is it possible to reorient the regulatory regimes in these communities in ways that protect the desire for order, safety, and sound investment without overly constraining individual freedom and access to public space or inequitably targeting the poor within them? Rethinking governance in these contexts away from regulatory paternalism and toward more participatory and inclusive engagement is one potential avenue (Chaskin, Khare, and Joseph 2012). A second is refocusing design to promote the integration of and access to public "civic" space and providing opportunities for more inclusive deliberation to establish collective norms about the use of such space (Chaskin and Joseph 2013). A third potential direction would focus greater attention on shaping the organizational infrastructure of these neighborhoods, moving beyond the overarching focus on housing to include attention to building and connecting residents to commercial, institutional, and organizational resources, such as stores, coffee shops, recreational facilities, and schools (Chaskin and Joseph 2013). Finally, more robust services and supports (education, training, job-placement assistance, case management) to help position relocated public housing and other low-income residents to participate more fully and effectively in these new communities and gain broader access to economic and social opportunities in the city are needed, but they are also limited in the broader context of shifting economic opportunity and other structural constraints that low-income people face and that policies like the transformation are fundamentally not designed to address. Local efforts focusing on human capital need to be promoted, along with a broader policy focus on structural barriers and inequality and on economic development, infrastructure, and institutional investment in education, technology access, and other foundational resources that are often either of inferior quality or out of reach for many low-income people (Chaskin 2013; Chaskin and Joseph, forthcoming).

Notes

1. A more extensive analysis of these dynamics can be found in Chaskin and Joseph (forthcoming).

2. On the relationship between disorder and perceptions of crime, see Lewis and Maxfield (1980), Skogan (1990), Taylor and Covington (1993), Perkins et al. (1993), and LaGrange, Ferraro, and Supancic (1992).

3. There are exemptions to this rule, for example, for those who are employed, disabled, or elderly.

4. The CHA provided federal stimulus dollars to all mixed-income and traditional public housing developments for camera installation at this time. Developer owner entities at Oakwood Shores, Park Boulevard, and Westhaven Park contributed additional funds, and homeowners and business owners paid a special assessment fee for exterior cameras at Park Boulevard.

5. Community Alternative Policing Strategy (CAPS) meetings are specifically designed to foster collaborative crime-reduction responses between the community and the police.

6. These extra patrols, in response to pressure from organized homeowner groups, were funded in part by the CHA.

7. Similarly, although development professionals clearly recognize the difference between relocated public housing residents and other subsidized renters (in large part because they have different responsibilities toward each), the work of property management staff in monitoring rule compliance and responding to infractions, including through the tenants meetings described above, focused on both categories of resident. In stark contrast, there was a strong tendency on the part of tax-credit renters to distance themselves from relocated public housing residents. They were often as vociferous in complaining about public housing residents as higher-income residents, and were quick to dissociate themselves from the values and behaviors of former public housing residents. In part this may be a response to shared stigma and an effort to manage that stigma through distancing strategies. See, for example, Goffman (1963), Link and Phelan (2001), and McCormick, Joseph, and Chaskin (2012).

8. One reason for this is that there are a number of institutions providing financing to subsidize units (e.g., the CHA, the U.S. Department of Housing and Urban Development, and banks and state agencies holding low-income housing tax credits), each of which claims oversight privileges to frequently check on their investment; it is thus more than just property managers who are targeting relocated public housing residents for inspection.

References

Chaskin, Robert J. 2013. Integration and exclusion: Urban poverty, public housing reform, and the dynamics of neighborhood restructuring. *The ANNALS of the American Academy of Political and Social Science* 647 (1): 237–67.

Chaskin, Robert J., and Mark L. Joseph. 2013. "Positive gentrification," social control, and the "right to the city" in mixed-income communities: Uses and expectations of space and place. *International Journal of Urban and Regional Research* 37 (2): 480–502.

Chaskin, Robert J., and Mark L. Joseph. Forthcoming. *Integrating the inner city: The promise and perils of mixed-income public housing transformation*. Chicago, IL: University of Chicago Press.

Chaskin, Robert J., Amy T. Khare, and Mark L. Joseph. 2012. Participation, deliberation, and decision-making: The dynamics of inclusion and exclusion in mixed-income developments. *Urban Affairs Review* 48 (6): 866–908.

Freeman, Lance. 2006. *There goes the 'hood: Views of gentrification from the ground up*. Philadelphia, PA: Temple University Press.

Gephart, Martha A. 1997. Neighborhoods and communities as contexts for development. In *Neighborhood poverty, volume 1: Contexts and consequences for children*, eds. Jeanne Brooks-Gunn, Greg J. Duncan, and J. Lawrence Aber, 1–43. New York, NY: Russell Sage Foundation.

Goffman, Erving. 1963. *Stigma: Notes on the management of spoiled identity*. New York, NY: Simon & Schuster.

Hyra, Derek S. 2008. *The new urban renewal: The economic transformation of Harlem and Bronzeville*. Chicago, IL: University of Chicago Press.

Joseph, Mark L., Robert J. Chaskin, and Henry S. Webber. 2007. The theoretical basis for addressing poverty through mixed-income development. *Urban Affairs Review* 42 (3): 369–409.

Kelling, George L., and James Q. Wilson. 1982, March. Broken windows: The police and neighborhood safety. *Atlantic Monthly*.

LaGrange, Randy L., Kenneth F. Ferraro, and Michael Supancic. 1992. Perceived risk and fear of crime: Role of social and physical incivilities. *Journal of Research in Crime and Delinquency* 29:311–34.

Leccesse, Michael, and Kathleen McCormick. 2000. *Charter of the new urbanism*. New York, NY: McGraw-Hill.

Lewis, Dan A., and Michael G. Maxfield. 1980. Fear in the neighborhoods: An investigation of the impact of crime. *Journal of Research in Crime and Delinquency* 17:160–89.

Link, Bruce G., and Jo C. Phelan. 2001. Conceptualizing stigma. *Annual Review of Sociology* 27:363–85.

McCormick, Naomi, Mark L. Joseph, and Robert J. Chaskin. 2012. The new stigma of relocated public housing residents: Challenges to social identity in mixed-income developments. *City and Community* 11 (3): 285–308.

Newman, Oscar. 1972. *Defensible space: Crime prevention through urban design*. New York, NY: Macmillan.

Owens, Ann. 2015. Assisted housing and income segregation between neighborhoods in U.S. metropolitan areas. *The ANNALS of the American Academy of Political and Social Science* (this volume).

Pattillo, Mary. 2007. *Black on the block: The politics of race and class in the city*. Chicago, IL: University of Chicago Press.

Perkins, Douglas D., Abraham Wandersman, Richard C. Rich, and Ralph B. Taylor. 1993. The physical environment of street crime: Defensible space, territoriality and incivilities. *Journal of Environmental Psychology* 13:29–49.

Ruppert, Evelyn. 2006. Rights to public space: Regulatory reconfigurations of liberty. *Urban Geography* 27:271–92.

Sampson, Robert J., Jeffrey D. Morenoff, and Thomas Gannon-Rowley. 2002. Assessing "neighborhood effects": Social processes and new directions in research. *Annual Review of Sociology* 28:443–78.

Sampson, Robert J., and Stephen W. Raudenbush. 1999. Systematic social observation of public spaces: A new look at disorder in urban neighborhoods. *American Journal of Sociology* 105:603–51.

Skogan, Wesley G. 1990. *Disorder and decline: Crime and the spiral of decay in American neighborhoods*. New York, NY: Free Press.

Soss, Joe, Richard C. Fording, and Sanford Schram. 2011. *Disciplining the poor: Neoliberal paternalism and the persistent power of race*. Chicago, IL: University of Chicago Press.

Taylor, Ralph B., and Jeanette Covington. 1993. Community structural change and fear of crime. *Social Problems* 40:374–97.

Locational Attainment and Housing Insecurity

Achieving the Middle Ground in an Age of Concentrated Extremes: Mixed Middle-Income Neighborhoods and Emerging Adulthood

By
ROBERT J. SAMPSON,
ROBERT D. MARE,
and
KRISTIN L. PERKINS

This article focuses on stability and change in "mixed middle-income" neighborhoods. We first analyze variation across nearly two decades for all neighborhoods in the United States and in the Chicago area, particularly. We then analyze a new longitudinal study of almost 700 Chicago adolescents over an 18-year span, including the extent to which they are exposed to different neighborhood income dynamics during the transition to young adulthood. The concentration of income extremes is persistent among neighborhoods, generally, but mixed middle-income neighborhoods are more fluid. Persistence also dominates among individuals, though Latino-Americans are much more likely than African Americans or whites to be exposed to mixed middle-income neighborhoods in the first place and to transition into them over time, even when adjusting for immigrant status, education, income, and residential mobility. The results here enhance our knowledge of the dynamics of income inequality at the neighborhood level, and the endurance of concentrated extremes suggests that policies seeking to promote mixed-income neighborhoods face greater odds than commonly thought.

Keywords: mixed-income neighborhoods; inequality; young adulthood; life course

Increases in income segregation, combined with the apparent loss of middle-class and mixed-income neighborhoods, have generated considerable attention. Among scholars and the public at large, recent attention has been focused primarily on the pulling away of the very rich—the so-called 1 percent (Piketty

Robert J. Sampson is Henry Ford II Professor of the Social Sciences at Harvard University.

Robert D. Mare is a distinguished professor of sociology at UCLA.

Kristin L. Perkins is a PhD candidate in sociology and social policy at Harvard University.

NOTE: This article was supported in part by Grant #95200 from the John D. and Catherine T. MacArthur Foundation and a grant from the Hymen Milgrom Supporting Organization to the University of Chicago.

DOI: 10.1177/0002716215576117

2014). At the other end, a classic urban literature has produced a wealth of studies on concentrated poverty and the "truly disadvantaged" (Wilson 1987).[1]

Mixed-income housing has become a major policy paradigm that seeks to address such income extremes. Based on evidence that links concentrated poverty to compromised life outcomes, federal policy-makers in the United States have advocated moving poor people out of concentrated public housing and increasing the presence of higher-income neighbors through mixed-income redevelopment of high-poverty neighborhoods. Two examples of these arguments are the Department of Housing and Urban Development (HUD)'s "Moving to Opportunity" voucher experiment, which sought to move the poor out of concentrated poverty; and HOPE VI, a place-based intervention that sought to increase income mixing in housing developments. With Mayor Bill de Blasio in New York City as the foremost example, city leaders around the country have also taken aim at policies to reduce residential inequality.[2]

Although neighborhood income mixing has surfaced as a favored policy tool and is the subject of growing scholarly discussion, research evaluating its sources and consequences is sparse. As a result, less is known about the nature of middle-income and mixed-income neighborhoods, and the experience of individuals transitioning into and out of the middle, rather than the extremes. The evidence that does exist has produced conflicting results (Joseph and Chaskin 2012), and much of the writing about mixed-income neighborhoods is normative or aspirational rather than analytic (Cisneros and Engdahl 2009). There are also important untested assumptions and unanswered questions about neighborhood change (Joseph, Chaskin, and Webber 2007). For example, if a neighborhood is middle or working class and a mixed-income housing intervention brings lower-income residents in, it is an open question whether the intervention would lead over time to the out-migration of existing middle-class residents or decreases in the in-migration of the nonpoor. More generally, mixed-income policy implicitly assumes a kind of static equilibrium with regard to intervention effects. Equally important is the fact that proportionately few people live in planned mixed-income housing or HOPE VI neighborhoods. Like ethnically diverse communities (Ellen 2000), mixed-income neighborhoods that are "naturally occurring" are much more common, and yet we do not know much about their course of development—for individuals or at the neighborhood level.

Framework and Research Approach

This article focuses on the dynamics of what might best be characterized as "mixed middle-income" neighborhoods: areas that are more evenly balanced than those at the extremes of either concentrated poverty or concentrated affluence and that have a reasonable mix among income groups, especially exposure of the poor to the middle and upper classes. Studying individual sorting and population flows into and out of mixed middle-income neighborhoods is fundamental to understanding neighborhood income inequality and, we argue, analytically necessary prior to assessing the impact of such neighborhoods on outcomes, whether at the individual or neighborhood level (see also Bruch and Mare 2006; Sampson and Sharkey 2008). We therefore analyze variation in mixed middle-income neighborhoods at both the neighborhood and individual levels, placing our emphasis on patterns of stability and change.

We begin by defining and validating a measure of mixed-income neighborhoods; we then examine neighborhood-level transitions over the course of several decades. Although neighborhoods are constantly in flux, the evidence on the persistence of neighborhood poverty (Sampson 2012) leads to the hypothesis that mixed-income neighborhoods tend to occupy similar positions in the citywide distribution over time. However, gentrification and the demolition of public housing in cities such as Chicago may have altered this pattern. For example, are economically integrated neighborhoods stable, or merely in transition between low-income and high-income areas?

We focus on income mixing that is organic or "naturally occurring" in the sense that neighborhood income status is not defined based on local or federal housing policy interventions. There are good reasons for this analytic move. In Chicago, our study site and a city of more than 2.7 million residents and a million housing units, we estimate that less than 5 percent of city residents live in a census tract where a public housing project is located. If we restrict our concern to HUD's HOPE VI housing program, the total number of units in mixed-income developments under the city's "Plan for Transformation" is about 12,000, approximately 1 percent of all housing units. Moreover, the effort to provide new subsidized housing units has stalled (Moore 2013), and in fiscal year 2014, only about 2,600 mixed-income units were leased.

The prevalence and correlates of moving in and out of naturally occurring mixed-income neighborhoods therefore looms large and motivates the second and major part of our article. Here we focus on individual exposure to neighborhood income mixing, based on a new longitudinal survey of almost 700 adolescents originally living in Chicago in 1995 and followed wherever they moved in the United States, with the most recent data collection ending in 2013. The study consists of a birth cohort followed up to 17 years of age and three later cohorts (ages 9–15) that were studied into young adulthood. Our central aim is to describe trajectories of neighborhood attainment for the young adult cohorts. They are of primary interest because we know the income status of both the neighborhood that their parents chose when respondents were children and the neighborhood that our respondents chose, or ended up in, during the critical transition to young adulthood (ages 25–32), which allows for an intergenerational perspective on neighborhood attainment. By contrast, the birth cohort members have not left their parents' home.

We first present descriptive patterns on trajectories of neighborhood mobility—how much stability or change is there in exposure to mixed-income areas? We focus primarily on race, cohort, and immigrant status as conditioning factors. Based on prior research, we expect that blacks and Latinos are disproportionately exposed to poverty compared with whites. However, we do not know how different racial/ethnic groups are exposed to mixed middle-income neighborhoods over time. Nor do we know if race/ethnic differences are attributable to preexist-ing differences in factors such as economic status or homeownership, or to time-varying factors such as residential mobility. Our last set of analyses thus examines the predictors of transitioning into mixed middle-income neighborhoods over young adulthood using a core set of theoretically selected characteristics.

The Mixed-Income Project

The larger project in which this study is embedded is the Mixed-Income Project (MIP), which was designed to examine neighborhood context, residential mobil-ity, and mixed-income housing in Chicago and Los Angeles. The two anchor studies that formed the backbone of MIP are the Project on Human Development in Chicago Neighborhoods (PHDCN) and the Los Angeles Family and Neighborhood Survey (LAFANS). The PHDCN and LAFANS are widely recog-nized for rich longitudinal data on neighborhoods and on educational, health, and behavioral outcomes. For present purposes, we focus on the city of Chicago and the PHDCN.

Project on Human Development in Chicago Neighborhoods (PHDCN)

The PHDCN is a longitudinal cohort study of 6,207 children and their caregiv-ers based on a representative sample drawn from eighty neighborhood clusters (NCs) in Chicago in 1995. A two-stage sampling procedure was conducted. U.S. census data were first used to identify 343 NCs in the city of Chicago: groups of two to three census tracts containing approximately 8,000 people that were rela-tively homogeneous with respect to racial/ethnic mix, socioeconomic status, housing density, and family structure. From these, a random sample of 80 of the 343 NCs was drawn within twenty-one strata defined by racial/ethnic composi-tion (seven categories) and SES (socioeconomic status: high, medium, and low).

Second, within the sampled eighty NCs, children falling within seven age cohorts (0 [birth], 3, 6, 9, 12, 15, and 18) were sampled from randomly selected households based on a screening of more than 35,000 households. Dwelling units were selected systematically from a random start within enumerated blocks. Within dwelling units, all households were listed, and all age-eligible children were selected with certainty. Multiple siblings were thus interviewed within some households. At baseline, the resulting PHDCN sample was 16 percent European American, 35 percent African American, and 43 percent Latino, evenly split by gender, and representative of families living in a wide range of Chicago neighborhoods.

Extensive in-home interviews and assessments were conducted with the sam-pled children and their primary caregivers three times over a 7-year period, at roughly 2.5-year intervals (wave 1 in 1995–1997, wave 2 in 1997–1999, and wave 3 in 1999–2001). Participants were followed no matter where they moved in the

United States. Participation at baseline and retention at wave 3 were relatively high for a contemporary urban sample, 78 percent and 75 percent, respectively.

MIP research design

The MIP follow-up located and reinterviewed randomly sampled participants last contacted at wave 3 of PHDCN in the original birth cohort (now about age 16–17) and the age 9–15 cohorts (now 26–32). These cohorts were selected to maximize variation in life-course experiences and because the age 18 cohort had the highest attrition rate at wave 3, and the MIP pilot test indicated that the age 3 and 6 cohorts were most difficult to locate. The Chicago field operation engaged in a multimethod tracking effort using electronic, phone-based, and in-person methods (e.g., knocking on doors). The majority of interviews were carried out in person (almost 60 percent), but phone interviews were allowed if preferred by respondents or easier to implement. Despite the long time that elapsed since last contact at wave 3 and the contemporary big-city setting, MIP achieved a final response rate in the Chicago Main Study of 63 percent of eligible cases overall, yielding 1,057 respondents (40 percent Latino, 37 percent black, and 19 percent white).

For this article, we examine the 226, 236, and 217 respondents in the 9-, 12-, and 15-year old cohorts, respectively. These respondents spent their early years of development in an era that included the violence epidemic and severe urban challenges in Chicago, entered young adulthood during the widespread crime decline, and then experienced their late 20s and early 30s in the era of the Great Recession. To capture exposure to neighborhood income mixing over this time span, we geo-coded addresses of the MIP sample to census tract boundaries and merged them with waves 1–3 of the PHDCN. Each individual was thus linked to a census tract for each of the four waves of the combined PHDCN-MIP survey.[3] We then integrated census data across three decades and the American Community Survey (ACS) data from 2005–9, 2006–10, 2007–11, and 2008–12.

Mixed in the Middle

A clear definition or consensus on measures of mixed-income neighborhoods is noticeably absent in past research, in part because income mixing and income inequality within neighborhoods are closely related yet distinct concepts. One way to think about income mixing is to focus solely on measures of income dispersion, but high values of inequality can be generated at the extremes of the neighborhood income distribution. For example, the Gini Index of income inequality, available in the ACS for 2007–11, is highest in the poorest and richest census tracts, defined according to the lowest and highest quintiles of median income. This is not surprising, as there are very low and high incomes that stretch the distribution. But having a lot of variation at the high end is not what we would typically consider a desired outcome, at least in the world of mixed-income housing. A pragmatic mixed-income policy seeks to expose the poor to middle-class society—going from poverty to the "Upper East Sides" of the world is neither realistic nor necessarily desirable. It is thus hard to achieve a meaningful measure of mixed income absent a simultaneous focus on the level of income or where in the distribution a neighborhood is located.

After exploring multiple indicators of inequality and diversity, in addition to more traditional indicators of concentrated poverty and median income, we follow Massey (2001) by focusing on simplicity and clear metrics in defining the Index of Concentration at the Extremes (ICE) as $\frac{A_i - P_i}{T_i}$, where A is the number of affluent residents in neighborhood i, P is the number of poor residents in neighborhood i, and T is the total number of residents in neighborhood i. The ICE can theoretically range from –1 (where all residents are poor) to 1 (where all residents are affluent). Relatively greater income mixing, in the form of an even balance of poor and affluent residents, is centered at 0. Operationally, we calculate the ICE using the national upper- and lower-income quintiles of family income as the cutoffs for affluent and poor families, respectively, assigning an ICE score for each year from 1990 to 2010 (using interpolation) at the census tract–level in the Chicago area and for all neighborhoods in the United States. In addition to its clearly interpretable definition, the ICE controls for shifting income distributions over time.

The ICE classification is desirable because it focuses directly on income extremes, but it has potential limitations in distinguishing mixed-income areas from purely homogenous middle-class areas. We therefore explored additional definitions of mixed-income neighborhoods using the Herfindahl index of income diversity, the Gini Index of income inequality, and the Interquartile Range (IQR) of income for each tract based on household income bins. The Herfindahl index is defined as $1 - \sum_{i}^{q} p_i^2$, where income bins were based on quintiles. The Herfindahl index ranges from 0, indicating everyone living in the neighborhood is in just one income category, to 1, indicating complete evenness across categories. The range of the index is limited by the number of categories used: with 5 categories, the index's maximum value is 0.8.

Analysis of the Herfindahl index shows that the highest level of income diversity is in the mixed-income category of the ICE, indicating convergent validity. Also, as expected, the IQR is much higher in the middle category of the classification (where the ICE centers on 0) than in the low-income category, whereas the largest IQR (i.e., the largest absolute difference in 25th and 75th percentile of incomes) is driven by the extremely high incomes found in the upper 20 percent of the ICE, which denotes concentrated affluence. As noted earlier, the Gini Index is highest in the upper and lower fifths of neighborhood median income (.46) compared with the middle category (.41).[4] At least in American society, it thus appears that the greatest exposure to mixed-income populations—especially among the poor or near poor—is found in the middle of the neighborhood ICE distribution, which also favors more equality.

To further explore this argument, we examined the observed amount of income mixing in our individual-level MIP sample according to the ICE classification. To do so, we defined each MIP respondent in wave 1 by the quintile of *household* income and then compared the distribution of the sample across the quintiles of *neighborhood* ICE. The greatest income mix whereby individual poor residents are exposed to upper-income residents in roughly equal proportions is in our category of mixed middle-income, compared with more affluent or poverty areas. For example, 19 percent of MIP residents in the middle ICE category are low-income (lowest category of household income), compared with 17 percent affluent (highest category of household income)—a gap of just 2 percent. Every other ICE category has larger gaps—especially, not surprisingly, the lowest and highest categories of the ICE, where the absolute gaps are 37 and 43 percent, respectively. But even in ICE categories 2 and 4, the gaps are considerably higher: 26.2 and 9.1, respectively. We also calculated Herfindahl indices. Consistent with the above analysis, the highest level of income diversity is in category 3 of the ICE, at .79. Like the neighborhood level analysis for the United States and for Chicago Cook County, then, our individual-level sample validates the mixed middle-income classification.

Neighborhood-Level Results

Figure 1 shows the ICE distribution in Chicago's Cook County, where more than 80 percent of our sample remained during the follow-up. The map is based on quintiles, corresponding to the five labeled neighborhood types, ranging from "homogenous low income" to "homogeneous high income." The ACS estimates from 2008–12 reveal that 20 percent of family households earned less than $27,000 per year, and 81 percent earned less than $116,000 per year (in 2011 dollars). We used these income thresholds to define poor and affluent households for computing the ICE for an examination of current income distributions. In Cook County, the ICE ranges from –0.85 to 0.70. The map illustrates the spatial variability in income mixing across neighborhoods. Some 18 percent of tracts in Cook County are mixed income, and approximately 16 percent of the overall Chicago sample lives in the mixed middle-income category.

Are economically integrated neighborhoods stable or merely in transition between relatively homogenous high- (or low) and low- (or high) income areas? The process of gentrification suggests a directional shift from lower to higher income areas, whereas a neighborhood decline narrative suggests the reverse. An inspection of Figure 1 suggests that mixed-income areas in Chicago are more proximate to homogenously poor areas on the city's South and West sides than to homogenously affluent areas, suggesting the potential fragility of mixed-income neighborhoods over time.

We assess neighborhood dynamics in Figure 2 by displaying the relationship between the ICE indicator in 1990 and 2005–9 for all census tracts in both Chicago/Cook County and the rest of the United States ($N = \sim 64,000$). There is

FIGURE 1
Mixed-Income Classification in Chicago and Cook County, 2008–12

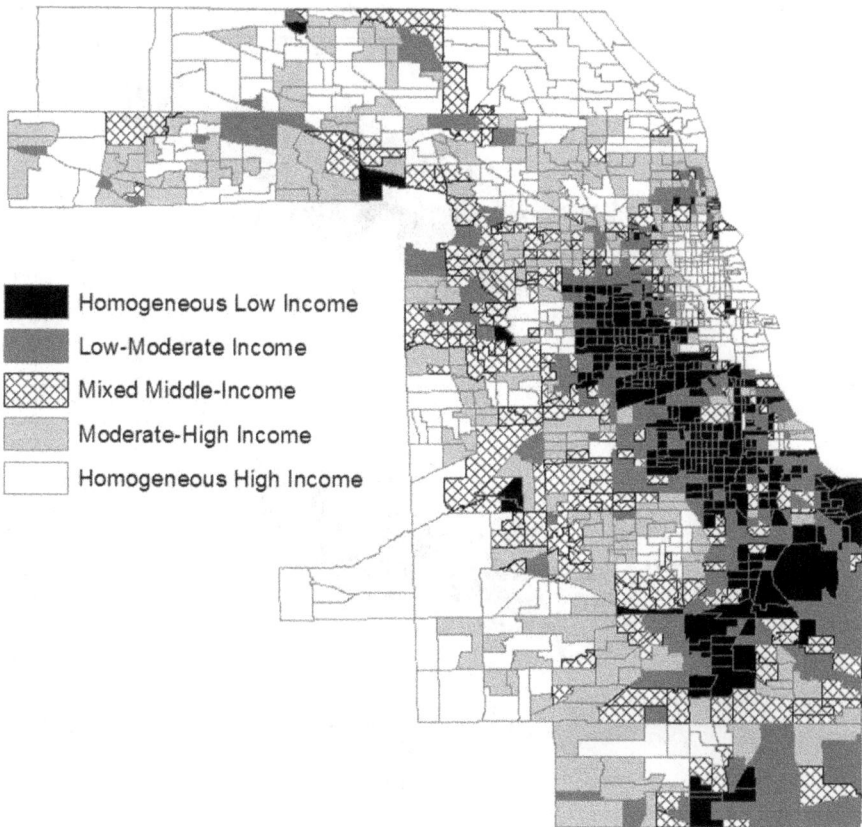

a high degree of persistence revealed in both graphs, especially at the high end, with areas of concentrated affluence largely remaining affluent. Chicago is thus not unique, and if anything, the persistence of income extremes is higher outside the Chicago area. For non-Chicago/Cook County tracts, the Pearson's and Spearman's rank-order correlations are very high—.87 and .86, respectively. Chicago comes in lower but still high, at.77 and .78. There is also suggestive evidence of gentrification in the Chicago area—note the areas above the regression line in the upper left quadrant of Figure 2, which represent poor neighborhoods that became more mixed income or affluent over time. Hwang and Sampson (2014) analyzed gentrification trajectories in Chicago and found that among neighborhoods that showed initial signs of gentrification in 1995 or that were nearby one another, trajectories of gentrification through 2009 were lower in areas with higher shares of blacks and Latinos—even after accounting for characteristics such as crime, poverty, and proximity to amenities.

FIGURE 2

ICE by Year, Chicago/Cook County and the Rest of the United States ($N = 63,709$ Census Tracts with 500 or More Population in 1990 and 2000)

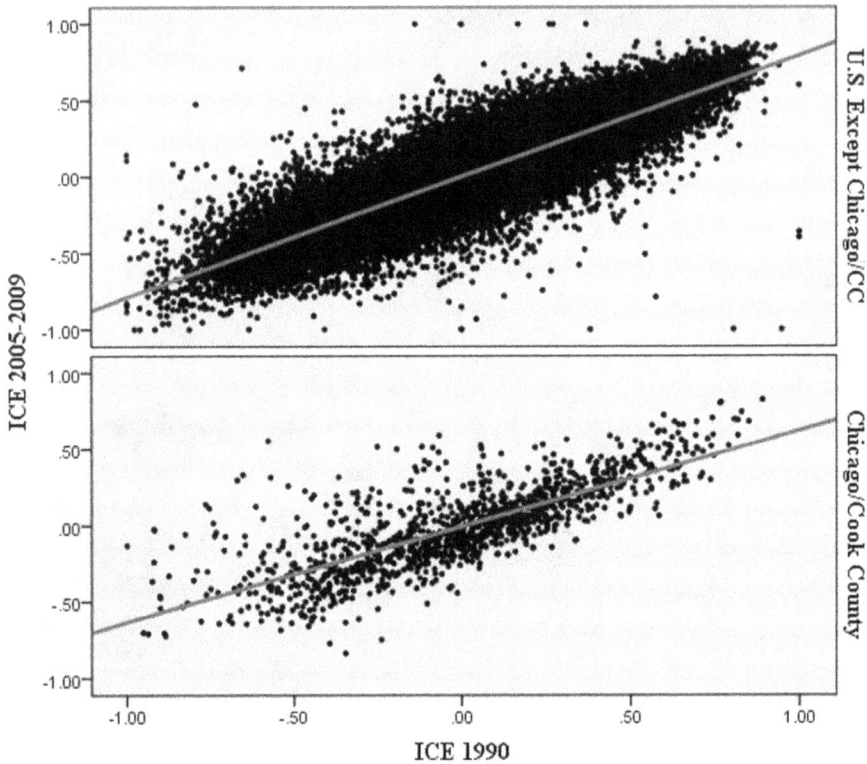

Although there is some evidence of upgrading and probably gentrification in areas that were largely poor, the overall correlations and transition matrices indicate that "stickiness" is the general rule. Indeed, the correlation of ICE over a 19-year time span exceeds .75 in both Chicago and the entire country, and in both settings the majority of neighborhoods that are in the top or bottom quintiles of income remain there. These patterns exist despite that inequality has been increasing at the top end (Reardon and Bischoff 2011)—the increase in the prevalence of rich neighborhoods does not override the strong inertial tendency of tracts to remain in a similar relative position over time.[5]

In Table 1 we look at population data from the Chicago area another way in the form of a transition matrix from 1990 to 2005–9 for quintiles of neighborhood ICE. Again we see evidence of greater persistence over time among neighborhoods at the extremes, with 65 percent or more of poor (lowest quintile) and affluent (highest quintile) neighborhoods remaining in the same category, compared with less than 30 percent of mixed middle-income neighborhoods. For the United States as a whole not counting the Chicago area, persistence is even

TABLE 1

Neighborhood-Level Transitions in the ICE, Chicago/Cook County (N = 1,264 Census Tracts, 1990 to 2005–9)

2005–9	1990 ICE Quintiles					
	1	2	3	4	5	Total
ICE Quintiles						
1	201	47	21	3	0	272
	66.34	30.72	10.66	1.02	0	21.52
2	45	49	76	28	1	199
	14.85	32.03	38.58	9.52	0.32	15.74
3	24	34	59	101	14	232
	7.92	22.22	29.95	34.35	4.42	18.35
4	23	12	27	133	94	289
	7.59	7.84	13.71	45.24	29.65	22.86
5	10	11	14	29	208	272
	3.3	7.19	7.11	9.86	65.62	21.52
Total	303	153	197	294	317	1,264
	100	100	100	100	100	100

greater—for example, 75 percent of affluent neighborhoods in 1990 remain affluent in 2007 and almost 70 percent of poor areas remain so.

These aggregate-level findings suggest that mixed-income neighborhoods are closely connected to larger structures of urban inequality. They also suggest that within the current structures that we observed, inducing large changes in exposure to neighborhood income mixing will likely result from individual mobility into and out of neighborhoods, rather than from changes within neighborhoods for persons who do not move. This is not to say neighborhoods do not change; clearly they do, and gentrification is real. But when we do not select on change and instead observe the universe of all neighborhoods either locally or nationally, we see strong evidence of the relative persistence of neighborhood position with respect to income.

Individual-Level Transitions and Trajectories in Chicago

The preceding analyses lead us to now consider individual-level trajectories of neighborhood income attainment. We ask, for example, how much change is there over time in the exposure of study participants to different income contexts, such as the transition from a poor neighborhood to a mixed-income or even affluent neighborhood? What is the profile of individuals who move into mixed middle-income neighborhoods over the life course of young adulthood? Is living in mixed-income areas a transitory state?

TABLE 2
Individual-Level Transitions in the ICE, Chicago Young-Adult Sample ($N = 671$
Individuals, 1995–2013)

Wave 4	Wave 1 ICE Quintiles					Total
	1	2	3	4	5	
ICE Quintiles						
1	140	50	30	6	4	229
	60.56	34.42	18.89	5.66	11.08	34.14
2	53	35	46	18	2	155
	23.17	24.36	29.27	17.5	6.24	23.06
3	23	36	33	15	3	109
	9.96	24.65	20.76	14.79	7.76	16.27
4	9	14	26	35	12	97
	3.99	9.73	16.69	34.61	33.53	14.46
5	5	10	23	28	15	81
	2.32	6.84	14.39	27.45	41.39	12.07
Total	231	145	157	101	37	671
	100	100	100	100	100	100

We begin with a transition matrix of individual exposure to concentrated poverty, concentrated affluence, and mixed middle-income neighborhoods over roughly an 18-year period (1995 to early 2013). Because we focus on the 9-, 12-, and 15-year-old cohorts, in effect we examine the transition from one's neighborhood in adolescence (of parental choice) to the destination neighborhood in young adulthood (the study participant's choice). As before, the neighborhood income measure is based on quintiles, but because study participants were free to move anywhere outside Chicago, we use national income distributions from 2012 to define the ICE quintiles.[6] These prevalence data and all analyses to follow are weighted to account for the stratified sample design at wave 1 and for potential attrition bias over time in the follow up.[7]

The results in Table 2 demonstrate that there is considerable inertia in the residential exposure of our sample at the low and high values of the ICE. Indeed, despite that there are five income categories, fully 61 percent of respondents who live in neighborhoods in the lowest ICE quintile at wave 1 remain in neighborhoods in the lowest ICE quintile almost two decades later at wave 4: in this sense, concentrated poverty exposure is durable. At the affluent extreme, just over 41 percent of respondents who live in neighborhoods in the top 20 percent of the ICE distribution at wave 1 remain there at wave 4 and another 34 percent end up in the next lowest category; thus only about a quarter of this group was significantly downwardly mobile.

By contrast, considerable fluidity characterizes the middle of the distribution. Based on our earlier analysis, we define mixed-income neighborhoods as those falling between the second and third quintile of the ICE. Higher shares of respondents living in mixed-income neighborhoods at wave 1 experience a change in neighborhood income mix by wave 4 than do respondents who live in more homogenous lower or higher income neighborhoods at wave 1. We see this pattern from the fairly even distribution of values in the column corresponding to residence in ICE category 3 at wave 1. The values in this column are more similar to each other than the values in other ICE quintile columns. In fact, only 21 percent of respondents who live in mixed-income neighborhoods at wave 1 remain in mixed-income neighborhoods at wave 4. This percentage is considerably lower than the share of respondents who start and end in more homogenous lower- or higher-income neighborhoods; continuous exposure to mixed-income areas is relatively rare—only 5 percent of our respondents live in mixed-income neighborhoods at both waves 1 and 4, whereas 65 percent of our respondents do not live in mixed-income neighborhoods at either wave. The balance, nearly a third of the sample (30 percent), transitions into or out of mixed-income neighborhoods between 1995 and 2012, once again suggesting fluidity in exposure to mixed-income neighborhoods.[8]

It is important to note that in Table 2 we used the national distribution of the ICE to determine the quintile cut points. Based on the national distribution, our mixed-income category consists of neighborhoods with ICE scores between –0.08 and 0.06 in 1995 and –0.06 and 0.10 in 2012. To test whether the results are sensitive to this definition, we examined ICE cut points based on the weighted distribution of sample respondents by the neighborhoods they lived in at waves 1 and 4. In this more local specification that is dominated by Chicago and its suburbs, the mixed-income category of neighborhoods includes those neighborhoods with ICE scores between –0.17 and –0.04 in 1995 and –0.19 and 0.03 in 2012. However, this definition does not alter our main findings about exposure to mixed-income neighborhoods. By both the national distribution and sample-based definitions, 5 percent of respondents lived in mixed-income neighborhoods at both waves 1 and 4. While using the national distribution shows that 21 percent of respondents who start in mixed-income neighborhoods end up in mixed-income neighborhoods, the corresponding statistic using the sample-based distribution is 24 percent. The basic pattern of exposure to mixed-income neighborhoods that we report above is thus robust to different cut points of the ICE variable.

Although revealing, the aggregate transition matrix in Table 2 conceals potential heterogeneity in exposure to poverty, affluence, and mixed-income areas by the race and ethnicity of the study participants. We thus examined exposure to our mixed-income classification of neighborhoods across the 18 years separately for Latinos, non-Latino blacks, and non-Latino whites. To simplify the data when looking at race and ethnicity, we examined a two-category classification that collapses the ICE quintiles into a mixed-income category versus all else. The results show that almost a third (29 percent) of young-adult Latinos who lived in mixed-income neighborhoods at wave 1 remained in mixed-income neighborhoods at wave 4, compared with only 18 percent of blacks and 13 percent of whites living in mixed-income neighborhoods at wave 1. Among respondents who do not live

FIGURE 3
Chicago ICE Trajectories of Young-Adult Sample by Race/Ethnicity, Adjusting for
Baseline Homeowner Status, Immigrant Generation, and Age (95 Percent CI)

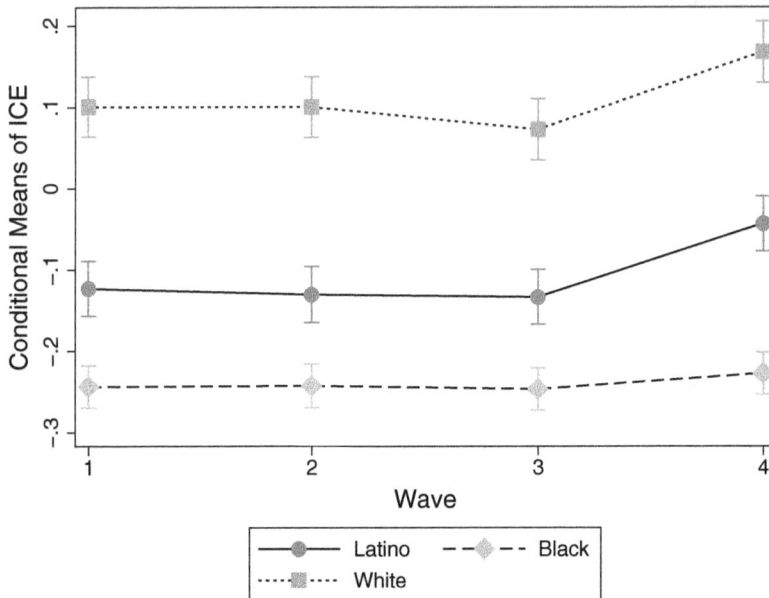

in mixed-income neighborhoods at wave 1, Latinos also have the highest rate of living in mixed-income neighborhoods at wave 4: 23 percent, compared with 12 percent and 11 percent for blacks and whites, respectively. Exposure to mixed-income neighborhoods is lowest among blacks, 70 percent of whom did not live in mixed-income neighborhoods at either wave 1 or 4 (table not shown). The equivalent statistics for Latinos and whites are 56 and 66 percent. Based on these results, Latino-Americans, at least in Chicago, are much more exposed to mixed-income neighborhoods.[9]

Another way to look at race differences is to examine the average trajectories of each group across the four waves of the MIP-Chicago study. Figure 3 presents the weighted trajectory results for ICE by race/ethnicity, with 95 percent confidence intervals shown for each group. In estimating these trajectories, we adjust for age cohort, homeowner status at wave 1, and parental immigrant status. As seen in Figure 3, the conditional mean differences among race/ethnic groups are quite stark at each wave of the study. Whites are much more likely to be exposed to higher values of concentrated affluence, and blacks to concentrated poverty (values of ICE below −.20). Latinos are again in the middle, with the final neighborhood attainment at nearly the middle (or .00) of the ICE distribution.

What is not yet clear is whether the consistent difference in experience we have observed so far among Latinos is a result of initial differences in socioeconomic status or different trajectories of residential mobility over time. In addition, we have not yet explored direct models of change in mixed-income exposure that simultaneously take into account demographic, socioeconomic, and residential mobility status. It is to these issues we now turn.

Who transitions to a mixed middle-income neighborhood?

Table 3 displays a series of models that predict whether our sample of young adults lives in a mixed-income neighborhood in 2012, controlling for baseline residence type in 1995. We are thus estimating a change model—that is, a model of who is most likely to transition into mixed-income residence over the course of the study. We begin with a logistic regression model that uses race, age, and residence in a mixed-income neighborhood at wave 1 to predict residence in a mixed-income neighborhood at wave 4. The results in model 1 show that living in a mixed-income neighborhood at wave 1 is positively related to living in a mixed-income neighborhood at wave 4 (model 1, OR = 1.40), but the coefficient is not significant at conventional levels. Controlling for age and residence in mixed-income neighborhood at wave 1, however, Latinos are substantially (2.7 times) more likely than whites to live in a mixed-income neighborhood at wave 4. Blacks are no more likely than whites to live in a mixed-income neighborhood at wave 4, and being in age cohort 12 compared with cohort 9 is only marginally associated with mixed-income neighborhood residence at wave 4.

Model 2 adds immigrant generation of the respondent's parents to the model, capturing the immigrant context of the household in which the respondent was an adolescent. Compared with respondents with third-generation parents and controlling for race, age, and residence in a mixed-income neighborhood at wave 1, respondents with first- and second-generation parents are no more or less likely to live in a mixed-income neighborhood at wave 4. Adding immigrant generation to the model also does not change the main result from model 1: Latinos compared to whites are significantly more likely to live in a mixed-income neighborhood at wave 4, but controlling for generational status actually increases the odds ratio for Latinos to 2.8.

In model 3, we add indicators of parental home ownership at wave 1 and whether the respondent lived in public housing at wave 2 (circa 1997). Neither growing up in a house that was owned nor living in public housing significantly predicts residence in a mixed-income neighborhood at wave 4, controlling for living in a mixed-income neighborhood at wave 1, race, age, and parent immigrant generation. The odds ratio for being Latino compared with being white remains substantial (3.2) after accounting for public housing and homeownership, while residence in a mixed-income neighborhood at wave 1 continues to be insignificant.

Model 4 is restricted to respondents who at wave 1 lived in homes that were rented. Although this specification reduces our sample size considerably, the model is justified because public housing residents were by definition not owners. In this restricted model, public housing is still not significantly associated with ending up in a mixed-income neighborhood at wave 4.

TABLE 3
Logistic Models Predicting Mixed Middle-Income Status at Wave 4

Predictors	Models					
	(1)	(2)	(3)	(4)	(5)	(6)
Mixed-income, wave 1	1.402 [0.797, 2.469]	1.381 [0.782, 2.439]	1.376 [0.700, 2.704]	2.471 [0.757, 8.069]	1.317 [0.733, 2.364]	1.182 [0.642, 2.174]
Latino	2.694°° [1.314, 5.523]	2.851°° [1.422, 5.719]	3.187°° [1.378, 7.372]	1.845 [0.417, 8.159]	2.927°° [1.380, 6.209]	3.413°° [1.545, 7.541]
Black	1.165 [0.532, 2.548]	1.147 [0.474, 2.774]	0.889 [0.317, 2.497]	0.652 [0.0809, 5.260]	1.142 [0.455, 2.866]	1.227 [0.479, 3.146]
Cohort 12	1.807+ [0.984, 3.318]	1.760+ [0.956, 3.239]	1.384 [0.692, 2.770]	0.820 [0.305, 2.201]	1.706+ [0.929, 3.133]	1.773+ [0.953, 3.298]
Cohort 15	0.922 [0.478, 1.779]	0.914 [0.472, 1.769]	0.728 [0.338, 1.565]	0.798 [0.286, 2.225]	0.881 [0.454, 1.709]	0.898 [0.446, 1.809]
Parent 1st generation		0.920 [0.415, 2.037]	0.718 [0.270, 1.911]	2.143 [0.412, 11.15]	0.922 [0.403, 2.108]	0.870 [0.374, 2.020]
Parent 2nd generation		0.903 [0.327, 2.489]	0.674 [0.215, 2.112]	0.703 [0.114, 4.327]	0.911 [0.333, 2.490]	0.945 [0.357, 2.502]
Homeowner, wave 1			1.253 [0.662, 2.373]		1.263 [0.696, 2.295]	1.313 [0.639, 2.698]
Public housing, wave 2			0.332 [0.0695, 1.589]	0.302 [0.0462, 1.969]		
Moved in Chicago					0.769 [0.434, 1.364]	
Moved outside of Chicago					0.383° [0.160, 0.916]	
Parent education						1.042 [0.766, 1.418]
Household income: $10,000–$19,999						0.547 [0.174, 1.724]

(continued)

TABLE 3 (CONTINUED)

			Models			
Predictors	(1)	(2)	(3)	(4)	(5)	(6)
$20,000–$29,999						1.852
						[0.696,
						4.929]
$30,000–$39,999						1.290
						[0.426,
						3.906]
$40,000–$49,999						1.103
						[0.270,
						4.515]
> $50,000						1.273
						[0.397,
						4.080]
N	645	638	498	277	627	602

NOTE: Exponentiated coefficients; 95 percent confidence intervals in brackets.
$+p < .10.$ $°p < .05.$ $°°p < .01.$

We next created indicators of neighborhood mobility status in the first three waves of the PHDCN: stayers, or those who did not move out of their baseline neighborhood over the first three waves (approximately seven years); movers within Chicago in the same period; and movers outside of the city proper again over the same period. Approximately 60 percent of the adolescent sample stayed in their family neighborhood as they transitioned to young adulthood, just under a third moved between neighborhoods but stayed within Chicago, and 11 percent moved out of the city. When we add the two indicator variables for moving within Chicago and out of Chicago (with the reference category being stayers) to model 5 in Table 3, we see that mobility status out of the city is associated with 62 percent lower odds of living in a mixed-income neighborhood compared with stayers; moves within the city are not significantly associated with mixed-income destinations. It appears that moving out of Chicago—mostly to the suburbs, but including regions around the country—is associated with increased odds of living in concentrated affluence and reduced odds of living in a mixed-income neighborhood.

The last model in Table 3 takes into account the relationship between family socioeconomic status during adolescence and residence in a mixed-income neighborhood at wave 4. To estimate this relationship, we include parental education and a series of dichotomous indicators for five categories of family income (with the reference group as those making less than $10,000 a year). Controlling for mixed-income neighborhood at wave 1, race, age, parent immigrant generation, and both homeownership and income status at wave 1, model 6 shows that none of the income indicators are significant.[10] Nor does parental education predict residence in a mixed-income neighborhood at wave 4. Including the socioeconomic status of the family of origin, therefore, does not alter our consistent findings that Latinos are three to four times more likely to live in a mixed-income neighborhood than are whites.

One question remains. Because it is known that Latinos are disproportionately likely to live in neighborhoods with a greater Latino composition, do our findings then reflect a contextual compositional effect? For example, are Latinos simply more attracted to Latino areas? Building from the full specification in model 6, we reestimated our results controlling first for percent Latino at baseline (1995) and then in the destination neighborhood (ACS 2008–12). In neither case was percent Latino significant, but the individual-level odds ratios for Latinos compared with whites remained large and significant (2.96 and 3.4, respectively). In addition, we compared Latino composition by income quintiles for the United States and Chicago. In both cases, the percentage Latino is actually higher or the same in the first two quintiles. In fact, for the United States as a whole, percent Latino is twice as high in the bottom quintile compared with the mixed middle income. Like socioeconomic status, contextual ethnic composition thus does not explain the strong pattern of Latino transitions to mixed middle-income neighborhoods.

Conclusion

Increases in income segregation have generated considerable debate, especially the separation of the very rich from everyone else. But our analysis shows that separation at the top is expressed in stark spatial terms and that this has been the case for some time. Indeed, about 75 percent of all neighborhoods in the United States in the highest quintile of income in 1990 were still there almost two decades later. The concentration of poverty is also very persistent, with more than 70 percent of poor neighborhoods remaining poor over time; for the Chicago area, two-thirds of both poor and affluent areas retained their position. Mixed-income neighborhoods that are naturally occurring are comparatively more unstable; there appears to be more churning or turnover in the middle of the neighborhood income distribution rather than the rich and poor neighborhoods changing places.

How do individuals fare against this strong pattern of structural inequality? Surprisingly, much less is known about exposure to neighborhood income mixing across the life course. To correct this gap, we examined 671 adolescents in Chicago who were followed over an 18-year span as they transitioned to young adulthood. The results show that the concentration of income extremes is persistent in the lives of individuals almost as much as in neighborhoods. Exposure to mixed middle-income neighborhoods is more infrequent and unstable.

The persistence of concentrated income extremes suggests that policies seeking to promote mixed-income neighborhoods face greater odds than commonly thought. Yet one pattern deserves consideration in thinking about the stability of mixed-income areas. When we considered characteristics of the individuals in our sample, Latino-Americans were much more likely than African Americans or whites to be exposed to mixed middle-income neighborhoods in the first place and to transition into them over time, adjusting for immigrant status, education, income, and residential mobility. Interestingly, this pattern is not explained by Latinos simply moving from or to neighborhoods with more Latinos.

Although our results call for more probing, the influx of immigration from Latin American countries may be not only creating a more diverse society but also inducing more income-mixing in the middle of the income distribution, perhaps offsetting what would otherwise be larger losses in the middle class as income inequality and the spatial separation of the poor and affluent increase. Immigrant and ethnic diversity thus need to be central to any debate over income mixing; the "black and white" frame that has dominated the urban sociology literature for decades will not suffice to capture the neighborhood context of income mixing now.

Notes

1. For a review of trends and an analysis of the increasing separation of poverty and affluence at the neighborhood level, see Reardon and Bischoff (2011).

2. Mixed-income housing is not without controversy. For a discussion of conflict in New York City's efforts, see http://nyti.ms/1pDg4G5.

3. The mobility of the sample reached well beyond the neighborhood clusters (themselves made up of census tracts) of the original PHDCN sampling design. We thus use census tracts to capture the income status of destination neighborhoods, assigning 2000 census tract boundaries for waves 1–3 and 2010 boundaries for wave 4. This strategy comports with past research and reflects the most accurate measure of the neighborhoods in which the participants were living at the time of each wave of data collection. For the neighborhood-level analyses we use 2000 boundaries to track neighborhood change from 1990 to 2009.

4. The entropy index is another candidate, but it gives the highest value to neighborhoods that are 50 percent poor and 50 percent affluent, which is a very rare context. Of census tracts in Chicago with more than 40 percent affluent, for example, none have a percentage poor greater than 40 and just a handful have greater than 25. The middle of the ICE distribution captures 50/50 splits but also the more realistic splits closer to 20 or 30 percent poor and 20 to 30 percent affluent.

5. We also decomposed the ICE into variability between census tracts and over time for all neighborhoods in the United States for the independent census years 1990, 2000, and 2005–9. The results revealed that 88 percent of variation is between neighborhoods. The same general pattern holds in Chicago. Thus the variability in concentrated income extremes is predominantly between neighborhoods rather than over time, which means that, consistent with Figure 2, relatively few neighborhoods are switching positions in the income distribution.

6. To maximize the temporal match to the main MIP follow-up interview year in 2012, we use the ACS, 2008–12. Because the Chicago area is similar to the national distribution of income, the results are similar if we use the local (county-level) quintile distribution.

7. The sampling weight is designed to adjust for the original stratification of the PHDCN by neighborhood SES and racial composition, along with the age-cohort selection and a post-stratification of population weights to census estimates of the age, gender, and race/ethnicity distribution of children in the City of Chicago in 1995. The attrition weight is defined as the inverse of the probability of being interviewed at wave 4 conditional on being in the study at wave 3. To model the probability of attrition at wave 4, missing data from waves 1 through 3 were first multiply imputed using chained regression equations. Attrition weights were then calculated by estimating a logit model for the probability of attrition at wave 4, based on individual- and household-level measures (e.g., socioeconomic status and family structure) as well as

neighborhood-level measures of demographic composition and social processes (such as collective efficacy and perceived violence). The inverse of each subject's probability of response was then calculated and standardized by the mean to yield the final attrition weights. The stratification and attrition weights were multiplied to produce the final weight. We examined results separately using the baseline stratification weights and attrition-based weights, but the patterns were very similar; therefore, we present data based on the final weights.

8. We also examined the Herfindahl index of income diversity defined earlier. Consistent with our typology, income diversity is significantly higher in the mixed-income category (.75) than the non-mixed-income category (.68) in our wave 4 sample (t-ratio of difference = −9.81; $p < .01$).

9. We also examined the immigrant status of parents among Latinos; there was a similar level of consistent exposure to mixed-income neighborhoods when comparing the first generation to the second and third generations. There is a somewhat bigger difference in mixed-income at wave 4 conditional on wave 1, with first-generation immigrants having a higher prevalence of living in mixed income at wave 4 conditional on wave 1.

10. We also examined an alternative model where we substituted welfare (TANF) receipt by the parent at wave 1 for income. But like income, welfare did not predict living in a mixed middle-income neighborhood. In addition, although most covariates in Table 3 have only modest levels of missing data, we reestimated all models using multiply imputed data to assess the robustness of results. We obtained substantively similar results.

References

Bruch, Elizabeth E., and Robert D. Mare. 2006. Neighborhood choice and neighborhood change. *American Journal of Sociology* 112:667–709.

Cisneros, Henry G., and Lora Engdahl, eds. 2009. *From despair to hope: HOPE VI and the new promise of public housing in America's cities.* Washington, DC: Brookings Institution Press.

Ellen, Ingrid Gould. 2000. *Sharing America's neighborhoods: The prospects for stable racial integration.* Cambridge, MA: Harvard University Press.

Hwang, Jackelyn, and Robert J. Sampson. 2014. Divergent pathways of gentrification: Racial inequality and the social order of renewal in Chicago neighborhoods. *American Sociological Review* 79:726–51.

Joseph, Mark L., and Robert J. Chaskin. 2012. Mixed-income developments and low rates of return: Insights from relocated public housing residents in Chicago. *Housing Policy Debate* 22 (3): 377–405.

Joseph, Mark L., Robert J. Chaskin, and Henry S. Webber. 2007. The theoretical basis for addressing poverty through mixed income development. *Urban Affairs Review* 42:369–409.

Massey, Douglas S. 2001. The prodigal paradigm returns: Ecology comes back to sociology. In *Does it take a village? Community effects on children, adolescents, and families*, eds. Alan Booth and Ann Crouter, 41–48. Mahwah, NJ: Lawrence Erlbaum Associates.

Moore, Natalie. 2013. CHA slows down on mixed-income housing. Available from http://www.wbez.org.

Piketty, Thomas. 2014. *Capital in the twenty-first century.* Cambridge, MA: Harvard University Press.

Reardon, Sean F., and Kendra Bischoff. 2011. Income inequality and income segregation. *American Journal of Sociology* 116:1092–1153.

Sampson, Robert J. 2012. *Great American city: Chicago and the enduring neighborhood effect.* Chicago, IL: University of Chicago Press.

Sampson, Robert J., and Patrick Sharkey. 2008. Neighborhood selection and the social reproduction of concentrated racial inequality. *Demography* 45:1–29.

Wilson, William Julius. 1987. *The truly disadvantaged: The inner city, the underclass, and public policy.* Chicago, IL: University of Chicago Press.

Immigrant Context and Opportunity: New Destinations and Socioeconomic Attainment among Asians in the United States

By
CHENOA FLIPPEN
and
EUNBI KIM

Immigrant-origin populations, once overwhelmingly concentrated in a handful of receiving gateways, have dispersed in recent decades to scores of new destinations throughout the United States. This pattern and its implications for immigrant incorporation have received a great deal of attention, but the vast majority of research has focused on Hispanics. This article examines the relationship between settlement patterns and socioeconomic attainment (income, occupational status, and homeownership) among Asians. Drawing on individual- and metro-level information from the 2009 to 2011 American Community Survey, results suggest that Asians in new destinations face an important trade-off between income and homeownership, and that differences across contexts are largely attributable to metropolitan labor and housing market conditions, rather than the ethnic context per se. However, there are important differences in outcomes among Asians by national origin and sex, and a comparison with whites suggests that inequality differs across new and more established Asian settlement areas.

Keywords: new destinations; Asian; immigration; status attainment

Social scientists have long been concerned with the impact of context on individual outcomes, including socioeconomic attainment. Immigration scholarship, in particular, has a lengthy tradition of considering how the context of reception shapes the adaptation of immigrant origin groups. Most of this research has focused on broad, macroeconomic and

Chenoa Flippen is an associate professor of sociology at the University of Pennsylvania. She has published extensively on the link between social context and ethno-racial stratification in the United States, Hispanic immigrant incorporation, and internal migration.

Eunbi Kim is a graduate student in sociology at the University of Pennsylvania. Her interests center on globalization, transnational corporations, and Asian immigration to the United States.

NOTE: Authors are listed alphabetically, reflecting equal contribution to the work.

DOI: 10.1177/0002716215577611

macrosocial aspects of context, such as the implications of deindustrialization and increasingly punitive immigration policies on successive immigrant-origin cohorts (Alba and Nee 2003). However, recent shifts in immigrants' geographic settlement patterns have invited more attention to how the conditions of local areas shape variation in outcomes within groups.

The geographic dispersion of U.S. immigrant-origin populations has grown rapidly since 1980. Once overwhelmingly concentrated in a handful of established metropolitan areas, immigrant-origin populations are now living in scores of new destinations, particularly throughout the Southeast and Midwest (J. Wilson and Singer 2011). This pattern and its implications for immigrant incorporation have received a great deal of attention, but the vast majority of the research has focused on Hispanics, with relatively little research on Asians (Desai and Joshi 2013). The relative dearth of research on new Asian destinations is problematic for several reasons. First, the Asian population is large and growing. According to the 2010 U.S. Census, 5.6 percent of the U.S. population is of Asian origin, with a large proportion foreign-born and recently arrived. Second, Asian populations were even more highly concentrated historically than Hispanics, and their geographic dispersion has been even more dramatic (Frey 2011). Third, Asians average relatively high socioeconomic status among both natives and immigrants. Most of the literature on the contextual influences on immigrant adaptation implicitly assumes immigrant disadvantage; studies are often at least partially motivated by the question of whether low-skill and nonwhite immigrants and their descendants are better able to overcome the challenges of labor market disadvantage and discrimination in contexts with sizable coethnic communities to draw on for support. How context shapes the adaptation of groups that are relatively advantaged has not received adequate consideration.

Accordingly, this article examines the link between Asian settlement patterns and their socioeconomic attainment. We pay particular attention not only to differences between traditional and new areas of destination but also to these areas' interaction with national origin and gender in explaining variation in outcomes within the Asian group. We do so by first constructing a typology of traditional, new, and other areas of Asian destination. We then examine differences among Asians' labor and housing market outcomes according to metropolitan context. That is, we analyze whether residents of new and traditional areas experience different levels of income, occupational status, and homeownership after controlling for individual demographic, human capital, and immigration characteristics. Income and occupational status are two of the most common barometers of socioeconomic attainment, while homeownership is often considered a hallmark of middle-class status, central to the process of wealth accumulation, and for immigrants an indicator of incorporation into U.S. society. We further explore the mechanisms behind contextual influences by assessing the extent to which the differences observed across Asian destinations remain after controlling for local labor and housing market conditions. And finally, we also examine the link between immigrant destinations and Asians' position relative to non-Hispanic whites (hereafter "whites"). Examining the interaction between race and metro context provides important insight into the potential impact of context on inequality.

Theoretical Background

Classical assimilation theory tended to frame the outcomes of immigrant-origin groups as a function of the characteristics inherent in the groups themselves. Individuals and national origin groups were expected to advance, both over time and across generations, in conjunction with acculturation and greater resemblance to the American "mainstream" (Gordon 1964). However, growing diversity in both the origins and outcomes of immigrant-origin populations has prompted more attention to variation across national origin groups. Segmented assimilation theory, in particular, argues for the need to systematically incorporate both the individual resources immigrants bring from abroad, or context of origin, and larger conditions in the receiving society, or context of reception (Portes and Borocz 1989).

Greater recognition of variation in the context of origin is especially salient for understanding the trajectories of Asian groups. One of the limitations of classical assimilation theory was the assumption of immigrant disadvantage. The theory for the most part was formulated to explain the incorporation of low-skilled immigrants and the conditions that would promote their advancement. The possibility of immigrant advantage relative to natives was largely unanticipated, or at least not discussed. Segmented assimilation theory, in contrast, emphasizes that variation in sending contexts, recruitment practices, and selection mechanisms affects who arrives to the United States and their position within the stratification system. While the literature tends to emphasize ethnic differentials, such as those separating Hispanics and Asians, the perspective can also accommodate national origin and gender differences within ethnic groups.

More importantly, segmented assimilation theory also asserts the importance of the context of reception in shaping immigrant-origin outcomes. Factors such as governmental response, legal context, and degree of native hostility vary across groups in ways that shape a host of outcomes, over and above the impact of individual attributes. A central dimension of the context of reception for our purposes is the coethnic community—its size and nature. Ethnic communities enhance social networks and access to information, which can boost labor and housing market prospects for immigrants and their descendants (Zhou 1997). In certain cases, ethnic communities are large enough to form residential and business enclaves. The relationship between ethnic enclaves and immigrant incorporation has been the subject of considerable debate (Xie and Gough 2011). On one hand, enclaves offer recent immigrants and their children comfort, solidarity, and ready access to employment even without English language skills and knowledge of U.S. society, potentially augmenting employment opportunities and offering informal training in return for cheap labor (K. Wilson and Portes 1980). On the other hand, enclave wages are highly polarized, with business owners faring relatively well and laborers often earning less than similarly skilled coethnics working outside the enclave (Sanders and Nee 1987). While the impact of enclave economies on labor market outcomes may be open to debate, large coethnic communities may shape the behavior of local institutions, such as schools, making them better equipped and more adept at servicing immigrant-origin communities. In the realm of housing, in particular, established coethnic communities arguably help minorities to gain access to the information necessary to navigate lending and real estate industries, facilitating homeownership (Flippen 2010). Thus coethnic communities can be an important source of infor-mation and support and facilitate adaptation in a wide array of outcomes.

New destinations and the context of reception

The vast majority of writing on segmented assimilation and the context of reception has focused on cross-national comparisons, to the relative neglect of how more localized variation in context can shape disparate outcomes within groups. A large part of the neglect of cross-locale variation in adaptation has been due to the fact that immigrants have historically settled in large cities in the Northeast or West that not only offered numerous employment opportunities but also had established ethnic communities. However, in recent decades the geo-graphical pattern of ethnic settlement has become more varied; while traditional gateways remain major receivers of new immigrants, approximately one-third of immigrants now reside outside of traditional settlement areas, and new destina-tions throughout the South and Midwest are experiencing some of the fastest immigrant growth rates in the nation (Frey 2011). The settlement patterns of Asians closely mirror those of the immigrant population at large. Indeed, new Asian immigrant destinations and new Asian destinations are generally one and the same, both because such a large share of the Asian population is foreign born and because native and foreign-born populations have shown similar patterns of dispersal over time.

These patterns present new opportunities for examining the impact of context on the incorporation of immigrant-origin populations (see Lichter, Parisi, and Taquino, this volume). New destinations differ from more traditional gateways in at least two important respects. First, they by definition lack a coethnic commu-nity. This implies that the institutions and social networks that are well developed in traditional gateways are absent or nascent in new destinations. The lack of ethnic networks could hinder the incorporation of Asians into the labor and hous-ing markets. On the other hand, competition with coethnics is likely to be less intense, particularly in the occupational niches in which Asians concentrate, potentially enhancing incorporation and social position relative to whites.

Second, relative to traditional areas, new destinations average radically differ-ent economic and housing structures. A long literature outside of immigration studies links population movements to regional and local variation in opportunity structures (Greenwood 1997). New Asian destinations are found throughout the country, but are disproportionately located in the South and Midwest. While considerable convergence has occurred over time across major U.S. regions, the

South, in particular, still averages lower wages and occupational prestige than do other regions. At the same time, housing market conditions (especially costs) are generally more favorable in many new destinations. It is thus crucial to distinguish between the impact of immigrant context and economic context when assessing the importance of new destinations.

Asian settlement patterns in historical perspective

Asian immigration to the United States dates back to at least the 1840s, when Chinese laborers were attracted to the West Coast by the Gold Rush. Migrants from China and Japan were important sources of labor in the West for much of the latter half of the nineteenth century, as were Koreans and Filipinos to a lesser extent. While the West Coast was the point of entry for most Asians at the time, exclusion and mistreatment encouraged many to try their fortunes in large urban areas in the East (Zhou 1992). Several decades of restrictive immigration legislation, including the Chinese Exclusion Act of 1882, 1908 Gentlemen's Agreement with Japan, and National Origins Quota acts of the 1920s, dramatically restricted Asian entry into the United States. Many Asian communities survived during this period but did not generally expand. The 1965 Hart Celler Immigration Act, however, marked a major turning point in Asian immigration history. The act removed the national origins quotas and explicit bans on Asian entry, and Asian immigration rose 600 percent between the early and late 1960s. The national origins of the U.S. Asian population also further diversified with an influx of migrants from Vietnam, Cambodia, and Laos following the Vietnam War, as well as skilled and unskilled labor migrants from other parts of the region.

While each national origin group has its own unique immigration history, which is reflected geographically, most of the post-1965 wave of Asian immigrants settled in the large cities in the West and Northeast that had longstanding Asian populations (Logan, Alba, and Zhang 2002; Min 2006). Thus, while some groups are heavily overrepresented in particular areas (e.g., Chinese, Filipinos, and Koreans in Los Angeles; Chinese and Indians in New York), the largest Asian groups, namely Chinese, Indians, Filipinos, Vietnamese, Koreans, and Japanese, are fairly evenly distributed in the top Asian destinations.

Overall, Asian geographic concentration remains pronounced; more than one-third live in just three metro areas (Los Angeles, New York, and San Francisco) and more than half live in the top ten destinations (Frey 2011). Absolute gains in Asian population have also been largest in traditional gateways. Nevertheless, in recent decades Asian dispersion to new destinations has been explosive; there are now more Asians residing in Texas than Hawaii, and the number in Atlanta now rivals that of San Francisco (Desai and Joshi 2013). Asians are in many ways following the same Sunbelt migration stream as Hispanics, whites, and African Americans. Several of the fastest growing Hispanic and Asian metros overlap, such as Richmond, VA, Raleigh, NC, and Indianapolis, IN. However, Asian movement into the South has been even larger than for other groups, and Asian populations are also growing in parts of the Southwest and south Florida that had large preexisting Hispanic populations (Schachter 2003). At the same time, metro areas in the

West such as Phoenix, Las Vegas, and Riverside have also attracted a large number of Asians from nearby traditional destinations (Frey 2011).

The significance of new destinations for Asian incorporation

In recent years a number of studies have compared educational, labor market, and housing outcomes across new and traditional destinations (Flippen 2010; Kochhar, Suro, and Tafoya 2005; Stamps and Bohon 2006). However, the vast majority of these studies focused on Hispanics, and new Asian destinations have remained seriously understudied (Waters and Jimenez 2005). There are a number of reasons why the contrast between new and traditional destinations may be important for Asians. First, like Hispanics, Asians are leaving labor markets where competition among similarly skilled coethnics is intense, for regions with very different labor market structures. However, even more than Hispanics, Asians are leaving high-wage, high occupational status, and high cost-of-living metro areas and settling in more affordable but lower wage areas. The net effect of these differences in contexts is important to assess.

Second, much of the work on immigrant incorporation is predicated on the assumption of immigrant disadvantage and how context may help newcomers to overcome the simultaneous challenges of low skill and minority status. However, Asians are unique among U.S. ethno-racial minorities in that they average higher levels of educational attainment than native whites, even among the foreign born. While some have argued that Asians are victims of a glass ceiling, suffering an earnings penalty relative to whites with similar education and skills, more recent research has challenged this idea, arguing that when factors such as foreign educational credentials are taken into consideration there is no evidence of systematic disadvantage (Sakamoto, Goyette, and Kim 2009; Zeng and Xie 2004). Advantaged immigrants may benefit less from the flow of information and cultural and linguistic buffers that large coethnic communities imply. And they may suffer more from the heightened competition with coethnics that comes from concentration in highly skilled occupations in traditional gateways. Diversity within the Asian population allows for further exploration of these relationships. The pan-ethnic "Asian" category not only encompasses numerous diverse ethnic groups with distinct cultural and linguistic backgrounds, it also represents considerable diversity in the context of origin. Groups such as the Chinese, Japanese, Filipinos, and Indians average markedly higher socioeconomic status than Koreans, Vietnamese, and other Asian groups (Pew Research Center 2013).

Accordingly, this article seeks to compare the socioeconomic attainment of Asian men and women in new immigrant destinations with their counterparts in more traditional receiving areas. As a first step, we refine a definition of new and traditional destinations and explore differences across destinations in both the average characteristics of residents and in larger metro-level conditions. Second, we examine whether residence in a new settlement area predicts Asian immigrants' income, occupational status, and homeownership once individual and household characteristics are controlled. Third, we examine whether immigrant context continues to exert a unique effect on Asian status attainment once the characteristics of local labor and housing market conditions are taken into account. That is, we explore whether earnings and homeownership differentials across locales are a simple function of metro-level wages and housing costs, or whether a unique penalty or benefit can be attributed to immigrant context itself. Fourth, we compare Asian status attainment outcomes with those of whites and evaluate whether residence in new vs. traditional destinations is beneficial for Asians in terms of shaping the socioeconomic gap with whites. Finally, in all of these analyses we explore differences by both sex and national origin, comparing relatively advantaged and disadvantaged groups.

Data and Methods

Data for the analysis come from a three-year sample (2009–11) of the American Community Survey (ACS; Ruggles et al. 2010). We restrict the sample to the noninstitutionalized, nonstudent single-race Asian and non-Hispanic white population aged 25 to 64 to capture prime working ages and exclude settlement patterns that might be connected with retirement. We likewise restrict the sample in the models of income and occupational status to those who were working and reported at least $1,000 in annual income. And finally, we restrict the sample to residents of the 130 metro areas with at least 5,000 nonstudent Asians in 2010, and to those where fewer than 25 percent of Asians are enrolled in school (to ensure that college towns, which represent unusual labor markets, do not influence the results). This yields a total sample of 77,741 Asian men, 73,594 Asian women, and 92,428 Asian households. Sample sizes for whites are 806,745, 703,531, and 1,090,525 for men, women, and households, respectively.

Model specification

Our three dependent variables—income, occupational status, and homeownership—are key indicators of status attainment. Income is defined as the total personal earned annual income reported in the year prior to the survey, logged. Occupational status is measured with the Duncan Socioeconomic Index (SEI), which assigns a prestige score to occupations based on educational attainment and income (Duncan 1961). Scores range from 0 to 100, with higher scores indicating greater prestige. Finally, homeownership is a household-level dummy variable that equals 1 if a household owns their residence and 0 otherwise.

The main explanatory variable of interest is type of Asian settlement area. Combining data from the 5 percent sample of the 1990 decennial census with the 2010 ACS, we distinguish between three types: established Asian settlement areas, new Asian destinations, and other areas (see online appendix Table A1 for the list of metros in each category).[1] Our definition combines three dimensions: absolute size, relative share of the total population, and growth between 1990 and 2010. We argue that both absolute and relative sizes are important for differentiating the ethnic context of reception. In large metros, even sizable Asian populations capable of generating a sense of community may represent a small share of the total population. Inversely, in smaller metro areas, smaller absolute numbers may nevertheless represent a large enough share of total population to affect the ethnic character of the area. These two dimensions, absolute size and relative share, also change over time, potentially giving rise to new destinations.

Accordingly, we define traditional destinations as metros that either had more than 50,000 Asians in 1990, or had between 10,000 and 50,000 Asians in 1990 and exceeded the national average for percent Asian (2.8 percent) by at least 1.5 times. The category thus includes areas such as San Francisco, where the more than 424,000 Asian residents in 1990 represented more than 15 percent of the total population, as well as Stockton, CA, where the 30,000 Asian residents in 1990 represented more than 10 percent of the population. It also includes metros such as Philadelphia, where the Asians' share of the total population was slightly below average in 1990, but which nevertheless had nearly 67,000 Asian residents in that year.

New areas of destination are defined as nontraditional areas that had an Asian growth rate of more than 200 percent between 1990 and 2000 and in which the share Asian in the total population reached at least half of the national average in 2010 (5.6). This category includes metros such as Raleigh, NC, whose Asian population grew more than 533 percent between 1990 and 2010, from under 9,000 to nearly 47,000, or 4.7 percent of the total population. It also includes Phoenix, AZ, where the 25,000 Asian residents made up 1.9 percent of the population in 1990, while the more than 92,000 Asian residents in 2010 represented 4.0 percent of the population.

We classify all other metros that do not qualify as either traditional or new destinations but had at least 5,000 nonstudent Asians in 2010 as "other" Asian settlement areas. This category includes metros such as Miami, FL, where the 28,000 Asian residents in 2010 accounted for 1.8 percent of the population. It also includes Cincinnati, OH, whose Asian population grew 265 percent between 1990 and 2010 (from 7,000 to 19,000) but still remained a relatively modest 2.0 percent of the total 2010 population.

In addition to the metro typology, the models also control for the demographic, human capital, and immigration characteristics that predict status attainment. These characteristics include age, measured continuously from 24 to 65, as well as age squared. Educational attainment is captured by four mutually exclusive dummy variables: high school graduate or lower (reference), some college, college graduate, and advanced degree. To capture immigration status, we construct four mutually exclusive dummy variables: the native born (reference), immigrants who arrived prior to 1980, immigrants who arrived between 1981 and 2000, and immigrants who arrived in 2001 or later. Because recent entrants to a local community may average different labor and particularly housing market outcomes, we also include a dummy variable indicating whether the respondent had moved within the previous year. Additional dummy variables identify those who are married and those who have minor children in the household. The models of homeownership also include dummies for whether the household head is male and employed, and total family income (logged).

As discussed above, because the Asian category is heterogeneous, we divided the Asian population into two groups: Asian A includes groups recognized as relatively advantaged, namely Chinese, Japanese, Indian, and Filipino origin populations, and Asian B includes less advantaged groups such as Koreans, Vietnamese, Cambodians, and other Asians (Pew Research Center 2013).[2] While models controlling for each group separately produced the same substantive results, this broader categorization is more parsimonious and allows us to test for the interaction between metro context and ethnicity in ways that are not possible for each national origin group individually.

And finally, we also include metropolitan-level indicators of local area conditions that are constructed by aggregating the individual and household information to the metropolitan level. These include total population size, occupational composition (the share of population employed in managerial or other professions, the share of the metro population that is minority (not non-Hispanic white), median total personal income (logged), and a dummy indicator of southern region (following the U.S. Census classification). The analysis of homeownership also includes controls for metropolitan homeownership rates, median housing values, and the share of housing that is new (i.e., built after 2000).

Our analytic strategy is first to assess the role of immigrant context in shaping socioeconomic outcomes within the Asian population. Doing so entails describing variation in both dependent and independent variables across contexts. We next estimate OLS models of income and SEI and logistic regression models of homeownership for the Asian population. In each of these models, we first estimate effects controlling only for individual- and household-level predictors, to assess the impact of context on socioeconomic attainment net of potential differences across contexts in the composition of the Asian population. We then add metropolitan-area controls to ascertain whether local labor and housing market conditions explain differences across immigrant destinations or whether ethnic context exerts a unique effect. These models are run separately by sex and also test for interactions between ethnic context and the two national origin types.

The second stage in the analysis assesses the link between context and the distance separating Asians from whites. Accordingly, we estimate the same models described above, pooling data for Asians and whites. To measure distance from whites, the models include a control for Asian origin group as well as interactions between Asian group and area of residence, again separately by sex. This also helps to assess the role of unmeasured aspects of metropolitan context in structuring observed relationships. Because the clustering of individuals and households within metropolitan areas violates the independence assumption in standard regression, we estimate robust standard errors.

Descriptive Results

Table 1 presents descriptive statistics of our dependent variables by metro typology, national origin group, and sex for Asians and whites. Results show that income is substantially lower, on average, outside of traditional areas of Asian settlement, while homeownership is higher. Differences in occupational status across contexts are relatively slight, but also favor traditional gateways. For instance, compared with residents of traditional Asian gateways, white men in new destinations average nearly $14,000 lower earnings but are 7 percentage points more likely to own a home. White men's SEI is lower in other areas relative to traditional gateways, but not in new destinations. While these generalities are also largely true for the two Asian origin groups, there are notable exceptions. First, Asian A men are the only group who do not average lower wages outside of traditional gateways. In fact, Asian A men's wages are highest in new destinations, in stark contrast to other groups. Likewise, Asian A men, and to a lesser extent women, also have modestly higher SEI outside of traditional areas, again in contrast to other groups. The homeownership advantage of residence outside of traditional settlement areas is essentially absent, however, among Asian A households. Second, Asian B men and women are the only groups whose SEI is substantially lower outside of traditional gateways. The disparity for Asian B women is particularly pronounced.

Before we can attribute the differences observed in Table 1 to the effects of context, we must also examine the extent to which individual characteristics vary across locales. If less educated or more recent immigrants disproportionately reside in new destinations, earnings and other disparities might reflect individual-level characteristics rather than context. Table A2 in the online appendix presents descriptive statistics for individual-level independent variables for the sample.[3] While age, sex, and family structure do vary appreciably across the metro types, there are marked differences in education and migration characteristics, which also vary across national origin groups. Among whites, residence outside of traditional Asian gateways is associated with lower educational attainment, but the disparities are mostly concentrated at the upper end, with fewer holders of advanced degrees in new and other areas. Among Asian A groups, in contrast, educational attainment is higher, on average, outside of traditional gateways; for both men and women, professional degrees are more common in new and other areas than traditional gateways, and for men the least educated groups (high school or less) are also underrepresented in nontraditional areas. Among Asian B groups, in contrast, nontraditional settlement areas are negatively associated with educational attainment. Unlike whites, however, this is most visible in the lower end of the educational distribution, with higher shares of the least educated groups in new and other areas relative to traditional gateways. Migration characteristics also vary across locales. Among both whites and Asian B groups there are few differences across metro types in the share that is foreign born or recent migrants to the local metropolitan area. Among Asian A groups, however, residents of new destinations are disproportionately recent entrants to both the United States and the local community.

It is also worth noting that the Asian A and Asian B individuals in our sample mirror the national figures reported above. The Asian A group is considerably more educated than both the Asian B and white samples. Aside from its greater

TABLE 1
Descriptive Statistics of Dependent Variables by Metro Typology, National Origin Group, and Sex

	Asian A			Asian B			White		
	Traditional	New	Other	Traditional	New	Other	Traditional	New	Other
Men									
Income (mean)	$76,270	$79,468	$76,550	$60,597	$52,765	$49,121	$84,772	$70,913	$62,801
SEI (mean)	54.0	59.0	58.5	48.4	44.4	42.5	50.4	49.6	46.7
N	39,599	9,928	3,589	16,597	6,058	1,970	307,632	293,871	205,242
Women									
Income (mean)	$56,531	$53,132	$50,644	$45,195	$36,363	$34,591	$53,813	$44,863	$39,824
SEI (mean)	53.2	54.3	53.6	47.3	40.7	39.5	54.2	53.1	51.4
N	38,249	8,920	3,330	15,371	5,796	1,928	261,464	259,438	182,629
Household heads									
Own home (%)	67.2	69.8	67.9	58.7	68.7	66.7	69.0	75.9	76.5
N	46,437	11,797	4,240	20,141	7,419	2,394	402,303	403,104	285,118

NOTE: Asian A refers to Chinese, Japanese, Indian, and Filipino origins; Asian B refers to Korean, Vietnamese, and other Asian origin groups.

immigrant representation, the Asian B group is not markedly more disadvantaged than the white population.

It is also important to understand how local labor and housing markets vary across new and traditional destinations. Table A3 in the online appendix reports descriptive statistics for contextual-level independent variables.[4] Traditional areas have more favorable labor market conditions while new settlement areas have more favorable housing market conditions. Specifically, traditional destinations have greater representation of people in managerial and professional occupations (8.1 percent) than do new (7.3 percent) and other (6.3 percent) destinations. The same pattern pertains to median personal income, which is higher in traditional gateways ($36,035) than in new ($32,108) and other areas ($28,941). Homeownership, in contrast, is higher in new and other destinations (71 percent) than in traditional Asian gateways (65 percent), likely reflecting the dramatically lower average cost of housing in new and other destinations ($200,091 and $164,421, respectively) compared with traditional areas ($355,381). New and other areas also average a larger supply of new housing relative to traditional areas. New and other destinations are also far more likely to be located in the South.

Multivariate Results

Table 2 presents results from models estimating the relationship between destination type and socioeconomic attainment for the Asian sample. For each dependent variable, the first column includes only individual socioeconomic controls while the second adds metro-level contextual predictors. The reference category is Asian B in traditional settlement areas. Focusing on the effect of the typology on the income and occupational status of the Asian B group, results show that both Asian B men and women pay a penalty for residing outside of traditional gateways. Even after accounting for individual background characteristics Asian B men's income in new and other destinations is 7 and 12 percent lower, respectively, than their counterparts in traditional gateways. The penalty is similar among women. Likewise among Asian B, average SEI is also lower in nontraditional areas, particularly for women. Asian B women in new and other destinations average 3.4 and 3.8 point lower SEI scores, respectively, than their peers with comparable human capital in traditional gateways. Thus, consistent with the view that ethnic communities are a resource, there is a clear labor market penalty to living outside of traditional gateways for Asian B men and women.

However, it is noteworthy that for Asian B men, all of the negative effects associated with residence in nontraditional areas are explained by the labor market conditions of those areas. That is, when we add controls for metro-level labor market context, both earnings and occupational status are indistinguishable across metro types. Thus, the beneficial effects of residence in established areas for the Asian B group reflect the greater concentration of high-quality jobs in these contexts rather than the protective effect of a well-established coethnic

TABLE 2
Coefficients from OLS Model Predicting Income and SEI, and Logit Models of Homeownership: Asians

	Income		SEI		Homeownership Household Heads	
	Men	Women	Men	Women		
Metro type (ref = Asian B in traditional settlement areas)						
New destination	-0.07**	-0.06**	-1.52	-3.39**	0.76**	0.15†
	(0.02)	(0.02)	(0.97)	(0.57)	(0.15)	(0.08)
Other destination	-0.12**	-0.11**	-1.30	-3.77**	0.75**	0.08
	(0.02)	(0.03)	(1.06)	(0.63)	(0.15)	(0.10)
Asian A	0.09**	0.13**	0.25	1.14**	0.29**	0.35**
	(0.01)	(0.01)	(0.43)	(0.25)	(0.06)	(0.06)
Asian A × new	0.06*	-0.02	3.59**	3.82**	-0.31**	-0.35**
	(0.03)	(0.02)	(0.72)	(0.65)	(0.10)	(0.09)
Asian A × other	0.04	-0.04	4.40**	3.86**	-0.37**	-0.39**
	(0.03)	(0.03)	(0.85)	(0.76)	(0.12)	(0.11)
Individual-level controls						
Age	0.07**	0.05**	0.05	0.11	0.11**	0.11**
	(0.00)	(0.00)	(0.12)	(0.08)	(0.01)	(0.01)
Age squared	0.00**	0.00**	0.00*	0.00**	0.00**	0.00**
	(0.00)	(0.00)	(0.00)	(0.00)	(0.00)	(0.00)
Male	—	—	—	—	0.10**	0.11**
					(0.02)	(0.02)
Education (reference: high school)						
Some college	0.31**	0.34**	10.93**	11.82**	0.33**	0.34**
	(0.02)	(0.02)	(0.36)	(0.28)	(0.05)	(0.03)
College	0.73**	0.72**	27.90**	22.65**	0.42**	0.43**
	(0.03)	(0.03)	(0.46)	(0.29)	(0.04)	(0.03)

(Continued)

TABLE 2 (CONTINUED)

	Income				SEI				Homeownership Household Heads	
	Men		Women		Men		Women		Household Heads	
Advanced degree	1.17**	1.16**	1.09**	1.08**	41.12**	40.81**	36.35**	36.17**	0.21**	0.19**
	(0.03)	(0.03)	(0.02)	(0.02)	(0.60)	(0.50)	(0.63)	(0.62)	(0.06)	(0.07)
Migration status (reference: native-born)										
Pre-1980	0.06**	0.07**	0.13**	0.12**	1.41**	0.96*	0.38	0.12	0.79**	0.80**
	(0.01)	(0.01)	(0.02)	(0.01)	(0.44)	(0.43)	(0.50)	(0.59)	(0.04)	(0.03)
1981–2000	-0.16**	-0.16**	-0.11**	-0.11**	-3.71**	-4.24**	-5.20**	-5.51**	0.25**	0.27**
	(0.01)	(0.01)	(0.02)	(0.01)	(0.45)	(0.37)	(0.38)	(0.48)	(0.01)	(0.02)
After 2001	-0.43**	-0.43**	-0.40**	-0.41**	-6.80**	-7.30**	-10.38**	-10.66**	-1.04**	-1.07**
	(0.02)	(0.02)	(0.02)	(0.02)	(0.57)	(0.45)	(0.43)	(0.48)	(0.04)	(0.04)
Migrated w/in prev. year	-0.05**	-0.05**	-0.08**	-0.07**	-0.45	-0.42	-0.14	-0.13	-1.16**	-1.19**
	(0.01)	(0.01)	(0.01)	(0.01)	(0.28)	(0.28)	(0.21)	(0.21)	(0.08)	(0.08)

	Income				SEI				Homeownership	
	Asian Men		Asian Women		Asian Men		Asian Women		Asian household heads	
Married	0.16**	0.16**	-0.02†	-0.02*	1.93**	1.87**	1.37**	1.39**	0.46**	0.45**
	(0.01)	(0.01)	(0.01)	(0.01)	(0.26)	(0.26)	(0.16)	(0.16)	(0.02)	(0.03)
Child in household	0.07**	0.07**	-0.02**	-0.02**	-0.16	-0.15	-0.75**	-0.73**	-0.28	-0.28
	(0.01)	(0.01)	(0.01)	(0.01)	(0.23)	(0.23)	(0.24)	(0.24)	(0.38)	(0.38)
Employed	—	—	—	—	—	—	—	—	0.18**	0.17**
									(0.03)	(0.03)
Family total income	—	—	—	—	—	—	—	—	0.00**	0.00**
									(0.00)	(0.00)
Metro-level contextual variables										
Median personal income (log)	0.45**		0.38**		—		—		—	
	(0.06)		(0.04)							

(Continued)

TABLE 2 (CONTINUED)

	Income		SEI				Homeownership	
	Men	Women	Men	Men	Women	Women	Women	Household Heads
% Employed in managerial/prof.	—	—	—	0.27	—	0.20**	—	—
				(0.19)		(0.06)		
Total population (logged)	-0.02*	0.00	—	0.75*	—	0.63**	—	-0.06**
	(0.01)	(0.01)		(0.36)		(0.11)		(0.02)
South	-0.03	-0.04**	—	1.35	—	0.69†	—	0.03
	(0.02)	(0.01)		(0.91)		(0.35)		(0.05)
% Minority	0.00	0.00	—	-0.01	—	0.02	—	0.01**
	(0.00)	(0.00)		(0.02)		(0.01)		(0.00)
% Homeownership	—	—	—	—	—	—	—	0.04**
								(0.00)
Median housing value	—	—	—	—	—	—	—	0.00**
								(0.00)
% New housing stock	—	—	—	—	—	—	—	0.01**
								(0.00)
Year 2010 (ref = 2011)	0.03**	0.03**	0.45*	0.43*	0.09	0.09	0.05	0.04
	(0.01)	(0.01)	(0.20)	(0.20)	(0.16)	(0.16)	(0.03)	(0.03)
Year 2009	0.04**	0.01	0.47*	0.48**	0.24	0.25†	0.05	0.03
	(0.00)	(0.01)	(0.18)	(0.17)	(0.15)	(0.15)	(0.03)	(0.03)
Intercept	8.70**	4.81**	35.90**	22.77**	40.13**	28.12**	-4.40**	-5.80**
	(0.09)	(0.43)	(2.49)	(6.40)	(1.79)	(2.69)	(0.16)	(0.48)
n	77741	73594	77741	77741	73594	73594	92428	92428
Adjusted R^2	0.31	0.25	0.43	0.43	0.39	0.39	0.22	0.23

**$p < .01$, *$p < .05$, †$p < .1$

community per se. While the same is true among women with respect to income, the lower average SEI of Asian B women outside of traditional areas remains even net of economic differences across contexts.

A very different pattern is evident among the Asian A group. Overall, Asian A men and women average higher incomes and SEI than their Asian B counterparts in traditional gateways (the only exception being Asian A men's SEI, which is not significantly higher than that of Asian B men). This pattern remains even after accounting for local area conditions. More interesting for our purpose is the significant interaction between ethnic context and national origin type. This interaction signifies that the advantage that the Asian A group enjoys over their Asian B counterparts in traditional areas is even larger in new and other destinations. It also indicates that among the Asian A group, SEI is higher outside of traditional gateways. To illustrate, Asian A men and women in new destinations average 3.4 and 3.8 points higher SEI, respectively, than comparable Asian A peers in traditional gateways. Not only are these effects opposite of those found among the Asian B group, they also remain significant even after accounting for variation across context in local labor market conditions. The same applies to residence in other destinations. Interestingly, only Asian A men translate better occupational standing into higher earnings; results show that Asian A men in new destinations earn 6 percent more net of socioeconomic background than comparable men in traditional areas. Asian A women, however, do not get the same income boost from residence in new destinations.

The intersection between ethnicity and gender is also highlighted in the association between socioeconomic background and income and SEI. Not surprisingly, those with more education average higher income and SEI than those with lower education. However, the returns to advanced education are lower among women. Likewise, marriage and the presence of children in the household are associated with higher earnings among Asian men but lower earnings among Asian women. The presence of children is also associated with lower occupational standing among women, but not among men.

Turning to the connection between metro context and homeownership among Asians (the final two columns of Table 2), results again point to important differences in the impact of context across national origin groups. For Asian B households, residence in new or other destinations is associated with higher homeownership rates (0.76 and 0.75, respectively). The homeownership advantages disappear, however, once we control for local housing market conditions, suggesting once again that for this group, locational differences stem mainly from the local socioeconomic context rather than the ethnic community. Once again, however, results differ for the Asian A group. Net of socioeconomic resources, Asian A households in traditional destinations are significantly more likely to own a home than comparable Asian B households, and the effect remains after controlling for local housing market conditions (0.35). However, the Asian A homeownership advantage relative to Asian B disappears in new and other destinations. Results show that among Asian A households, residence in new or other destinations reduces the log-odds of homeownership by –0.35 and –0.39, respectively, effectively cancelling their advantage observed in traditional destinations. Thus, even though their homeownership rates remain relatively high, Asian A households in nontraditional destinations face obstacles in translating their higher socioeconomic status into homeownership.

Disparities with whites

The next set of analyses, presented in Table 3, investigates the relationship between ethnic context and the socioeconomic positions of Asians relative to whites. The sample includes Asians and whites, and the models predict income, SEI, and homeownership separately by sex, including interaction terms between Asian race, separately by national origin group, and the metro typology. Results indicate significant differences by both sex and national origin group in the earnings gap between Asians and whites by settlement area.

Focusing first on the role of context in shaping men's wages, results show that similar to Asians in Table 2, white men's wages are on average 11 and 18 percent lower in new and other Asians destinations, respectively, relative to white men in traditional Asian gateways. As was the case for Asians, most of that deficit is attributable to the lower-wage context in new and other destinations. Both Asian A and Asian B groups earn less than whites, on average, in traditional areas, though the gap is smaller among the former Asian group. However, the interaction terms indicate that the gap between Asian A and white men is significantly smaller outside of traditional areas of settlement, in both new and other destinations. In fact, after accounting for metro economic context, the gap between Asian A and white men is essentially eliminated in new (–0.04 + 0.10) and other (–0.05 + 0.10) destinations. The same is not true for Asian B groups, among whom the gap with whites is equally significant across contexts.

A very different pattern is evident for Asian women. White women, like white men, average lower wages in new (–0.10) and other (–0.17) destinations than their peers in traditional Asian areas. However, the pay gap between Asian and white women in traditional areas is only significant among Asian B women. Also in contrast to men, the relative position of Asian and white women does not vary by ethnic context. Once again, controlling for metropolitan characteristics reduces the size of the coefficients but does not change the pattern of results.

Findings for the models predicting SEI show a number of differences from those predicting income. While average SEI for white men and women are not lower in new Asian destinations, those in the other destinations do average lower occupational prestige than their white counterparts in traditional Asian gateways. Once again men and women in the Asian A and Asian B groups, but particularly the latter, average lower occupational prestige than whites with similar human capital characteristics in traditional gateways. However, there are again important interactions by both national origin group and sex in the link between context and the gap between Asians and whites. Once again, the race gap is smaller in new and other destinations than in traditional gateways, but this narrowing is again present only among men and advantaged (i.e., Asian A) national origin groups. With SEI, as with income, the gap between Asian A and white men that is significant in traditional gateways is eliminated in nontraditional areas. This effect is not

TABLE 3

Coefficients from OLS Model Predicting Income and SEI, and Logit Models of Homeownership: Asians and Whites

	Income Men	Income Men	Income Women	Income Women	SEI Men	SEI Men	SEI Women	SEI Women	Homeownership Household Heads	Homeownership Household Heads
Metro type (ref = Whites in traditional settlement areas)										
New destination	-0.11** (0.03)	-0.04* (0.02)	-0.10** (0.02)	-0.01 (0.01)	0.31 (0.31)	0.36 (0.33)	-0.09 (0.31)	-0.26 (0.30)	0.67** (0.15)	0.06 (0.05)
Other destination	-0.18** (0.03)	-0.05* (0.02)	-0.17** (0.02)	-0.03* (0.01)	-1.10** (0.30)	-0.50 (0.38)	-0.93** (0.30)	-0.79* (0.36)	0.76** (0.15)	0.04 (0.06)
Asian A	-0.20** (0.03)	-0.21** (0.02)	0.02 (0.02)	0.01 (0.01)	-1.07** (0.36)	-0.77** (0.35)	-2.76** (0.26)	-2.51** (0.27)	-0.26** (0.09)	-0.05 (0.07)
Asian B	-0.30** (0.02)	-0.30** (0.03)	-0.12** (0.02)	-0.12** (0.01)	-1.47* (0.66)	-1.23† (0.66)	-4.46** (0.51)	-4.28** (0.56)	-0.46** (0.07)	-0.30** (0.08)
Asian A × new	0.10** (0.03)	0.10** (0.03)	0.02 (0.02)	0.02 (0.02)	1.68* (0.70)	1.40* (0.67)	0.60 (0.70)	0.38 (0.66)	-0.23* (0.12)	-0.39** (0.11)
Asian A × other	0.09** (0.04)	0.10** (0.03)	0.01 (0.03)	0.02 (0.02)	3.16** (0.69)	3.00** (0.70)	0.81 (0.62)	0.63 (0.66)	-0.39** (0.13)	-0.53** (0.12)
Asian B × new	0.03 (0.03)	0.02 (0.03)	0.02 (0.02)	0.01 (0.02)	-1.98* (0.75)	-2.41** (0.75)	-4.18** (0.61)	-4.49** (0.63)	0.16† (0.09)	0.01 (0.09)
Asian B × other	0.05 (0.03)	0.04 (0.03)	0.03 (0.03)	0.03 (0.02)	-1.34 (0.82)	-1.52† (0.83)	-4.11** (0.74)	-4.29** (0.77)	0.06 (0.10)	-0.05 (0.10)
Individual-level controls										
Age	0.09** (0.00)	0.09** (0.00)	0.07** (0.00)	0.07** (0.00)	0.54** (0.04)	0.53** (0.04)	0.40** (0.03)	0.40** (0.03)	0.08** (0.01)	0.09** (0.01)
Age squared	0.00** (0.00)	0.00** (0.00)	0.00** (0.00)	0.00** (0.00)	-0.01** (0.00)	-0.01** (0.00)	0.00** (0.00)	0.00** (0.00)	0.00** (0.00)	0.00** (0.00)
Male	—	—	—	—	—	—	—	—	0.02 (0.01)	0.02* (0.01)

(Continued)

TABLE 3 (CONTINUED)

	Income				SEI				Homeownership Household Heads	
	Men		Women		Men		Women			
Education (reference: High school)										
Some college	0.30** (0.01)	0.30** (0.01)	0.30** (0.01)	0.29** (0.01)	12.31** (0.26)	12.28** (0.26)	8.33** (0.34)	8.34** (0.34)	0.31** (0.03)	0.33** (0.02)
College	0.69** (0.02)	0.67** (0.02)	0.60** (0.02)	0.59** (0.02)	27.34** (0.24)	27.13** (0.25)	18.41** (0.33)	18.30** (0.33)	0.54** (0.02)	0.57** (0.02)
Advanced degree	0.99** (0.02)	0.97** (0.02)	0.91** (0.02)	0.89** (0.02)	38.29** (0.32)	38.00** (0.36)	27.62** (0.38)	27.48** (0.38)	0.40** (0.03)	0.43** (0.03)
Migration status (reference: Native-born)										
Pre-1980	0.18** (0.01)	0.19** (0.01)	0.20** (0.01)	0.21** (0.01)	4.12** (0.34)	4.24** (0.34)	4.88** (0.39)	4.96** (0.38)	0.78** (0.07)	0.80** (0.05)
1981–2000	-0.11** (0.01)	-0.11** (0.01)	-0.07** (0.01)	-0.07** (0.01)	-3.37** (0.18)	-3.38** (0.19)	-3.76** (0.19)	-3.78** (0.20)	-0.06† (0.03)	0.00 (0.03)
After 2001	-0.07** (0.01)	-0.08** (0.01)	-0.13** (0.01)	-0.14** (0.01)	-0.75** (0.18)	-0.86** (0.19)	-1.12** (0.39)	-1.19** (0.38)	-0.72** (0.07)	-0.80** (0.05)
Migrated w/in previous year	-0.08** (0.00)	-0.08** (0.00)	-0.08** (0.01)	-0.08** (0.01)	-0.96** (0.14)	-0.95** (0.14)	-1.20** (0.09)	-1.20** (0.09)	-1.63** (0.05)	-1.64** (0.05)
Married	0.25** (0.00)	0.25** (0.00)	-0.05** (0.00)	-0.05** (0.00)	2.95** (0.13)	2.91** (0.13)	2.26** (0.09)	2.23** (0.10)	0.87** (0.03)	0.87** (0.03)
Child in household	0.08** (0.00)	0.08** (0.00)	-0.13** (0.00)	-0.13** (0.00)	-0.36** (0.11)	-0.31** (0.12)**	-1.39** (0.09)	-1.35** (0.10)	0.17** (0.03)	0.18** (0.03)
Employed	—	—	—	—	—	—	—	—	0.22** (0.02)	0.21** (0.02)
Total family income	—	—	—	—	—	—	—	—	0.00** (0.00)	0.00** (0.00)

(Continued)

TABLE 3 (CONTINUED)

	Income				SEI				Homeownership	
	Men		Women		Men		Women		Household Heads	
Metro-level contextual variables										
Median personal income (log)	—	0.60** (0.05)	—	0.52** (0.03)	—	0.60** (0.13)	—	0.36** (0.10)	—	-0.01 (0.01)
Total population (logged)	—	0.00 (0.00)	—	0.00 (0.00)	—	1.41** (0.28)	—	1.38** (0.20)	—	0.00 (0.03)
South	—	0.00 (0.01)	—	0.00 (0.01)	—	-0.03** (0.01)	—	-0.02* (0.01)	—	0.00 (0.03)
% Minority	—	0.00 (0.00)	—	0.00** (0.00)	—	-0.03** (0.01)	—	-0.02* (0.01)	—	0.00† (0.00)
% Employed in managerial/prof. occupations	—	—	—	—	—	0.34** (0.08)	—	0.18** (0.05)	—	—
% Homeownership	—	—	—	—	—	—	—	—	—	0.05** (0.00)
Median housing value	—	—	—	—	—	—	—	—	—	0.00** (0.00)
% New housing stock	—	—	—	—	—	—	—	—	—	0.00 (0.00)
Year 2010 (ref=2011)	0.04** (0.00)	0.04** (0.00)	0.04** (0.00)	0.04** (0.00)	0.43** (0.05)	0.42** (0.05)	0.41** (0.06)	0.41** (0.06)	0.07** (0.01)	0.07** (0.01)
Year 2009	0.05** (0.00)	0.05** (0.00)	0.05** (0.00)	0.05** (0.00)	0.78** (0.07)	0.77** (0.07)	0.75** (0.05)	0.74** (0.05)	0.13** (0.01)	0.12** (0.01)
Intercept	8.21** (0.03)	2.02** (0.47)	8.55** (0.03)	3.06** (0.27)	19.48** (1.04)	9.28** (2.05)	32.47** (0.78)	26.41** (1.49)	-3.99** (0.24)	-6.28** (0.39)
n	897542	897542	789251	789251	897542	897542	789251	789251	1E+06	1E+06
Adjusted R^2	0.2601	0.2663	0.1632	0.1675	0.3536	0.3553	0.2498	0.2511	0.2464	0.2604

**$p < .01$, *$p < .05$, †$p < .1$

present among Asian A women or Asian B of either sex. In fact, for both Asian B men and women, the gap with whites is actually larger in new and other destinations relative to traditional gateways.

The complexity of the connection between immigrant context and socioeconomic standing also extends to the analysis of racial disparities in homeownership, reported in the final columns of Table 3. As was the case with the Asian models, whites are more likely to own a home outside of traditional destinations, though only because the housing markets in those areas are more conducive to ownership. Overall, Asians are less likely to own a home than are whites, but among Asian A groups this is only because they tend to live in more expensive housing markets than whites. When local conditions are controlled, only Asian B groups are less likely than whites to own a home in traditional gateways. Once again, however, the interaction terms show important differences in the relationship between context and inequality with whites across national origin groups. That is, while Asian A groups are not disadvantaged relative to whites in traditional gateways, they are less likely to own their homes than are whites in new and other destinations. That is, the homeownership gap for the Asian A group and whites exists only outside of traditional areas of settlement. Thus, there is some indication that, at least in the realm of housing, there are relative disadvantages to residence in new destinations for advantaged Asians, as was previously found for Hispanics and African Americans (Flippen 2010). For the less advantaged Asian B group, disadvantage relative to whites is constant across immigrant contexts.

Conclusion

The dramatic dispersal of immigrant-origin groups outside of traditional gateways and the rise of new destinations have brought renewed attention to the extent to which context shapes the incorporation of minorities into broader society. The overwhelming majority of this work has focused on Hispanics, to the serious neglect of Asian origin groups. Asians as a group are relatively unique among ethno-racial minorities in the United States. Highly selective migration flows and a high-achieving second generation have translated into a population that averages better educational and labor market outcomes than the general U.S. population. Thus, whether and how Asians may benefit from contexts with large coethnic communities—and how these benefits differ from less advantaged groups—remain open and important questions.

This study examined these issues drawing on data from the 2009–11 ACS. After defining traditional, new, and other areas of Asian settlement, we examined whether status attainment was related to metropolitan ethnic context. The main question was whether Asians paid a penalty for leaving traditional settlement areas, and if so, how this penalty varied within the Asian population (i.e., by sex and across more and less advantaged national origin groups). To answer the question, we compared absolute income, occupational status, and homeownership in traditional and nontraditional settlement areas and examined whether differences remain after accounting for variation across contexts in local labor and housing market conditions. We also considered not only differences across locales within the Asian population but also differences across contexts in the extent of inequality with whites.

The results show considerable complexity in the relationship between context and Asian status attainment, with variation between labor and housing market outcomes, as well as by sex and national origin group. For the less advantaged Asian-origin group (namely, Korean, Vietnamese, and other Asians), results indicate an important tradeoff between labor and housing market outcomes. Wages and occupational status are both lower outside of traditional gateways, net of human capital and demographic characteristics. Homeownership, on the other hand, is higher. However, these disparities emanate from differences in labor and housing market conditions across destinations rather than from the influence of ethnic context per se. When metro-wide labor and housing contexts are taken into consideration, there appears to be no particular advantage or disadvantage among the Asian B group to living outside of traditional areas of settlement.

For the more advantaged national origin groups (Chinese, Japanese, Indians, and Filipinos), results are profoundly different. For the Asian A group, residence outside of traditional gateways is associated with higher wages and occupational status, particularly among men. These patterns do not extend to homeownership, however, as the advantaged national origin group averages lower rates of homeownership in new and other destinations than in traditional gateways.

When we consider not only Asians relative to one another according to metro context, but also inequality between Asians and whites, the picture is likewise markedly different for labor market and housing outcomes, men and women, and national origin groups. First, for Asian B men and women, the gap with whites is only marginally related to ethnic context. While the gap in occupational status between Asian B men and women and their white counterparts is larger outside of traditional destinations, the gap in income and homeownership does not vary across contexts. For the Asian A group, in contrast, the gap with whites varies across contexts on all three of our measures of status attainment, at least for men. Thus, Asian A men's deficit in earnings and occupational status relative to whites is lower in nontraditional metros. These benefits are offset, however, by a wider Asian A disadvantage in homeownership in new and other destinations, once metro-level housing conditions are controlled. The labor markets of traditional Asian receiving areas are often saturated with workers in the highly skilled niches in which the Asian A group concentrates. In this environment, Asian A men pay a wage penalty relative to white men that does not appear to be present in nontraditional destinations. In housing, on the other hand, the benefits of greater access to information and greater institutional outreach to minority communities make homeownership relative to whites easier in traditional than in nontraditional areas.

The results also hint at important gender differences in the link between context and socioeconomic outcomes. While the race gap is generally smaller among women than among men, Asian women do not get the same boost relative to whites from residence in new and other destinations that men do. Thus, as with the unequal payoff to education, marriage, and childbearing, Asian women may receive a lower return to internal migration than do their male peers. Further research is needed to explore potential differences by marital status; it is possible that the greater tendency of wives to relocate in accordance with their husbands' careers could undermine the benefits from leaving immigrant niche areas.

It is also worthwhile to consider what these findings portend for national-level inequality between Asians and whites. Here the results are somewhat contradictory. On one hand, Asian dispersal to lower wage environments will tend to narrow the national pay advantage held by Asians today. On the other hand, movement out of niche labor markets into areas with fewer Asians seems to lower the wage penalty faced by comparably educated Asians (at least for men from relatively advantaged national-origin groups). Taken together, these two trends could contribute to growing convergence in Asian and white occupational and housing outcomes, for both better and worse.

Finally, it is important to recognize that these patterns pertain to the period following the 2007 economic recession. Emerging evidence on the differential impact of the recession on the Asian population is somewhat mixed. On one hand, there is evidence that Asian homeownership was not affected more than other groups, despite their heavy concentration in the West (especially California), where the effects of housing collapse were keenly felt (Painter and Yu 2012). On the other hand, there are indictors that long-term unemployment may have been higher among highly educated Asians than whites with similar skills, with regional concentration playing an important role in the disparities (Kim 2012). It is important for future work to explore how both economic change and the continual evolution of the composition of the Asian population over time influence the link between context and status attainment.

Notes

1. See http://ann.sagepub.com/supplemental.

2. In many respects Koreans occupy a middle ground in Asian socioeconomic attainment. On one hand, their average educational attainment is similar to that of U.S. Chinese, Filipino, and Japanese populations, indicating relative advantage. On the other hand, their individual and household incomes are starkly lower and poverty rates markedly higher than these other groups and even slightly worse than those of the Vietnamese, indicating relative disadvantage (Pew Research Center 2013). As such, we included Koreans in the disadvantaged category in this analysis. An alternative specification, including Koreans in the relatively advantaged category, produced the same substantive results.

3. See http://ann.sagepub.com/supplemental.

4. Ibid.

References

Alba, Richard, and Victor Nee. 2003. *Remaking the American mainstream: Assimilation and contemporary immigration*. Cambridge, MA: Harvard University Press.

Desai, Jigna, and Khyati Joshi, eds. 2013. *Asians in Dixie: Race and migration in the South*. Urbana, IL: University of Illinois Press.

Duncan, Otis. 1961. A socioeconomic index for all occupations. In *Occupations and social status*, ed. Albert J. Reiss, 109–38. New York, NY: Free Press.

Flippen, Chenoa. 2010. The spatial dynamics of stratification: Metropolitan context, population redistribution, and black and Hispanic homeownership. *Demography* 47 (4): 845–68.

Frey, William. 2011. *The new metro minority map: Regional shifts in Hispanics, Asians, and blacks from Census 2010*. Washington DC: The Brookings Institution. Available from http://www.brookings.edu.

Gordon, Milton. 1964. *Assimilation in American life*. New York, NY: Oxford University Press.

Greenwood, Michael. 1997. Internal migration in developed countries. In *Handbook of population and family economics*, eds. Mark R. Rosenzweig and Oded Stark, 1–12. New York, NY: North Holland.

Kim, Marlene. 2012. *Unfairly disadvantaged? Asian Americans and unemployment during and after the Great Recession*. Washington DC: Economic Policy Institute. Available from http://www.epi.org.

Kochhar, Rakesh, Roberto Suro, and Sonya Tafoya. 2005. *The new Latino South: The context and consequences of rapid population growth*. Washington, DC: Pew Hispanic Center. Available from http://www.pewtrusts.org.

Lichter, Daniel T., Domenico Parisi, and Michael C. Taquino. 2015. Spatial assimilation in U.S. cities and communities? Emerging patterns of Hispanic segregation from blacks and whites. *The ANNALS of the American Academy of Political and Social Science* (this volume).

Logan, John R., Richard D. Alba, and Wenquan Zhang. 2002. Immigrant enclaves and ethnic communities in New York and Los Angeles. *American Sociological Review* 67 (2): 299–322.

Min, Pyong Gap, ed. 2006. *Asian Americans: Contemporary trends and issues*. Thousand Oaks, CA: Pine Forge Press.

Painter, Gary, and Zhou Yu. 2012. *Caught in the housing bubble: Immigrants' housing outcomes in traditional gateways and newly emerging destinations*. Los Angeles, CA: Center for the Study of Immigrant Integration.

Pew Research Center. 2013. *The rise of Asian Americans*. Washington DC: Pew Research Center. Available from http://www.pewsocialtrends.org/files/2013/04/Asian-Americans-new-full-report-04-2013.pdf.

Portes, Alejandro, and Jozsef Borocz. 1989. Contemporary immigration: Theoretical perspectives on its determinants and modes of incorporation. *International Migration review* 23 (3): 606–30.

Ruggles, Steven, Trent Alexander, Katie Genadek, Ronald Goeken, Matthew Schroeder, and Matthew Sobek. 2010. Integrated public use microdata series: Version 5.0. Minneapolis, MN: University of Minnesota.

Sakamoto, Arthur, Kim Goyette, and ChangHwan Kim. 2009. Socioeconomic attainments of Asian Americans. *Annual Review of Sociology* 35:255–76.

Sanders, Jimy M., and Victor Nee. 1987. Limits of ethnic solidarity in the enclave economy. *American Sociological Review* 52:745–73.

Schachter, Jason. 2003. *Migration by race and Hispanic origin: 1995–2000*. Census 2000 Special Report. Washington, DC: U.S. Census Bureau.

Stamps, Katherine, and Stephanie Bohon. 2006. Educational attainment in new and established Latino metropolitan destinations. *Social Science Quarterly* 87:1225–40.

Waters, Mary, and Tomas Jimenez. 2005. Assessing immigrant assimilation: New empirical and theoretical challenges. *Annual Review of Sociology* 31:105–25.

Wilson, Jill, and Audrey Singer. 2011. *Immigrants in 2010 metropolitan America: A decade of change*. Washington, DC: The Brookings Institution. Available from http://www.brookings.edu.

Wilson, Kenneth L., and Alejandro Portes. 1980. Immigrant enclaves: An analysis of the labor market experiences of Cubans in Miami. *American Journal of Sociology* 86:295–319.

Xie, Yu, and Margaret Gough. 2011. Ethnic enclaves and the earnings of immigrants. *Demography* 48:1293–1315.

Zeng, Zhen, and Yu Xie. 2004. Asian-Americans' earnings disadvantage reexamined: The role of place of education. *American Journal of Sociology* 109:1075–1108.

Zhou, Min. 1992. *Chinatown: The socioeconomic potential of an urban enclave*. Philadelphia, PA: Temple University Press.

Zhou, Min. 1997. Segmented assimilation: Issues, controversies, and recent research on the new second generation. *International Migration Review* 31 (4): 975–1008.

The Great Risk Shift and Precarity in the U.S. Housing Market

In this article, we propose that metropolitan areas represent differential "risk contexts" to the people who live within them and argue that growing insecurity in U.S. metropolitan areas arises out of cross-cutting economic weaknesses that are too often seen in isolation. The housing crisis that led up to the Great Recession was a moment in which the underlying vulnerabilities in our markets and institutions were laid bare. The crisis also occurred in the context of the "great risk shift" in American society—where individuals are increasingly responsible for managing the ordinary risks of life in a modern economy. The multiple sources of precarity in the housing market highlight the complex nature of insecurity that many Americans face. We look at metropolitan variability in foreclosures to identify conditions that contributed to the housing crisis. We build on prior research by showing different sources of vulnerability to the housing crisis in metropolitan areas—including labor market insecurity and housing market insecurity—and find that some of the metropolitan areas that fared the worst faced problems in both markets before the crisis.

Keywords: housing; inequality; insecurity; labor market; urban areas

By
RACHEL E. DWYER
and
LORA A. PHILLIPS LASSUS

Rising inequality in U.S. society has been accompanied by rising insecurity across a range of domains. American families face increasing risks of job loss, income decline, and housing insecurity (Kalleberg 2011; Hacker

Rachel E. Dwyer is an associate professor of sociology at Ohio State University. She studies the causes and consequences of rising economic inequality and insecurity in U.S. housing, job, and credit markets. She has published on these issues in Social Problems, Social Forces, Gender & Society, *and the* American Sociological Review.

Lora A. Phillips Lassus is a PhD candidate in the Department of Sociology at Ohio State University. Her research focuses on inequality and mobility, particularly as these processes are contextualized geographically, politically, and historically. In one recent project she examined job loss among older workers during the Great Recession.

DOI: 10.1177/0002716215577612

2006). The poor face the most severe threats to security, but precarity increasingly afflicts families across a range of economic circumstances (Western et al. 2012). Decades of economic restructuring, deregulation, and government retreat from social insurance have culminated in what Jacob Hacker (2006) calls the "great risk shift," where individuals are increasingly responsible for planning for retirement, health crises, and the ordinary risks of life in a modern economy. At the same time, corporations provide fewer benefits, wages have stagnated, and there are fewer opportunities for secure long-term employment, resulting in scant resources for individuals and families to manage and plan for these myriad risks (Kalleberg 2011). In other words, the great risk shift devolved responsibility to individuals just as risks became less individually manageable.

The contours of the great risk shift were put into sharp relief by the 2008 housing crisis. Americans had become accustomed to the idea that housing is a commodity, and they viewed homeownership underwritten by a long-term mortgage as a relatively stable strategy for wealth accumulation (Pattillo 2013). This ethos was most thoroughly developed during the expansion of homeownership in the 1950s and 1960s. The mortgage markets that undergirded that midcentury expansion were very different from the deregulated and securitized markets of the first decade of the 2000s, however. By the turn of the twenty-first century, mortgage markets had become far more treacherous than many realized (Immergluck 2009). This insecure housing market was a peril not just for the poor and disadvantaged minority populations, but also for a large range of white middle-class suburban families who were traditionally believed to be immune from this kind of calamity.

In this article, we argue that the housing crisis holds significant lessons for understanding growing insecurity in twenty-first century America, and these lessons have significant implications for ongoing policy debates. The crisis represents a moment of exposed risk, where the underlying insecurity of mortgage markets that developed during the 2000s was suddenly in full sight. We study metropolitan variability in foreclosures to identify the conditions that led to greater precarity in housing markets during and after the 2008 housing crisis. We build on prior research on the housing crisis by identifying different sources of precarity in metropolitan areas, including both housing market insecurity and labor market insecurity. Investing in homeownership becomes riskier in a context of high levels of labor market risk, characteristic of many metropolitan areas in the early 2000s. We argue that housing market risk and labor market risk are additive sources of precarity that intertwine but also create somewhat distinctive forms of vulnerability for residents of U.S. metropolitan areas.

Theoretically, we build on the long history of research on housing insecurity. Prior research on urban homelessness and residential instability in poor and minority segregated neighborhoods has identified causes and consequences of housing precarity among highly disadvantaged populations (Lee, Tyler, and

NOTE: This project is partly supported by the National Science Foundation Graduate Research Fellowship Program under Grant No. DGE-0822215. The opinions expressed here are solely those of the authors and do not represent those of the National Science Foundation.

Wright 2010). Lately, there has been a renewed focus on precarity for the urban poor, most notably in Matthew Desmond's work on evictions (Desmond 2012), as well as in studies that highlight the extreme levels of housing insecurity among ex-felons, worsened by laws that prohibit felons from living in subsidized housing and extensive police surveillance of residences for even minor parole violations (Goffman 2009). We contribute to this growing consensus that housing precarity is an important dimension of residential stratification in contemporary urban settings by considering the homeownership market and the sources of precarity for a broader set of populations within metropolitan areas, beyond the most disadvantaged. By studying foreclosures, we use an indicator of housing insecurity particularly likely to affect nonpoor families given the relatively low homeownership rates among the poor (Joint Center for Housing Studies 2012). Yet the strength or weakness of the homeownership market has significant consequences for entire communities, as the housing crisis brought into all too clear relief.

This study contributes to policy debates on the housing crisis and Great Recession and the rebounding effects of these events on inequality in America. We also contribute to developing theory and evidence on precarity in American life more broadly. Despite the key role of housing in the Great Recession, many studies of economic insecurity hardly touch on insecurities related to housing (Western et al. 2012). Here we take up Mary Pattillo's (2013) call for greater attention to housing as a commodity, arguing that what the Great Recession has revealed more generally are the risks of mortgaged homeownership in a time of precarity. At the same time, studies of the housing crisis often focus almost exclusively on weakness in mortgage markets to the exclusion of the role of labor market precarity. The pattern of foreclosures across metropolitan areas helps to uncover these overlapping fault lines.

Risk and Insecurity in U.S. Metropolitan Areas

Metropolitan areas are locations of opportunity and disadvantage that shape residents' life chances. Local labor markets and housing markets typically organize at the metropolitan scale and the municipalities within metro regions compete for jobs and residents through these markets. Racial and economic segregation are primarily metropolitan dynamics that take different shapes depending on the economic base and population composition of urban regions (Dwyer 2007, 2010; Rugh and Massey 2010). Wage and income inequality are also at least in part a function of metropolitan forces, shaped by the economic base of cities, by the character of competition among social groups in local labor markets (affected especially by relative group size), and by all the ways that segregated neighborhoods provide differential access and exclusion to racial and ethnic populations (McCall 2000). Of course, metropolitan areas are not the only types of settlements that produce segregation. Residential segregation occurs at larger and smaller spatial scales, including between subnational regions and within smaller towns and rural areas (Lobao 2004; Lichter, Parisi, and Taquino 2012).

Still, metropolitan characteristics shape residential segregation and economic inequality in part by influencing access to opportunities and exposure to harm at other geographic scales, especially within the cities, suburbs, and local neighborhoods that make up metropolitan areas. For example, the economic specialization of metropolitan areas affects the spatial distribution of jobs and transit opportunities, in turn shaping residents' proximity to jobs in different neighborhoods.

The same metropolitan dynamics that shape stratification structures should influence exposure to insecurity in local labor and housing markets. Local markets likely create different *risk contexts*, producing different levels of unemployment, housing difficulties, and foreclosure (see Hall, Crowder, and Spring, this volume, on foreclosure). There has been much less attention to the potentially variable sources of insecurity in metropolitan areas, however, than to inequality, poverty, and segregation. One possible explanation for this lack of research on metropolitan risk contexts is that scholars reasonably expect that deprivation and insecurity are fellow travelers. The poorest and most disadvantaged populations also face the highest levels of insecurity, and thus metropolitan areas that produce high levels of disadvantage also likely produce high levels of insecurity. Yet there is growing evidence that insecurity and deprivation are distinct dimensions of disadvantage that do not always clearly align. Deregulated housing and labor markets have increased the risk of experiencing an income- or wealth-destroying event for many Americans, and a weakening social safety net has increased the costs of experiencing such negative events (Hacker 2006). Deregulation and welfare state retrenchment mean that families face multiple sources of risk—including job loss and housing precarity—that can independently or jointly create a financial crisis, even for middle-class families.

One perhaps overlooked feature of the great risk shift is that the federal retreat from regulation and social insurance increasingly exposes individuals and families to their local context. Where federal rules and benefits may once have gone some way toward leveling life chances among places with different economies, housing regimes, and histories, now the consequences of uneven development unfold with little buffer. Even the government regulations and services that remain have frequently devolved from the federal level to state, county, and municipal levels, creating variable policy environments as governmental authority fragments across entities at different geographic scales (Lobao 2004). Just as devolving services have made Americans more susceptible to localized political environments, deregulated markets have made Americans more vulnerable to local economic conditions (Katz and Bradley 2013). These local variations become especially apparent during an economic downturn.

The Great Recession exposed underlying risks that had been building for years in the American political economy. Because the recession was instigated by a housing crisis, a great deal of attention has been given to risks in the housing market. Yet growing insecurity in the labor market significantly contributed to the crisis as well. The long shift toward deregulated and deinstitutionalized markets came to fruition in the 2000s in a calamitous combination of an overheating housing market and a lackluster job market. Here we study metropolitan variation in the housing and jobs crises and then analyze the overlapping versus cross-cutting nature of these crises. We focus on indicators of housing and labor market weakness and consider the policy and political variation discussed above as potentially contributing background factors. While we do not directly measure policy interventions, in the conclusion we return to policy issues as a crucial direction for future research.

Housing market risk

Housing market insecurity has long accompanied residential inequality in American communities. Prior research on urban homelessness and high levels of residential instability in poor and minority segregated neighborhoods catalogs the devastating consequences of housing precarity among highly disadvantaged populations (Lee, Tyler, and Wright 2010; Desmond 2012). The Great Recession raised new questions about precarity for homeowners as well. Homeowners tend to be more advantaged than the populations exposed to the most severe housing insecurity, but homeowners also face a particular set of risks associated with mortgage and property markets (Anacker, Carr, and Pradhan 2012). We focus on three types of housing market precarity that were important before and after the housing crisis but that varied significantly among metropolitan areas: the prevalence of high-interest loans, rapidly rising house prices, and racial residential segregation.

The clearest manifestation of deregulated mortgage markets in the 2000s was the rise of the subprime loan. Deregulated and securitized mortgage markets separated issuers of mortgages from the risks of default (at least in the short term), leading to an explosion in complicated mortgage instruments and, consequently, much wider access to mortgages than in prior decades (Immergluck 2009). Subprime loans were targeted toward supposedly higher-risk consumers and often contained other complex provisions for adjustable rates and fees for early payback. While these loans allowed a greater number of people to enter into the housing market, the complex instruments also increased the risk of taking on a mortgage. Lax underwriting standards made subprime transactions even riskier and opened up opportunities for abuse—especially by mortgage agents who faced large incentives to give out loans—but few deterrents because their companies expected to bundle and sell the loans. While subprime loans spread nationally, local regulatory and market conditions produced metropolitan differences in the rate of subprime loans (Immergluck 2009).

Another source of housing market precarity fueled by deregulation was speculative investment in residential property. While speculation was facilitated and sometimes pursued with the aid of subprime loans, the buying and selling frenzy associated with the housing boom also had distinct sources (Immergluck 2009). The sheer number of people looking to enter the real estate market made it easier for investors to quickly move property, and rising prices made it easier to get home improvement loans for those wishing to "flip" property. Not all metropolitan areas saw this level of investment activity, however. The market for new

housing construction, population growth, and local real estate actors all influenced the level of activity.

Finally, racial segregation represents a potential source of fragility within a
metropolitan area's housing market. Segregated populations experience higher
levels of housing insecurity (owners and renters) because they are subjected to
greater predation and unscrupulous housing industry behavior, and because
property values are suppressed by segregation due to depressed demand for
houses in those areas (Rugh and Massey 2010). Metropolitan areas with high
levels of segregation may also be more likely to experience speculation and price
competition among more affluent homeowners seeking to locate in separate
spaces from the less advantaged populations (i.e., an arms race effect).

Labor market risk

As William Julius Wilson powerfully argued in his 1987 classic *The Truly
Disadvantaged*, inequality and insecurity in American cities are strongly linked to
labor market conditions. Employment insecurity has since become even more
important, as American households rely more on jobs as other forms of social
insurance decline, as aid to single mothers requires work, and as more families
rely on two earners to make ends meet (Hacker 2006). At the same time, polari-
zation and precarity in labor markets became even more severe during the first
decade of the 2000s (Kalleberg 2011).The U.S. labor market never fully recov-
ered from the twin shocks of the technology sector bust in 2000 and the 2001
terrorist attacks and subsequent war economy. The 2001 recession appears mild
when compared with the deep crisis of 2007–9, but the recovery throughout the
2000s was anemic, with much lower job growth than during the previous two
periods of expansion (Wright and Dwyer 2003).

Job loss is one of the main causes of housing distress (Sullivan, Warren, and
Westbrook 2000), but discussions of the jobs crisis and the housing crisis largely
proceed separately.[1] We propose three sources of labor market risk that we
expect were significant for metropolitan variation in the severity of the housing
crisis in those areas: the weakness of the labor market before the crisis as indicated by unemployment during the 2000s expansion, the weakness of the labor
market after the crisis as indicated by job growth and unemployment, and the
economic specialization of the metropolitan area.

Job market conditions in a metropolitan area before the housing crisis set the
stage for problems during the recession. In places with weaker job markets and
higher levels of unemployment, residents struggled to save enough to weather
economic downturns and entered the recession with considerable stress. Some
metropolitan areas have had faster recoveries in their housing markets than others, and unemployment and job growth after the crisis was also significant for
whether a metropolitan area continued to have high foreclosures into the economic recovery after the official end of the recession.

The economic specialization of metropolitan areas also shapes the metropolitan risk context. One of the key sources for rising inequality and insecurity in the
American jobs structure has been economic restructuring from manufacturing to

service. Service industries are more likely to employ part-time workers, use temporary help, and have employment-at-will work environments than are other sectors, especially compared with unionized manufacturing and public sectors (Kalleberg 2011). Metropolitan areas with large service economies are also likely to have higher levels of inequality because workers in the high-end services that pay lucratively in finance and health care generate demand for low-level services such as food service and childcare that are low-paying and more precarious than middle-wage jobs (Sassen 2001; Moller and Rubin 2008).

Here we take first steps toward understanding how labor market insecurity and housing market insecurity are distributed across different types of urban places in the United States. How did weakness in the labor and housing markets during the 2000s affect the housing crisis that followed during the Great Recession? How did these two factors intersect and develop over time?

Data and Methods

We construct a dataset on insecurity in U.S. metropolitan areas using a range of data sources. We use the American Community Survey (ACS) data (1-year files) from 2005 to 2012 for demographic and economic variables (U.S. Census Bureau 2014). Unemployment rates come from the Bureau of Labor Statistics (BLS 2014). The Federal Housing Finance Agency provides data on the house price index as well as home loan data gathered by mandate of the Home Mortgage Disclosure Act (Federal Housing Administration 2012, 2013). From Brown University's American Communities Project we utilize several segregation indices (Brown University 2014). Finally, we use foreclosure data from the Local Initiatives Support Corporation (LISC) calculations, derived from Lender Processing Services Applied Analytics and Mortgage Bankers Association data on mortgage problems from large loan servicers and several other sources (LISC 2013). We include all 366 metropolitan areas for which foreclosure rates are provided in the LISC data (defined by 2008 metropolitan boundaries).[2] We conduct supplemental analyses including only the 100 largest metropolitan areas and find largely similar results; we report differences where relevant in the text.

We measure the depth of the *housing crisis* in metropolitan areas with foreclosure rates. We calculate these rates in two different time periods: first, in 2010 at the peak of the crisis and second, averaged from 2010 to 2013 in the slow recovery that followed. This gives us a view of the acute and chronic effects of the crisis.

We measure *housing market precarity* in the homeownership market with three main indicators. Change in the housing price index from 2000 to 2006 indicates the degree to which a speculative housing bubble developed in the metropolitan area. The mean percentage of subprime loans originated in the metropolitan area from 2004 to 2007 indicates how many high-risk mortgages were issued in the metropolitan area leading up to the crisis. We measure black-white and Hispanic-white racial segregation with the dissimilarity index for 2000.

We also have three main indicators of *labor market precarity*. The mean unemployment rate in the metropolitan area during the 2000s economic expansion before the crisis (2001–6) and during and after the recession (2007–10) indicates the level of precarious work in the metropolitan area. We also measure how much levels of employment declined during the 2000s recession as an indicator of labor market health. We measure economic specialization with the percent of employment in key industries including services, manufacturing, the public sector, and other industries as an indicator of the types of jobs that prevail in metropolitan areas.

Analytically we estimate variation in foreclosure rates across metropolitan areas as a function of these sources of precarity. We use graphical techniques, cluster analysis, and ordinary least squares regression analysis to study variation in foreclosures during the most acute phase of the housing crisis and in the slow recovery that followed the official end of the recession. Because we are among the first to attempt to chart the variability in the risk contexts of U.S. metropolitan areas, we start by considering the key factors that have received the most attention during the Great Recession: housing and jobs. Other sources of risk in metropolitan areas also likely played a role in the Great Recession. For example, rental markets vary significantly in levels of precarity through market dynamics, through differences in the ease with which owners can evict residents, and through variation in the foreclosure rate in rental housing given that even tenants current on their rent can be forced out by banks when property owners go into default (Immergluck 2009). We consider a smaller set of factors for now as an entry point for conceptualizing the risk contexts of metropolitan areas.

Housing and Labor Market Precarity in the Foreclosure Crisis

Our findings show that both housing and labor market precarity shaped metropolitan areas during the Great Recession, but there was variation in the pattern of precarity that was most significant. Housing speculation and subprime loans were important in all metropolitan areas with serious foreclosure crises, but the foreclosure crisis was most severe in places that also experienced significant jobs crises. Metropolitan areas with more secure labor markets were much better able to weather the foreclosure crisis even if they experienced a large number of subprime loans. There was also change over time in the relative role of these factors, from the initial onset of the housing crisis to the continuing foreclosure crisis that followed into the economic recovery. All metropolitan areas experienced the housing boom and bust of the 2000s, but some were particularly hard-hit, and others were relatively more insulated. While we cannot identify all the mechanisms that led to these differential experiences, our findings provide insight into the types of market weaknesses that beset the worst-impacted areas.

Both housing market precarity and labor market precarity are associated with variation in foreclosures among metropolitan areas. The top panel of Figure 1 shows scatterplots with fitted regression lines of the association among our three measures of housing precarity, all of which are positively associated

FIGURE 1

Scatterplots and Fitted Regression Lines for Association between Housing and Labor
Market Precarity and Foreclosure Rates in U.S. Metropolitan Areas

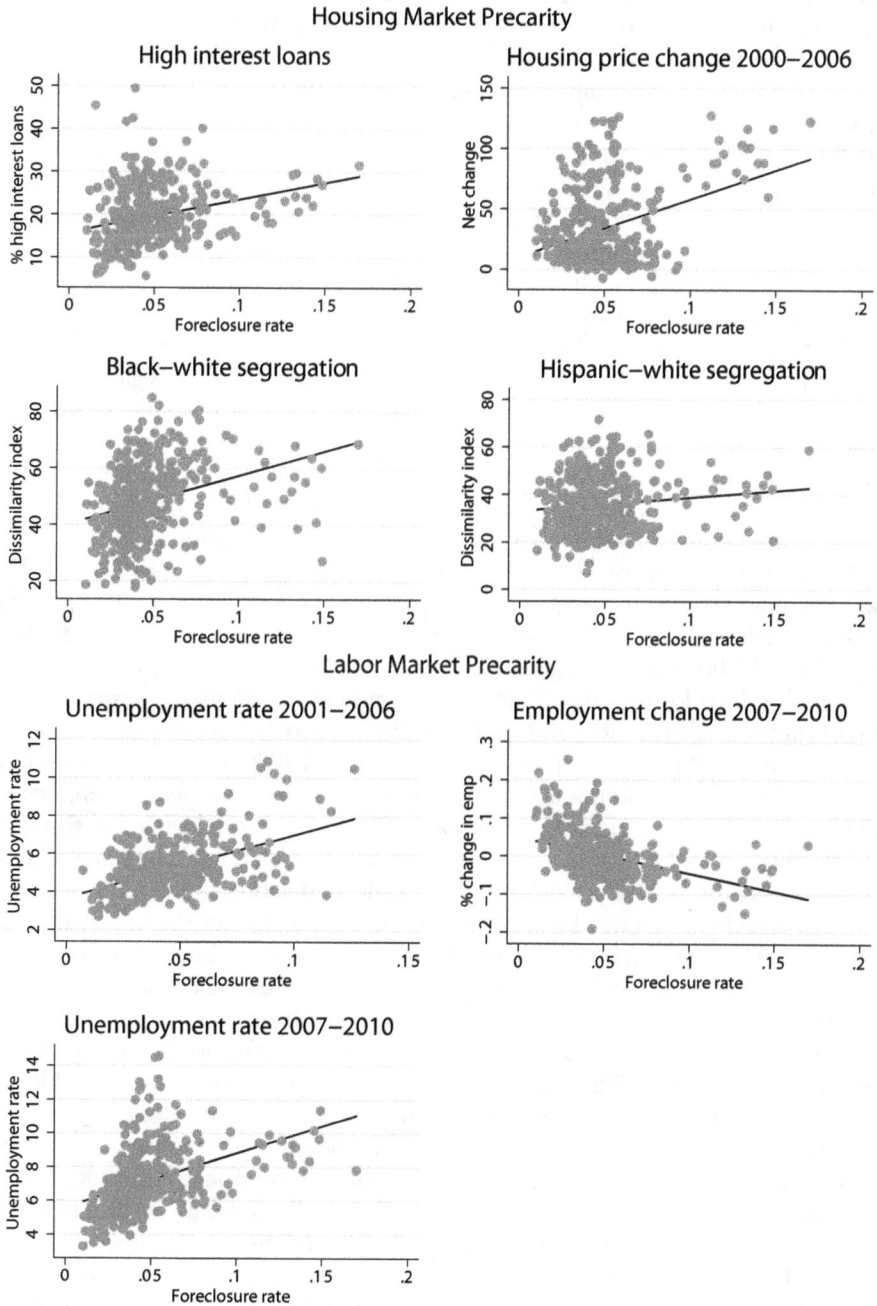

Housing Market Precarity

Labor Market Precarity

with foreclosures. Metropolitan areas with larger home price increases, greater percentages of high-interest loans, and higher levels of racial segregation experienced higher foreclosure rates. The association between the level of high-interest loans in a metropolitan area during the housing boom and the foreclosure rate in 2010–13 displays a relatively compact dispersion. The association between housing price changes and foreclosure rates is a little more complex, with more dispersion around the line, suggesting that there may have been several different processes shaping the relationship between price increases and housing precarity. For both high-interest loans and housing price changes, there is a somewhat distinct group of metropolitan areas that are high on both indicators. The greater dispersion for home price increases shows that there were metropolitan areas with significant housing booms that nevertheless saw a smaller foreclosure crisis than other similar places. One reason is likely differences in labor market health between places with similar home price booms.

Metropolitan areas with higher levels of racial residential segregation also have higher levels of foreclosure, consistent with arguments that segregation fosters housing insecurity. The association is particularly strong for black-white segregation, consistent with Rugh and Massey's (2010) findings for the early crisis period. The relationship between foreclosures and Hispanic-white segregation is a little weaker in comparison, but Hispanic-white segregation is lower overall and is less variable among metropolitan areas than black-white segregation.

Labor market precarity is also significantly associated with foreclosures, but here the most significant factor was unemployment in the years running up to the housing crisis, as shown in the bottom panel of Figure 1. Metropolitan areas that experienced labor market weakness before the crisis hit were also more likely to have higher foreclosure rates. This association is not particularly surprising, but it is often overlooked in accounts that treat the jobs crisis as a consequence of the housing crisis rather than as a long-run structural feature of the U.S. economy, which varies in its contours across metropolitan areas. The percent change in employment after the recession (indicating job growth or decline) is similarly negatively associated with foreclosures, showing that stronger economies fared better during the housing crisis. We found little association of the industrial sector with foreclosure in any time period, however (results not shown). It may be that the industries used here are too aggregated to capture the kinds of precarity that may impact housing security. Or perhaps much of the relevant metropolitan variation in labor market insecurity occurs within industries.

Next we evaluated the joint distribution of labor market and housing market precarity. Did metropolitan areas with high foreclosure rates typically evidence both job and housing market precarity, or did the housing crisis look different in different places? First, we conducted a cluster analysis using the most significant indicators of precarity, including the percent of subprime loans, the unemployment rate, and the foreclosure rate. Cluster analysis finds groups in data by identifying similarities and dissimilarities between observations.[3] Our analysis identifies five clusters that fall into two main categories: metropolitan areas with roughly higher than average foreclosure rates and metropolitan areas with roughly lower than average foreclosure rates. We see these clusters as further differentiating tendencies among metropolitan areas rather than sharply delineating types.

Table 1 reports average levels on our indicators of labor market and housing market precarity as well as foreclosure rates for these clusters, and several interesting patterns emerge. First, there are two clusters for metropolitan areas with the highest foreclosure rates. These clusters are particularly distinctive from those with lower foreclosure rates in evidencing higher than average unemployment, not only after the recession, but before it as well. The cluster with the highest foreclosure rates has a significantly high unemployment rate before the housing crisis, and also has the highest percentage of subprime loans and the largest home price increases during the boom. These metropolitan areas were in very risky territory in both their housing and labor markets before the crisis, and the downturn has been particularly severe in those places. The other cluster with relatively high foreclosure rates looks similar but has a somewhat lower percent of subprime loans and foreclosures though still high unemployment rates. We find similar results when we restrict the cluster analysis to the 100 metropolitan areas with the largest populations. The only notable difference is that the cluster with the second-highest foreclosure rate in this restricted analysis has the lowest unemployment rate before the crisis in 2001–6, and then a much higher unemployment rate after the crisis. This is consistent with the argument that housing market precarity was the principal precursor to decline in some communities. Metropolitan areas in this cluster include Las Vegas, Phoenix, and Orlando. We explore this pattern further in the regression analysis below.

The clusters with average or below-average foreclosure rates are distinctive in being particularly low on some measure of precarity. One is lowest on the percent subprime loan measure, another is lowest on the home price increase measure, and the last is lowest in unemployment. In fact, the cluster that has the lowest foreclosure rates both in the immediate and longer term of the housing crisis has the lowest unemployment rates, even while having a reasonably large percentage of subprime loans. Of course, it is important to recall throughout this discussion that the foreclosure rates are still high even for the cluster with the lowest rate of homeowner distress.

The clusters are also geographically distinctive. Table 1 shows that metropolitan areas in the two clusters with the highest foreclosure rates are predominately located in the West and South. Within those regions, the cluster with the highest foreclosure rate is concentrated in the Pacific division, including San Diego and Fresno, and the South Atlantic division, including Miami and Palm Beach. Cluster 2 (with the second-highest foreclosure rate) has more metropolitan areas in the Mountain division, including Las Vegas and Tucson. Cluster 3, which is characterized by a higher housing price increase but the lowest percentage of subprime loans, is fairly evenly spread across the Northeast, South, and West, with the highest concentration in the Northeast. Cluster 4, with higher unemployment, is concentrated in the Midwest, including Toledo, Youngstown, and Detroit. Cluster 5, with the lowest foreclosure rate and distinctively low unemployment before the crisis is disproportionately in the South, and includes Nashville, Albuquerque, and Birmingham.

TABLE 1
Cluster Analysis of Labor and Housing Market Precarity and Foreclosure

	Cluster	Foreclosure rate 2010–13	Unemployment 2001–6	Unemployment 2007–10	Percent high-interest loans	Home price increase 2001–6[*]	Northeast	Midwest	South	West
Above-average foreclosure	1	7.4	6.4	10.0	23.1	186	4%	0%	30%	65%
	2	6.6	6.0	9.1	18.3	135	11%	0%	36%	53%
Below-average foreclosure	3	4.8	4.7	6.5	15.7	83	37%	6%	24%	33%
	4	4.5	5.4	7.2	20.7	14	7%	45%	42%	7%
	5	3.9	4.9	6.2	19.3	39	13%	16%	53%	19%

[*]In thousands of dollars.

TABLE 2

Ordinary Least Squares Regression of Labor and Housing Market Precarity on
Foreclosure Rates

	2000s precursors	Recession period
	Coefficient (standard error)	Coefficient (standard error)
Low subprime and High unemployment	.138 (.070)*	.318 (.065) ***
High subprime and Low unemployment	.355 (.071)***	.078 (.065)
High subprime and High unemployment	.294 (.058)***	.498 (.054) ***
Constant	−3.385 (.041)	−3.380 (.038)
R^2	.09	.21

NOTE: *Low* is defined here as the bottom half of percentages of subprime loans across metropolitan areas. *High* is defined here as the top half of percentages of subprime loans across
metropolitan areas
*$p \leq .05$. **$p \leq .01$. ***$p \leq .001$ (two-tailed tests).

To further explore the intersections of housing and labor market precarity, we
estimate an ordinary least squares regression of the interaction between high and
low levels of unemployment and subprime loans in U.S. metropolitan areas. One
advantage of this additional analysis is that we can more clearly isolate the influence of different types of precarity than in the cluster analysis, which produces
best matches that may in fact combine metropolitan areas with significant differences. We consider high and low rates above versus below the median for all
metropolitan areas. (We replicate the results with more conservative measures of
high and low with similar patterns.) In Table 2, we show a basic model with only
the precarity measures. (In supplemental analyses not shown to conserve space,
we estimate the model with control variables and find a similar pattern of results.)
The results are consistent with the cluster analysis, where the highest foreclosure
rates are in metropolitan areas with both high unemployment and high levels of
subprime purchase loans in the 2000s. Foreclosure rates are also elevated in
metropolitan areas high on either subprime loans or unemployment, though not
as high as in places with both. In the early crisis years, having a high percentage
of subprime loans was most significant, whereas in the later period, unemployment became more important in predicting foreclosure, as the housing crisis
worsened the jobs crisis in many places.

These contrasts are more vivid when we consider exemplar metropolitan areas
in each category of interaction between housing market and labor market precarity. Table 3 gives key examples. The places in the "high-high" category with the
highest foreclosure rate are locations that we already know to be extremely distressed, including Detroit, MI, and McAllen, TX. The "low-low" category

TABLE 3
Exemplar Metropolitan Areas

	Higher than average unemployment	Lower than average unemployment
Higher high-interest loans	Las Vegas, NV (in recession)	Las Vegas, NV (in 2000s)
	St. Louis, MO	Dallas, TX
	Detroit, MI	Phoenix, AZ
	McAllen, TX	Indianapolis, IN
Lower high-interest loans	Cleveland, OH	Columbus, OH
	Pittsburgh, PA	Minneapolis, MN
	Louisville, KY	Austin, TX

includes relatively more robust economies such as Minneapolis and Austin. Places with large levels of high-interest loans but lower unemployment include metros in the Southwest such as Dallas and Phoenix. Metropolitan areas with higher unemployment and fewer high-interest loans include older rust belt metros such as Cleveland and Pittsburgh. This latter category highlights a limitation of our subprime measure. It includes only loans for home purchases and not home equity loans. The predatory lending that occurred in places such as Cleveland and Pittsburgh was heavily focused on encouraging existing homeowners to take out loans against equity held in their home (Dillman 2013). Thus, one important form of housing precarity is still unmeasured in our analysis. However, the differences in these types of housing market risks are in fact related to the labor market trajectories in these different areas. Predatory lending in the older and more segregated industrial cities targeted minority homeowners more than any other group, producing a type of housing insecurity tied to the particular socioeconomic history in those places.

Finally, Las Vegas stands out (as it ever has during the crisis) as an illustrative case of the relationship between housing and labor market fragility. While it was in the low category of unemployment before the crisis, it developed higher than average unemployment into the 2000s recovery. The very extreme housing boom and bust in the Las Vegas metropolitan area brought the labor market down with it. This is a microcosm of what happened with the housing crisis nationally and globally as risks that built up in one market spread out to other markets. This kind of risk contagion is an important reason to better understand the sources of risk in our metropolitan economies. Only by identifying vulnerabilities can we develop the necessary policy instruments to prevent or stop crises before they become nationwide, and indeed worldwide, calamities.

Conclusion

The foreclosure crisis in the United States was rooted in both labor market weakness and increasingly risky mortgage markets that hit some metropolitan areas harder than others. The metropolitan areas that fared the worst faced problems in both markets before the crisis and showed the highest levels of unemployment and foreclosure into the postrecession period. Other metropolitan areas faced particular vulnerabilities on one or the other dimension: some areas experienced more housing market risk, whereas others had a more severe jobs crisis. In this article we argue that growing insecurity in U.S. metropolitan areas arises out of multidimensional and cross-cutting economic weaknesses that are too often seen in isolation. We think the analysis is particularly novel in drawing attention to the important role of labor markets in ongoing housing market problems in U.S. cities. A fundamental characteristic of the contemporary period is growing insecurity across distinct arenas that, at their worst, accumulate into extreme crises like the Great Recession. There may also be increasing inequality in risk contexts, as jobs, housing, and credit risk combine in different ways in different areas, exposing some populations to accumulating risk, whereas other areas can rely on security in one area (especially the labor market) as a buffer against insecurity on other dimensions.

Why are community risk contexts important above and beyond our more traditional measures of inequality, poverty, and segregation? In addition to the analytic reasons we have offered above, we believe that bringing attention to risk and insecurity in metropolitan areas can contribute to several important areas of policy and public debate. First, a number of the most pressing concerns in poverty policy involve managing insecurity. High levels of housing insecurity and eviction in many metropolitan areas significantly worsen life chances, especially for poor minority families living in predominantly minority neighborhoods. Suburban poor populations often face distinctive risks, partly because there are often less robust social services and greater transportation insecurity in suburban settings than in cities (Murphy and Wallace 2010). All of this is to say that neighborhoods and municipalities within metropolitan areas also represent differential risk contexts on a wide array of dimensions. In turn, metropolitan factors significantly shape the risk contexts of smaller places, not least through the labor and housing market dynamics we have discussed here.

Second, this analysis highlights the multiple dimensions of growing insecurity in American life. The jobs crisis and the housing crisis are experienced simultaneously by Americans looking for work, trying to pay mortgages, and struggling to save for an ever-more-uncertain future. These crises have similar origins and, as we have discussed here, they intersect in patterns that have significant consequences for life chances. There is also inequality across metropolitan areas in the degree to which Americans are exposed to labor market and housing market risk. To the extent that the federal government retreats from regulation and social insurance, its "leveling" capacity is diminished and locational attainment in America is shaped by the risk context of the communities in which we live.

Third, we need more research on how variable policy environments shape insecurity contexts. Lobao and collaborators' data on county governments is a very important step, demonstrating just how much variability there is across and between geographic scales (Lobao 2004; Lobao, Adua, and Hooks 2014). The Great Recession itself appears to have encouraged city and metropolitan policy

initiatives. Cities are investing in innovation districts designed to encourage economic growth, including high-precarity places such as Cleveland and Detroit and
places with already robust economics such as Minneapolis (Katz and Bradley
2013). Much of the focus of policy and research is on interventions to increase
job growth. We need more attention as well to the conditions and policies that
may shape housing market precarity, including policies that shape access to housing, banking density, and policing of mortgage fraud (Carswell and Bachtel 2009;
Nelson 2010; Lucio and Ramirez de la Cruz 2012). City governments may be
hampered in their efforts, however; even as authority has been devolved in some
areas, other policy arenas are seen to be the purview of higher levels of government, even if state and federal agencies take little action. In a striking example,
Cleveland's city government passed an ordinance that restricted predatory high-
interest mortgages only to see that ordinance repealed by an Ohio State Supreme
Court decision that the law inappropriately treaded on a "statewide concern"
(Xu 2014, 126).

Finally, we think "insecurity" may be in certain respects a more effective focus
for policy intervention than "inequality," both in terms of government capacity and
for reasons of political framing. There is energy around addressing inequalities in
cities today. Mayors elected in 2013 in New York, Minneapolis, Boston, and Seattle
campaigned on pledges to reduce disparities. Policy center programs such as the
Brookings Institution Metropolitan Policy Program promote the idea that cities can
take on the task of economic development. However, it is difficult for governments
to intervene directly into economic inequality and growth, with the exception of
certain targeted policies such as the minimum wage. Governments may be most
effective when they instead provide insurance against the risks inherent in capitalist
markets. Social insurance policies do redress inequality but do so by setting a floor
below which the most disadvantaged do not fall, and by putting constraints on the
degree to which capitalist organizations and investors can monopolize the fruits of
productivity. Developing a new language around the need for social insurance and
a challenge to the great risk shift would thus bolster efforts to redress inequality.
The political obstacles to such a strategy are quite significant, however. The bitter
and continuing debate over the Affordable Care Act shows only too well how difficult it is to initiate a new program of collective insurance, and even the longstanding insurance programs such as Social Security and Medicaid have come
under increasing attacks. The forces that led to the great risk shift will continue to
be arrayed against efforts to reverse the shift.

There may be a political framing advantage to shifting the conversation to
insecurity in addition to inequality, however. A conversation about insecurity may
provide greater potential of moving toward a "win-win" framing of governmental
intervention. Discussions of inequality in the U.S. context are all too easily
framed in a "win-lose," "us-versus-them" discourse blighted with only faintly
disguised racial animosity (if disguised at all). Insecurity after the Great Recession
may be harder to frame in these terms and thus could provide a firmer basis for
political mobilization. In cross-national studies, political scientists find that support for welfare state policies is strongest in countries where the population
exposed to risk is broader than the disadvantaged population. Where poverty and

insecurity overlap more strongly, there is less support for social insurance (Rehm, Hacker, and Schlesinger 2012). One unintended consequence of the great risk shift is to put Americans from diverse racial and social class positions in a similar position of exposure to insecurity and, perhaps as a result, widen the potential for coalition building. Certainly the Great Recession has changed the conversation around inequality and insecurity in America. Only time will tell whether that changed conversation will translate into changed institutions and improved life chances for Americans from all social backgrounds. Greater discussion of growing insecurity in American life may very well contribute to moving the conversation—and then the institutions—in the right direction.

Notes

1. Some analysts attempt to downplay the role of housing market actors in the Great Recession by arguing labor market weakness is more important, but we argue that the two forms of precarity are overlapping and linked as outcomes of deregulated neoliberal markets.

2. The LISC data are less rich than RealtyTrac data, but we use them because they are publicly available and show similar patterns to other sources. The data work for our purposes because they describe aggregate foreclosure rates at the metropolitan level. More detailed analyses of neighborhood risk contexts are desirable, but this would require the larger RealtyTrac database.

3. We use k-means clustering, with Euclidean distance defined as the similarity measure. We tested clusters with different numbers of groups and identified the best fit as the cluster with the largest Calinski-Harabasz pseudo-F index.

References

Anacker, Katrin B., James H. Carr, and Archana Pradhan. 2012. Analyzing foreclosure among high-income black/African American and Hispanic/Latino borrowers in Prince George's County, Maryland. *Housing & Society* 39 (1): 1–28.

Brown University. 2014. *Segregation indices*. American Communities Project. Available from http://www.s4.brown.edu/us2010/Data/Download1.htm.

Bureau of Labor Statistics (BLS). 2014. *1990–2013 local area unemployment statistics*. Washington, DC: BLS. Available from http://www.metrotrends.org/data-download-new.cfm.

Carswell, Andrew T., and Douglas C. Bachtel. 2009. Mortgage fraud: A risk factor analysis of affected communities. *Crime, Law, and Social Change* 52:347–64.

Desmond, Matthew. 2012. Eviction and the reproduction of urban poverty. *American Journal of Sociology* 118:88–133.

Dillman, Jeffrey D. 2013. Subprime lending in the city of Cleveland and Cuyahoga County. In *Where credit is due: Bringing equity to credit and housing after the market meltdown*, eds. Christy Rogers and john a. powell, 140–62. Lanham, MD: University Press of America.

Dwyer, Rachel E. 2007. Expanding homes and increasing inequalities: U.S. housing development and the residential segregation of the affluent. *Social Problems* 54:23–46.

Dwyer, Rachel E. 2010. Poverty, prosperity, and place: The shape of class segregation in the U.S. *Social Problems* 57:114–37.

Federal Housing Finance Agency (FHA). 2012. *1997–2012 Home Mortgage Disclosure Act*. Washington, DC: FHA. Available from http://www.metrotrends.org/data.cfm.

FHA. 2013. *2000–2013 House Price Index*. Washington, DC: FHA. Available from http://www.metrotrends.org/data-download-new.cfm.

Goffman, Alice. 2009. On the run: Wanted men in a Philadelphia ghetto. *American Sociological Review* 74:339–57.

Hacker, Jacob S. 2006. *The great risk shift: The assault on American jobs, families, health care, and retirement*. Oxford: Oxford University Press.

Hall, Matthew, Kyle Crowder, and Amy Spring. 2015. Variations in housing foreclosures by race and place, 2005–2012. *The ANNALS of the American Academy of Political and Social Science* (this volume).

Immergluck, Dan. 2009. *Foreclosed: High-risk lending, deregulation, and the undermining of America's mortgage market*. Ithaca, NY: Cornell University Press.

Joint Center for Housing Studies. 2012. *State of the nation's housing 2012*. Cambridge, MA: Harvard University.

Kalleberg, Arne L. 2011. *Good jobs, bad jobs: The rise of polarized and precarious employment systems in the United States, 1970s to 2000s*. New York, NY: Russell Sage Foundation.

Katz, Bruce, and Jennifer Bradley. 2013. *The metropolitan revolution: How cities and metros are fixing our broken politics and fragile economy*. Washington, DC: Brookings Institution Press.

Lee, Barrett A., Kimberly A. Tyler, and James D. Wright. 2010. The new homelessness revisited. *Annual Review of Sociology* 36:501–21.

Lichter, Daniel T., Domenico Parisi, and Michael C. Taquino. 2012. The geography of exclusion: Race, segregation, and concentrated poverty. *Social Problems* 59:364–88.

Lobao, Linda. 2004. Continuity and change in place stratification: Spatial inequality and middle-range territorial units. *Rural Sociology* 69 (1): 1–30.

Lobao, Linda, Lazarus Adua, and Gregory Hooks. 2014. Privatization, business attraction, and social services across the United States: Local governments' use of market-oriented, neoliberal policies in the post-2000 period. *Social Problems* 61:644–72.

Local Initiatives Support Corporation. 2013. *June 2010–September 2013 metropolitan delinquency and foreclosure data*. Available from http://www.foreclosure-response.org.

Lucio, Joanna, and Edgar Ramirez de la Cruz. 2012. Affordable housing networks: A case study in the Phoenix metropolitan region. *Housing Policy Debate* 22:219–40.

McCall, Leslie. 2000. Explaining levels of within-group wage inequality in U.S. labor markets. *Demography* 37:415–30.

Moller, Stephanie, and Beth A. Rubin. 2008. The contours of stratification in service-oriented economies. *Social Science Research* 37:1039–60.

Murphy, Alexandra, and Danielle Wallace. 2010. Opportunities for making ends meet and upward mobility: Differences in organizational deprivation across urban and suburban poor neighborhoods. *Social Science Quarterly* 91:1164–86.

Nelson, Kathryn O. 2010. Housing needs and effective policies in high-tech metropolitan economies. *Housing Policy Debate* 13:417–68.

Pattillo, Mary. 2013. Housing: Commodity versus right. *Annual Review of Sociology* 39:509–31.

Rehm, Philipp, Jacob S. Hacker, and Mark Schlesinger. 2012. Insecure alliances: Risk, inequality, and support for the welfare state. *American Political Science Review* 106:386–406.

Rugh, Jacob S., and Douglas S. Massey. 2010. Racial segregation and the American foreclosure crisis. *American Sociological Review* 75:629–51.

Sassen, Saskia. 2001. *The global city: New York, London, Tokyo*. Princeton NJ: Princeton University Press.

Sullivan, Theresa A., Elizabeth Warren, and Jay Lawrence Westbrook. 2000. *The fragile middle class: Americans in debt*. New Haven, CT: Yale University Press.

U.S. Census Bureau, American Community Survey. 2014. *2005–2012 American Community Survey*. http://factfinder2.census.gov.

Western, Bruce, Deirdre Bloome, Benjamin Sosnaud, and Laura Tach. 2012. Economic insecurity and social stratification. *Annual Review of Sociology* 38:341–59.

Wilson, William Julius. 1987. *The truly disadvantaged: The inner city, the underclass, and public policy*. Chicago IL: University of Chicago Press.

Wright, Erik Olin, and Rachel E. Dwyer. 2003. The patterns of job expansions in the United States: A comparison of the 1960s and 1990s. *Socio-Economic Review* 1:289–325.

Xu, Yilan. 2014. Does mortgage deregulation increase foreclosures? Evidence from Cleveland. *Regional Science and Urban Economics* 46:126–39.

Variations in Housing Foreclosures by Race and Place, 2005–2012

This study describes the spatial and racial variations in housing foreclosure during the recent housing crisis. Using data on the 9.5 million visible foreclosures (public auctions and bank repossessions) occurring between 2005 and 2012, we show that the timing and depth of the foreclosure crisis differed considerably across regions and metropolitan areas, with those located in the Mountain and Pacific West regions experiencing the highest foreclosure risks. The crisis was patterned sharply along racial/ethnic lines, with metros and neighborhoods with large black and Latino populations—as well as racially mixed neighborhoods—having high rates of foreclosure. Our analysis also highlights the particular vulnerability of Latino households, who not only had very high individual risk of foreclosures but tended to reside in areas hit hardest by the crisis. The race-stratified geographic patterns of foreclosure revealed here are substantially more complicated than a narrative that depicts only the unique disadvantage of black households during the crisis, and likely reflect some level of specific targeting of minority populations and neighborhoods by predatory and subprime lenders.

Keywords: race/ethnicity; foreclosures; neighborhoods; metropolitan areas

By
MATTHEW HALL,
KYLE CROWDER,
and
AMY SPRING

The bursting of the housing bubble in the first decade of the 2000s was one of the most cataclysmic housing-related events in a century. Over the past decade, Americans had seen the value of their homes rise to unprecedented levels, only to have a large portion of

Matthew Hall is an assistant professor in the Department of Policy Analysis and Management and a faculty affiliate of the Cornell Population Center at Cornell University. His research centers on racial/ethnic inequality in housing and labor markets, and immigrant incorporation in new destination areas.

Kyle Crowder is a professor in the Department of Sociology and a faculty affiliate of the Center for Studies in Demography and Ecology at the University of Washington. His research focuses on the causes and consequences of residential segregation, neighborhood stratification, and environmental inequality.

DOI: 10.1177/0002716215576907

their equity erased when the housing market crashed. The contraction of housing prices, and the subsequent economic recession, contributed heavily to an explosion in the number of homeowners unable to afford their mortgages. Since 2006, approximately 16.2 million homes have entered foreclosure (RealtyTrac 2015). The scale of the crisis, thus, has been colossal, directly affecting nearly one in six American households. Yet estimates of the incidence of foreclosure completion miss the full scope of the crisis—while an enormous number of homeowners saw their homes enter the foreclosure process during the crisis, an even higher number experienced foreclosure vicariously through their extended families and friends, and an untold number dealt with the repercussions of foreclosure in their neighborhoods and broader communities. The full scale of the crisis has thus stretched far beyond individual homeowners undergoing housing-related stress.

Despite the wide reach of the foreclosure epidemic, media reports and commercial data outlets point to the strong likelihood that the impact of the crisis was unevenly distributed across the American residential landscape, with some locales seeing foreclosures accumulate rapidly and others hardly witnessing any. The geographic clustering of foreclosures means not only that people differentially experienced housing distress, but that the populations residing in these places were also unevenly exposed to housing foreclosures. In particular, the uneven distribution of racial/ethnic group populations across the United States presents the possibility that foreclosure and its consequences were racially variant. Furthermore, the high rates of subprime and other risky mortgages among racial minorities—driven not just by their relatively lower position on the socioeconomic ladder, but also by financially diminished family support networks and explicit targeting of black and brown neighborhoods by predatory lenders—suggests that minority homeowners may have been uniquely harmed by the housing crisis and experienced especially high rates of foreclosure, independent of their regional location.

A growing body of work is directed at documenting and explaining the consequences of the foreclosure crisis, including its effects on wealth accumulation (e.g., Pfeffer, Danziger, and Schoeni 2013), mental and physical health (e.g., Houle 2014), neighborhood crime (e.g., Ellen, Lacoe, and Sharygin 2013), and neighborhood instability (Hall, Crowder, and Springer 2014). While this work has been crucial for expanding our understanding of the consequences of housing foreclosure, we still lack basic knowledge of how the foreclosure crisis differentially affected communities and populations.

Information about which areas, and which populations, have been most affected by the foreclosure crisis is crucial to the development of policy responses capable of stabilizing vulnerable neighborhoods and remediating the individual- and community-level effects of the crisis. For example, the effectiveness of the federal Neighborhood Stabilization Program and similar state and local programs designed to interrupt the effects of foreclosure concentrations on housing decay,

Amy Spring is an assistant professor in the Department of Sociology at Georgia State University. Her research is directed at understanding spatial inequality, residential mobility, and population aging.

crime, and disorder hinges on a solid understanding of the types of neighborhoods most at risk of experiencing such concentrated effects (Immergluck 2012).
Similarly, mortgage default counseling, loan modification programs, and related
state and local programs are likely to be much more impactful if effectively
designed for, and targeted to, the populations and neighborhoods most heavily
affected by the foreclosure crisis (Collins, Lam, and Herbert 2011; Collins and
Schmeiser 2013; Mayer et al. 2012). More generally, variations in foreclosure
patterns across populations and geographic areas provide important clues about
the underlying forces that may precipitate the next crisis.

In this study, we seek to build on this basic knowledge by describing geographic and demographic variation in foreclosures since the onset of the crisis. In
doing so, we take a geographically and temporally expansive view, examining
foreclosures across virtually all metropolitan areas during the run-up to the crisis,
during its peak, and into the recovery (2005–12). Working our way down the
geographic scale, we document temporal patterns of foreclosure accumulation
across U.S. regions and metropolitan areas, explore whether foreclosure concentrations in metropolitan areas, their cities, their suburbs, and the constituent
neighborhoods were patterned along racial lines, and show how individual households faced different risks of experiencing foreclosures themselves and in their
neighborhood settings. To describe these patterns, we combine population-level
data on nearly all foreclosure events in the United States between 2005 and 2012
with data on individual households from the Panel Study of Income Dynamics.

To preface our findings, we demonstrate here that the timing and depth of the
foreclosure crisis varied significantly across the regions of the country, and that
the concentration of foreclosures was especially pronounced in metropolitan
areas—and their constituent cities and suburbs—containing relatively large
shares of African Americans and Latinos. Moreover, foreclosures were highly
clustered in metropolitan neighborhoods with large minority populations, a finding that stands in sharp contrast to the popular depiction of high rates of foreclosures in white suburbs. Finally, our data indicate that these spatial variations
corresponded with individual-level variations in the likelihood of foreclosure in
ways that reinforce the disadvantages experienced by some racial-ethnic groups;
in comparison to whites, individual African Americans and, especially, Latinos
were considerably more likely to experience foreclosure during (and following)
the crisis period.

Background

According to conventional wisdom, the foreclosure crisis that hit the United
States in 2007 was precipitated by a pernicious combination of unsustainable
lending practices, irresponsible borrowing, and unrealistic expectations about the
appreciation of real estate (Been, Chan, et al. 2011; Engel and McCoy 2011). In
the years leading up to the crisis, housing prices rose steadily in most metropolitan areas, and astoundingly so in some. These housing-price increases reflected

not only general economic prosperity during the last years of the twentieth century, but the assumption by homeowners, speculators, and lenders that real estate prices would continue to rise.

This expectation of ever-increasing equity helped to justify the increasing use of risky financing strategies, all enabled by years of federal deregulation of the banking industry (Immergluck 2008). With the growing popularity of zero-down mortgages, loan-to-value ratios soared and an increasing share of homeowners had little or no equity even at the origination of their loans. These problems were exacerbated by the proliferation of predatory lending practices that left millions of homeowners vulnerable to the financial burden of large balloon payments and unsustainably high interest rates. Moreover, even owners who had been in their homes for years were tempted by low interest rates and rising prices to borrow against the value of their homes. The end result was that by 2007, more than one in five homeowners had negative equity in their homes, owing more than the value of the property (CoreLogic 2010).

With the general slowing of the economy, this negative-equity epidemic fueled the foreclosure crisis. Declining economic activity and growing long-term unemployment left millions of homeowners—especially those with unfavorable mortgage terms—in properties they could no longer afford (Gerardi, Ross, and Willen 2011). In a normal housing market, such homeowners would be able to reduce their financial hardship by selling their homes. But for millions, negative equity and falling home prices made such an escape impossible, leading to high rates of default and foreclosure (Been, Chan, et al 2011). Even for many who did not experience job loss, making payments on a high-cost loan for a property that was no longer expected to appreciate lost its financial imperative. The result was millions of homeowners in default on their mortgages and a huge increase in the number of housing foreclosures. Between 2007 and 2010, the estimated number of annual foreclosures increased by more than fourfold, from about 650,000 to 2.9 million (RealtyTrac 2011).

Yet the underlying dynamics of the foreclosure crisis are likely to have varied dramatically across the country (see Dwyer and Lassus, this volume). For example, fast-growing metros of the Southwest saw aggressive development strategies and huge speculation-driven inflation of housing prices while other locations saw more modest increases (Gerardi, Ross, and Willen 2011). Similarly, subprime, zero-down, and other risky and high-cost loans were more common in some areas than in others, particularly in areas with highly inflated housing prices (Immergluck 2008). Finally, some areas were much harder hit than others by job losses (Kuehn 2011) and housing-price declines (Zillow 2011) brought about by the general economic recession. These geographic disparities are likely to have been exacerbated by variation in state policies related to the administration of foreclosures and lender strategies for dealing with defaulted loans and repossessed units (Collins, Lam, and Herbert 2011; Immergluck 2010, 2012).

Geographic patterns, along with sharp variations in the risk of foreclosure, highlight the highly racialized dynamics of the foreclosure crisis. In the years leading up to the foreclosure crisis, Latino and African American homeowners were substantially more likely than whites to hold high-interest or otherwise risky mortgages (Faber 2013; Cheng, Lin, and Liu 2014). Minorities were also hit hardest by layoffs when the recession began (Hoynes, Miller, and Schaller 2012) and in a weaker position than whites to weather the financial burden of these layoffs. The result has been substantially higher levels of foreclosure among ethnic minorities than among whites during the crisis (Bocian, Li, and Ernst 2010).

There is substantial evidence that these racial disparities were a function of—and exacerbated by—the explicit targeting of minority communities by subprime lenders (Engel and McCoy 2011). A U.S. Housing and Urban Development/U.S. Treasury report (2000) found that subprime loans were five times more frequent in predominately black neighborhoods than predominately white ones, and during the 1990s, the annual percentage of all subprime loans signed by those in minority neighborhoods increased from 2 to 18 percent. There is also consistent evidence that the disproportionate concentration of risky loans in minority-populated areas was facilitated by broader patterns of racial stratification, with residential segregation by race leaving whole communities vulnerable to predatory lending practices and the subsequent collapse of the housing market during the recession (Hyra et al. 2013; Rugh and Massey 2010). Moreover, the processes through which lenders have addressed the glut of repossessed properties since the end of the recession has been substantially slower in minority-populated communities than in areas occupied mostly by whites (Ellen, Madar, and Weselcouch 2014), leaving these same minority communities vulnerable to the prolonged impacts of weakened housing dynamics.

The role of race in the foreclosure crisis is particularly troublesome given what we know about the impacts of foreclosures on individuals, families, and communities. Most obviously, the foreclosure crisis has entailed substantial decreases in individual and family wealth and expanded the already large racial-wealth gap (Pfeffer, Danziger, and Schoeni 2013; Saegert, Fields, and Libman 2011). Moreover, the process of foreclosure by itself is associated with elevated levels of stress (Bennett, Scharoun-Lee, and Tucker-Seeley 2009; Nettleton and Burrows 1998) and often triggers other taxing life changes, including residential mobility, school change, and family disruption (see Been, Ellen, et al 2011; Stoll 2014). This increased stress is implicated in the apparent link between foreclosure and both physical and mental health at the individual level (see Houle 2014; Osypuk et al. 2012; Pollack and Lynch 2009).

Even for those not experiencing foreclosure, the concentration of foreclosures in a neighborhood has potentially important impacts on a variety of outcomes. Available research indicates important links between local foreclosures and public health (see Arcaya et al. 2014; Houle and Light 2014). In addition to foreclosures diminishing the value of surrounding properties (see Daneshvary and Clauretie 2012; Wassmer 2011), they are also likely to have profound impacts on neighborhood distress and instability, leading to a density of vacant, neglected, and abandoned properties; heightening the appearance of neighborhood deterioration; driving up crime; and increasing the likelihood of racial transition (see Capone and Metz 2003; Ellen, Lacoe, and Sharygin 2013; Williams, Galster, and Verma 2014). The concentration of foreclosures in neighborhoods may thus erode residential satisfaction and propel residents who are financially able to out-migrate. These changes in neighborhood social conditions emerging from high foreclosure concentrations may significantly alter patterns of neighborhood integration by differentially affecting patterns of in- and out-migration for members of different racial groups (Hall, Crowder, and Spring 2014).

Data and Methods

To explore our main research questions, we combined data from a variety of sources. Our main source of foreclosure data comes from RealtyTrac, which collects local foreclosure listings and documents from county assessor's offices across the United States. The complete database includes virtually all preforeclosure filings, public auction notices, and bank repossessions since 2005, which provide the universe of foreclosure filings for nearly all metropolitan counties. Most important for our purposes, these data include the physical addresses of all properties in the foreclosure process and the timing of the filings.

Utilizing the address-level RealtyTrac file, we compiled a complete listing of residential foreclosures between 2005 and 2012. With these data, we created a panel file of unique foreclosure events that tracked individual properties through the foreclosure process. To do so, we used a fuzzy-matching algorithm based on multiple fields—including address, tax parcel number, transaction and judicial case IDs—that identified unique properties, removed potential sources of redundancy, and imputed any incomplete information (e.g., property type recorded on the lis pendens but not on the Notice of Trustee Sale). Although our panel file includes all events in the foreclosure process, we restricted our analyses to cases representing the first visible sign of housing distress; a listing for public auction or bank repossession. The final file included 9.5 million visible foreclosures over the 2005 to 2012 period.

Geographic longitude and latitude of each record were determined using Bing Maps REST Services API. With these geocodes, we used GIS tools to attach census geography to each observation. Specifically, for each foreclosure event, we identified its census division, metropolitan area (using the 2010 CBSA definitions), and census block group. The metropolitan location—city or suburb—of each block group was determined by the U.S. Census Bureau's central city indicator. We used this information to aggregate the number of foreclosures in each census division, metropolitan area, central city, suburban ring, and block group between 2005 and 2012. Dividing the annual or cumulative number of foreclosures during this period by the number of housing units at the start of each interval produced our foreclosure rate measure (expressed in 100s of housing units). For some descriptive purposes, we also classified foreclosure rates into one of five categories: none; low (0–1 foreclosures per 100 homes); moderate (1–5); high (5–10); very high (10+). Demographic information on area racial compositions comes from Summary File 1 of Census 2000. To summarize the racial/ethnic structure of each block group, we used a modified version of the typology developed by Farrell and Lee (2011) who define neighborhood racial/

ethnic types consistently across metropolitan areas using information for five racial groups: Hispanics, non-Hispanic whites, blacks, and Asian and Pacific Islanders (see Hall, Crowder, and Spring 2014 for more information).

To describe foreclosure risk for individual households, we used data from the Panel Study of Income Dynamics (PSID). Started in 1968, the PSID collected detailed information of the economic and social well-being of households annually until 2001 and biennially since. The PSID is well suited for the analysis because starting in 2009, it incorporated a set of questions on recent experience with mortgage distress and housing foreclosure. Specifically, we used the 2009, 2011, and (preliminary) 2013 releases of the PSID data on mortgage distress, but were able to construct foreclosure histories back to 2001. We restricted our analytic sample to white, black, and Latino heads of PSID households.[1]

Foreclosure in the PSID is assessed via a set of questions on whether household members have faced difficulty in maintaining their mortgages. In 2009, 2011, and 2013, householders were asked if a bank had begun the foreclosure process on their current home; they were also asked about whether a foreclosure was initialized on any other residential property dating back to 2001. We used this information, along with the foreclosure start date and interview date, to determine whether a householder was ever in foreclosure between the two-year observation intervals since 2001. These data were used to calculate rates of foreclosure for PSID householders by dividing the number of foreclosures into the number of households who either owned their home or lost their home through foreclosure during the interval. We also used this information to calculate *cumulative* foreclosure rates by keeping a running indicator of whether a household ever experienced foreclosure.

Reflecting our interest in documenting the patterns of and trends in foreclosure, our analytic approach is strictly descriptive. We first summarized the temporal patterns of the foreclosure crisis, then documented regional and metropolitan variation in the levels and timing of the foreclosure crisis and explored whether foreclosure rates differed by metropolitan racial compositions. We considered similar processes of foreclosure in the central cities and suburban rings of metro areas and then moved down to the neighborhood level where we assessed how foreclosure concentrations were related to racial/ethnic structures of census block groups. We then moved down to the household level and considered how foreclosure risk evolved over time and differed by race.

Results

We start by documenting basic geographic patterns of foreclosures between 2005 and 2012. To illustrate the variation in foreclosure risk across the national landscape, we show cumulative and quarter-specific foreclosure rates in U.S. census divisions, respectively, in Figures 1 and 2. Differences across census divisions in the cumulative rates (Figure 1) show that the Mountain division (AZ, CO, NM, NV, UT) had the highest cumulative rate of housing foreclosures: by the middle

FIGURE 1

Trend in Cumulative Foreclosure Rates by Census Division, 2005–12

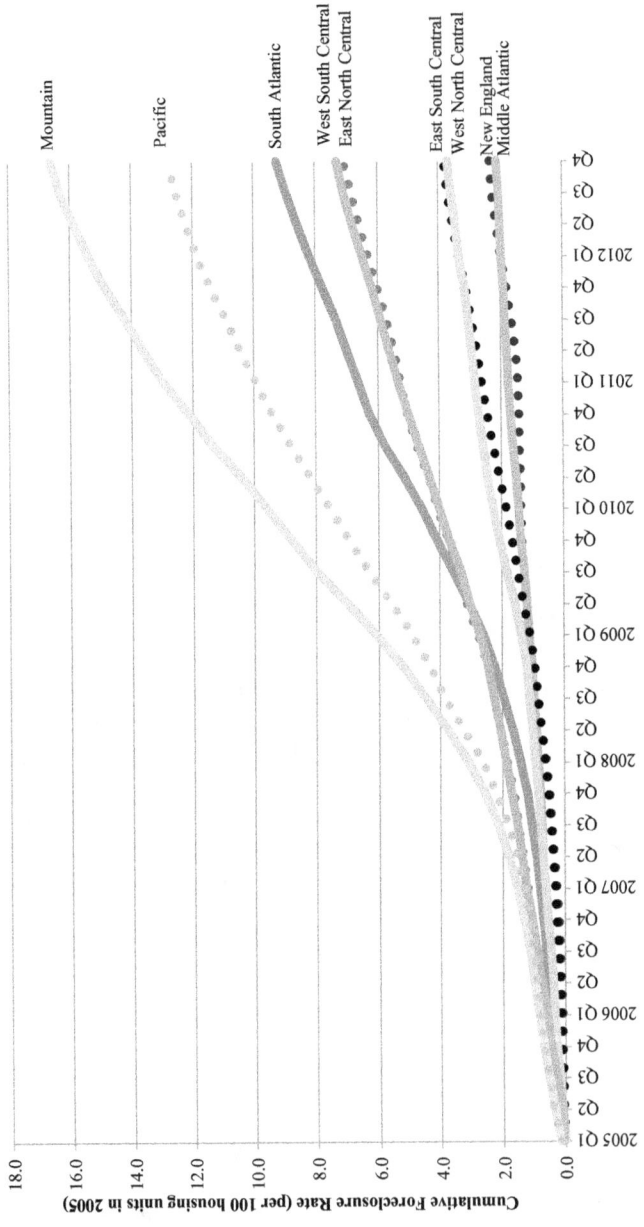

FIGURE 2

Trend in Quarterly Foreclosure Rates by Census Division, 2005–12

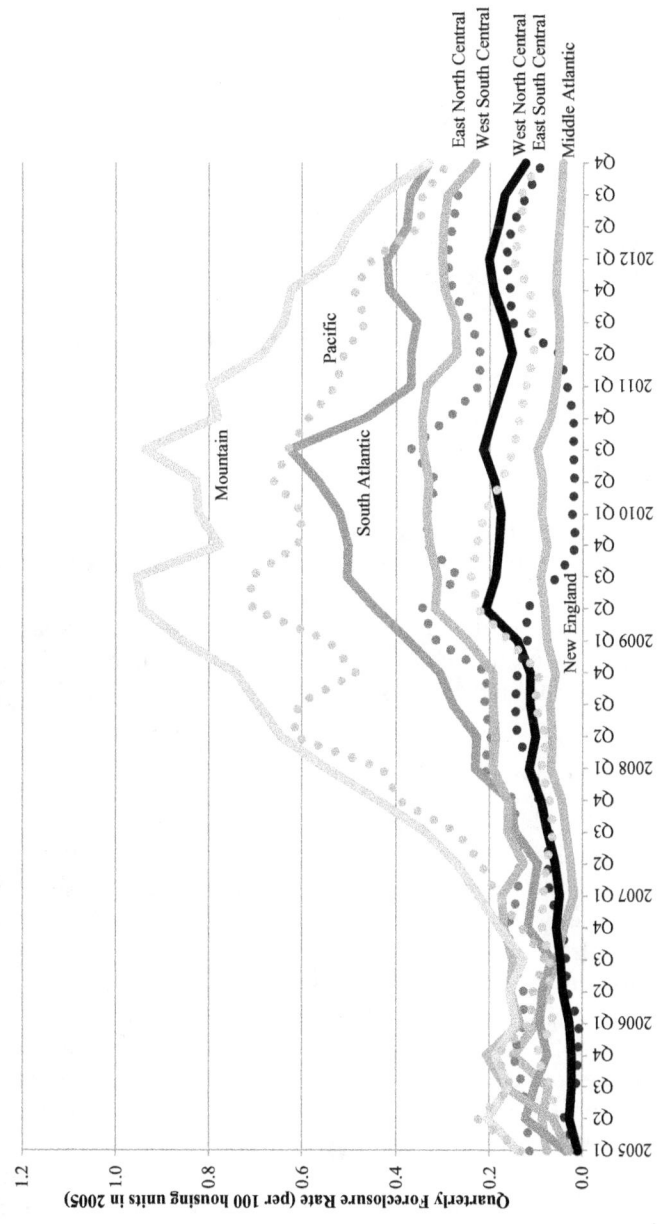

225

of 2010, more than one in ten homes in the Mountain division had been fore-closed on, and by the end of 2012, nearly one in six had. The exceedingly high rates of foreclosure in the Mountain division should not distract from the very high rates in other divisions: by the end of 2012, one in eight homes in the Pacific and one in eleven homes in the Southern Atlantic had been foreclosed on. By contrast, rates of foreclosure—while still inflated—were comparatively low in the Northeast, West North Central, and East South Central areas. The quarterly rates in Figure 2 better illustrate the variability across divisions in the timing of the crisis. Not surprisingly, most divisions hit their peak foreclosure levels between the end of 2009 and the middle of 2010. Yet the onset of the crisis was quite variable, with foreclosure rates starting to climb in the Mountain and Pacific divisions in 2006 but later in the South. In the Middle Atlantic division (PA, NJ, NY), foreclosure rates were actually higher in the precrisis years (2005–6) than during the crisis.

We now shift scales and describe patterns of foreclosure in metropolitan areas for the 291 metros with complete coverage of foreclosure data as of 2005.[2] Foreclosure rates for the ten metropolitan areas with the highest rates in 2005, 2007, 2009, and 2011 are shown in Table 1. Indianapolis posted the highest fore-closure rate in 2005, with 2.8 foreclosures for every 100 homes. Several other Midwestern metros—Anderson (IN), Dayton, Akron, and South Bend—were among the 2005 top ten. By the end of 2007, Western metros—including four in California—dominated the top ten. At the height of the crisis in 2009, the highest rates were seen in Phoenix, Las Vegas, and Atlanta, where about eight out of every 100 homes were foreclosed on. As the housing market started to recover in 2011, foreclosure rates declined modestly, but the general ranking of metropoli-tan foreclosure rates mostly persisted.

Racial context of foreclosure in metropolitan areas

We next consider the extent to which these variations in foreclosure rates across metropolitan areas were associated with the racial/ethnic composition of metropolitan areas. Specifically, Table 2 summarizes eight-year foreclosure rates—expressed as the sum of foreclosures in each metro between 2005 and 2012 divided by the number of housing units (in 100s) in 2005—across metro-politan areas with differing shares of blacks and Hispanics in 2000. Table 2 shows that during the crisis period, the typical metropolitan area had 6.26 foreclosures for every 100 housing units; it also shows that about half of the 291 metros fell in the "moderate" foreclosure range (between 1 and 5), a quarter in the high range (between 5 and 10) and about one in six in the very high range (above 10). Most importantly, average foreclosure rates differed by metropolitan racial shares, with the lowest rates observed in metros with very small black or Latino shares. The pattern, however, across areas with varying levels of percent black is not espe-cially pronounced; the foreclosures rates in the most-black metros are only slightly higher, on average, than those in the least-black metros. In contrast, foreclosure rates across metros with differing Hispanic populations are striking; metros with Hispanic shares exceeding 20 percent have average foreclosure rates

TABLE 1
Metropolitan Areas with Highest Foreclosure Rates, by Year

2005		2007	
Indianapolis-Carmel, IN	2.78	Merced, CA	3.60
Anderson, IN	2.05	Stockton, CA	3.26
Dayton, OH	1.91	Greeley, CO	3.06
Lake Havasu City-Kingman, AZ	1.89	Denver-Aurora-Broomfield, CO	2.75
Akron, OH	1.86	Modesto, CA	2.75
Salem, OR	1.83	Riverside-San Bernardino-Ontario, CA	2.74
South Bend-Mishawaka, IN-MI	1.71	Atlanta-Sandy Springs-Marietta, GA	2.69
Punta Gorda, FL	1.60	Las Vegas-Paradise, NV	2.59
El Centro, CA	1.60	Phoenix-Mesa-Glendale, AZ	2.43
Tucson, AZ	1.53	Pueblo, CO	2.30
2009		2011	
Phoenix-Mesa-Glendale, AZ	8.70	Atlanta-Sandy Springs-Marietta, GA	7.55
Las Vegas-Paradise, NV	8.50	Las Vegas-Paradise, NV	6.74
Atlanta-Sandy Springs-Marietta, GA	7.61	Gainesville, GA	6.47
Merced, CA	7.12	Phoenix-Mesa-Glendale, AZ	5.90
Riverside-San Bernardino-Ontario, CA	6.63	Riverside-San Bernardino-Ontario, CA	4.13
Stockton, CA	6.27	Stockton, CA	4.08
Modesto, CA	6.08	Modesto, CA	3.96
Cape Coral-Fort Myers, FL	6.00	Vallejo-Fairfield, CA	3.91
Gainesville, GA	5.98	Dalton, GA	3.89
Bend, OR	5.83	Savannah, GA	3.71

NOTE: Foreclosure rates expressed as annual number of foreclosures per 100 housing units.

(mean = 13.80) more than three times as high as metros where Hispanics make up less than 5 percent of the population (mean = 3.95). The foreclosure level profiles indicate that while just 5 percent of the least-Hispanic metros have "very high" foreclosure rates, nearly two-thirds of the most-Hispanic metros do. This is one of the first signs that Hispanics were particularly hard hit by the foreclosure crisis.

The popular story of the crisis often includes narratives of foreclosures in white Western suburbs (*Economist* 2011) and minority-heavy central cities in the Midwest (Haughney and Roberts 2009). We explore this portrayal descriptively in the two right columns of Table 2, which show average foreclosure rates over 2005–12 in metropolitan central cities and their suburban rings. On average, foreclosure rates were slightly higher in cities than suburbs during the crisis period, but there is a substantial regional patterning to this difference. In Northeastern and Midwestern metros, foreclosure rates were considerably higher in central cities than in suburban areas. In Metro Cleveland, for example,

TABLE 2
Foreclosures in 2005–12, by Metro Racial/Ethnic Composition (in 2000)

	Metropolitan Areas					Cities	Suburbs
	Foreclosure Rate	Foreclosure Levels (%)				Foreclosure Rate	Foreclosure Rate
		Low	Moderate	High	Very High		
All metropolitan areas	6.26	12.37	45.36	25.77	16.49	6.47	6.09
Division							
New England	2.14	7.14	85.71	7.14	.00	2.77	1.87
Middle Atlantic	1.81	34.48	65.52	.00	.00	2.73	1.52
East North Central	5.89	5.26	42.11	47.37	5.26	7.78	4.98
West North Central	2.14	33.33	55.56	11.11	.00	2.80	1.86
South Atlantic	7.13	7.41	42.59	31.48	18.52	6.12	6.93
East South Central	3.26	20.00	65.00	10.00	5.00	4.13	2.65
West South Central	2.79	26.92	50.00	19.23	3.85	2.80	2.95
Mountain	10.43	3.45	37.93	20.69	37.93	10.19	11.13
Pacific	12.22	.00	15.91	34.09	50.00	11.87	12.56
Area percent black							
<5%	5.95	16.92	43.85	23.08	16.15	6.08	5.96
5–10%	7.91	7.94	38.10	28.57	25.40	8.38	8.33
10–20%	5.09	10.00	51.67	33.33	5.00	6.60	3.97
>20%	6.38	7.89	52.63	18.42	21.05	5.94	6.29
Area percent Hispanic							
<5%	3.95	17.18	55.83	22.09	4.91	4.29	3.58
5–10%	6.24	5.17	44.83	39.66	10.34	6.04	7.35
10–20%	9.06	6.25	25.00	37.50	31.25	8.30	11.59
>20%	13.80	7.89	18.42	10.53	63.16	10.74	14.79

NOTE: $N = 291$ metropolitan areas with complete foreclosure data in 2005; "Area" refers to metro areas, central cities, and suburban rings.

its central city foreclosure rate of 14.0 was twice as high as its suburban rate (7.0). However, in Southern and Western metropolises, foreclosures were noticeably higher in suburban areas than in cities. While foreclosures in Phoenix-Mesa-Glendale were prevalent throughout the metro, the central city foreclosure rate (30.4) was a third smaller than its suburban foreclosure rate (45.5).

Table 2 also shows that the same racial/ethnic patterning of foreclosures observed at the metro levels existed at the city and suburban levels. While black concentrations were mostly unrelated to average foreclosure rates in cities and suburbs, they were strongly conditioned by Hispanic shares. In both cities and suburbs, foreclosure rates were considerably higher in areas with larger Hispanic populations, with the highest rates being observed in suburban areas that were more than one-fifth Hispanic.

Foreclosures and race in metropolitan neighborhoods

The concentration of foreclosures in particular metropolises, and their distribution across central-city and suburban zones of these metros, offers suggestive evidence of substantial group differences in exposure to heightened foreclosure levels in residential environments. However, even within these metropolitan segments, levels of foreclosure likely varied substantially across neighborhoods. We assess this possibility in the next set of tables, summarizing foreclosure rates across block groups with varying racial/ethnic structures. Table 3 shows that in the 291 metros analyzed, there were 8.5 foreclosures per 100 housing units in the average neighborhood during the 2005 to 2012 period. The categorical breakdown shows that less than 5 percent of block groups reported no foreclosures, about half were in the low to moderate range, one-fifth had high rates, and about one-fourth had very high rates. Average foreclosure rates, however, varied substantially by neighborhood racial composition. In all-white and Asian neighborhoods, there were fewer than 5 foreclosures for every 100 homes, and just 1 in 8 of such neighborhoods had foreclosure rates over 10 ("very high"). By contrast, mostly black and mostly Hispanic neighborhoods had foreclosures over 12.9 and 11.4, respectively, and nearly half of these neighborhoods had very high rates. For the most part, most neighborhood types, including a mix of whites and minority groups, fell somewhere between all-white and all-minority neighborhoods. The exceptions are Hispanic-white and integrated neighborhoods, which experienced especially high rates of foreclosure (14.0 and 15.1) and were very likely to fall in the "high" or "very high" foreclosure classification.

We break these patterns down further by metropolitan location in the two right columns of Table 3, showing average neighborhood foreclosure rates in central city and suburban areas. Despite that central cities, as a whole, had slightly higher foreclosure rates than did suburban zones (see Table 2), the average suburban block group had a modestly higher foreclosure rate than the average city block group. More importantly, the racial patterning to foreclosure levels is similar in city and suburban neighborhoods, with all-white and Asian neighborhoods having the lowest foreclosure rates in both locations. In other types of neighborhoods, it is generally true that foreclosure rates were higher in suburban

TABLE 3
Foreclosures in 2005–12, by Neighborhood Racial/Ethnic Structure (in 2000)

| | All metropolitan neighborhoods | | | | | | | Central city neighborhoods | Suburban neighborhoods |
| | | Foreclosure Levels (%) | | | | | | | |
	Foreclosure Rate	None	Low	Moderate	High	Very High		Rate	Rate
All neighborhoods	8.50	4.86	13.25	35.05	19.85	27.00		8.36	8.84
Asian-white	4.31	10.30	21.75	37.47	18.33	12.15		3.23	5.22
All white	4.89	5.58	16.95	46.26	18.67	12.54		5.07	4.85
White shared	7.07	4.26	13.59	37.07	22.72	22.37		6.31	7.59
Black-white	8.51	4.10	12.54	37.44	20.27	25.64		8.04	9.02
White mixed	10.93	4.91	11.69	26.99	19.07	37.34		8.99	13.23
All minority	10.97	8.66	14.44	21.90	15.04	39.96		10.07	14.02
Mostly Hispanic	11.42	5.14	9.99	23.27	19.99	41.61		10.18	13.07
Mostly black	12.92	2.94	7.41	24.23	21.32	44.10		12.07	17.21
Hispanic-white	14.00	2.25	7.33	22.79	20.71	46.93		12.57	15.17
Integrated	15.14	5.05	11.92	17.78	15.80	49.45		13.33	17.73

NOTE: Table 3 includes block groups with total populations of at least 20 in 1990, 2000, and 2010 that are located in the 291 metropolitan areas with complete foreclosure data in 2005.

than city neighborhoods with similar racial make-ups. In mostly black neighborhoods, for example, suburban neighborhoods had mean foreclosure rates that were 21 percent higher than their counterparts in central cities (17.21 percent vs. 12.07 percent). Overall, however, black- and Latino-populated neighborhoods—whether in the suburbs or a central city—tended to have higher rates of foreclosure than did similarly located neighborhoods occupied largely by whites and Asians. Thus, despite media accounts of the crisis primarily targeting suburban white and urban minority neighborhoods, the descriptive patterns in Table 3 suggest that white neighborhoods were mostly shielded from the worst of the foreclosure crisis, while black, racially mixed, and especially Hispanic neighborhoods were hit especially hard.[3]

We repeat these analyses separately by census division in Table 4 to assess whether the differences in neighborhood foreclosure rates across racial/ethnic types simply reflect the general regional concentration of foreclosures documented earlier. To facilitate interpretation of the table, the neighborhood type within each division that has the highest and second-highest foreclosure rate is highlighted in dark grey and light grey, respectively. (Foreclosure rates for types with fewer than fifty neighborhoods in a division have been suppressed.) Table 4 not only reinforces that there was strong regional clustering of foreclosures but also buttresses the racialized context of the crisis. Specifically, in all divisions, the lowest average foreclosure rates are observed in all-white or Asian neighborhoods. By contrast, racially mixed and solidly minority neighborhoods—e.g., mostly black and all-minority areas—consistently recorded some of the highest rates. Foreclosure rates among black-white and white-mixed neighborhoods in the Mountain division were in excess of one in four homes over the 2005 to 2012 period. The highest average rates were observed in Southern Atlantic integrated neighborhoods where one foreclosure for every three homes was logged. In all but one division where there were a sufficient number of block groups, mostly black neighborhoods had the highest or second-highest average foreclosure rate, followed by black-white neighborhoods, which ranked second-highest in four divisions. Mostly Hispanic and Hispanic-white neighborhoods also recorded exceptionally high rates of foreclosure in several divisions, including the Mountain West and South. A central point to take away from this analysis is that in each region, neighborhoods containing sizable shares of African American and Latino populations tended to be the most heavily burdened by foreclosures in almost every division of the country. Only in the South Central and Mountain divisions did white-dominated—or, more specifically, white-mixed—neighborhoods register as a category with especially high foreclosure concentration, and even there, black and Latino neighborhoods were not far behind.

Household-level rates of foreclosure

Metropolitan and neighborhood level analyses point to a strong racial patterning of foreclosures, with rates of foreclosure being substantially higher in minority neighborhoods, especially those with large concentrations of blacks or Latinos. These descriptive patterns raise several important questions about the extent to

TABLE 4

Foreclosures Rates in 2005–12, by Neighborhood Racial/Ethnic Structure (2000) and Census Division

	New England	Middle Atlantic	East North Central	West North Central	South Atlantic	East South Central	West South Central	Mountain	Pacific
All neighborhoods	2.6	2.4	9.5	4.3	10.6	5.7	4.0	18.7	13.1
All white	2.0	1.9	7.3	3.1	6.7	2.9	3.0	9.4	8.0
Mostly black	—	4.9	17.2	13.5	17.3	10.8	4.5	—	20.4
Mostly Hispanic	—	1.7	8.5	—	9.7	—	3.0	23.4	15.5
White shared	2.5	1.8	8.1	4.5	9.6	4.9	4.2	16.2	8.6
Black-white	2.3	4.0	12.1	7.4	10.0	6.5	3.0	27.1	9.6
Hispanic-white	4.3	2.2	9.4	4.9	12.5	—	6.1	22.0	16.3
Asian-white	1.6	1.0	4.9	—	9.1	—	—	—	5.9
White mixed	4.3	2.7	10.7	6.2	14.3	6.0	5.6	27.6	15.0
All minority	7.8	2.2	11.3	—	13.7	—	—	—	18.1
Integrated	—	2.7	7.6	—	30.0	—	—	—	19.8

NOTE: Cells with fewer than 50 block groups are omitted; includes block groups with total populations of at least 20 in 1990, 2000, and 2010 that are located in the 291 metropolitan areas with complete foreclosure data in 2005.

which the incidence of foreclosure differs by race. We explore this issue in even greater detail in the final stage of our analysis using household-level data from the PSID to calculate racial differences in household-level rates of foreclosure.

Table 5 summarizes foreclosure rates for non-Hispanic white, non-Hispanic black, and Hispanic householders in the PSID during the 2000s. The upper panel of Table 5 shows rates of foreclosure starts between successive interviews and thus refers to rates over each two-year interval (e.g., rates for 2009 refer to foreclosures between the 2007 and 2009 interviews). The lower panel in Table 5 shows cumulative rates over the 2001–13 period, reflecting, in a given year, whether homeowners (or previous homeowners) had entered foreclosure on a home at any time since 2001.

The values in Table 5 underscore the racialized nature of the foreclosure crisis: in every year, foreclosure rates for minority households exceeded those for whites, and as the crisis unfolded, the racial gaps expanded. In the 2005–7 period just preceding the crisis, 1.2 percent of whites, 1.7 percent of Hispanics, and 2.4 percent of blacks had experienced a foreclosure start. At the height of the crisis, blacks were more than twice as likely as whites to enter foreclosure and Hispanics nearly three times as likely, with about one out of every thirteen Hispanic homeowners entering the foreclosure process. Table 5 also shows that household foreclosure rates peaked for white and black owners in 2009 but reached their height for Latino owners in 2011. The most recent data, covering the 2011–13 period,

TABLE 5

Foreclosure Starts for PSID Householders, by Race and Year, 2001–13

Year	White owners	Black owners	Hispanic owners
Interval Incidence			
2001	.000	.002	.000
2003	.003	.007	.006
2005	.006	.013	.018
2007	.012	.024	.017
2009	.026	.057	.076
2011	.025	.037	.080
2013[†]	.023	.054	.040
Cumulative Incidence			
2001	.000	.002	.000
2003	.003	.008	.006
2005	.009	.023	.023
2007	.021	.047	.039
2009	.049	.107	.114
2011	.069	.133	.182
2013[†]	.079	.151	.190

[†]preliminary 2013 data.

indicate that foreclosure rates have declined modestly but remain at quite high levels. Indeed, foreclosure rates for black owners appear to have rebounded, potentially reflecting the resolution of so-called zombie foreclosures—units with mortgages in default for an extended time that lenders were slow to take action on.

Arguably more striking than the racial differences in the interval rates are the cumulative rates summarized in the lower portion of Table 5. The cumulative numbers underscore the enormity of the crisis as well as the substantial racial variation is foreclosure risk. By 2013, about one in thirteen whites had experienced a foreclosure start, more than one in six black homeowners did, and nearly one in four Hispanic homeowners had gone through a foreclosure.

Conclusion

Using a combination of data from individual foreclosure filings, household records of mortgage stress, and census data on the neighborhood and metropolitan characteristics, this article provides a comprehensive view of variations in the timing and depth of the housing crisis across multiple levels of aggregation. Several key conclusions emerge from this examination. First, the foreclosure crisis that began in 2007 clearly played out quite differently across the country, in terms of both the timing of the housing crash and the depth of the crisis. Foreclosures began to rise much earlier in the Midwest region than in other parts of the country, but the crisis was ultimately more intense in southern and western parts of the country, especially in the Mountain, Pacific, and South Atlantic divisions. Yet even within these regions, there was substantial metropolitan-level variations in both the timing and magnitude of the foreclosure crisis, presumably reflecting variations across markets in the specific ingredients of the crash: prerecession inflation of housing prices; the popularity of risky mortgage vehicles; the prevalence of negative equity; the magnitude of the general economic slowdown; and variations in the legal processes related to the processing of foreclosures and lenders' strategies for dealing with repossessed homes.

Most apparent in the data examined here, however, are the important roles of race and ethnicity in shaping patterns of the foreclosures during the crisis. Racialized patterns of foreclosure are observed at all levels of aggregation: concentrations of foreclosures were substantially higher in metropolitan areas—as well as their cities and suburbs—with the highest concentration of blacks. Most importantly, in comparison to whites, individual black householders were much more likely to have experienced foreclosure in their own home and, because of their residential location, were also likely to have been exposed to much higher levels of foreclosure at the neighborhood level.

Less appreciated in the discourse on stratified foreclosure processes is the even more extreme burden borne by the Latino population during the foreclosure crisis. Individual Latino householders had the highest cumulative risk of foreclosure during the crisis, and neighborhoods and metropolitan areas where Latino populations were present in significant numbers recorded some of the highest rates of foreclosure. This Latino disadvantage has been apparent in both suburban and central city neighborhoods of metropolitan areas across virtually all regions of the country. Thus, geographic patterns of foreclosure during the crisis, as well as the underlying patterns of stratification that fed these disparities, were substantially more complicated than is indicated by depictions of unique disadvantages for black households.

The racial and ethnic stratification revealed here likely reflects, at least in part, the specific targeting of minority populations and neighborhoods by predatory and subprime lenders, and it has potentially profound implications for broader patterns of racial inequality. We know, for example, that personal experiences of foreclosure typically entail significant loss of wealth-building capacity, and often precipitate residential mobility, family instability, and stress, all with important repercussions for health. And those living in high-foreclosure neighborhoods, regardless of their own housing situation, face the prospect of declining property values, rising crime, and growing disorganization in their neighborhood. Thus, by disproportionately impacting minority individuals and communities of color, the foreclosure crisis has likely affected substantial changes to racial disparities in well-being along a variety of dimensions. The downstream impacts of these effects are likely to unfold slowly, potentially reshaping patterns of racial stratifications for decades.

By delineating these racially stratified patterns, the research presented here highlights the importance of several key areas for policy intervention. First, it is clear that programs aimed at preventing individual foreclosures should be aggressively targeted to and—given evidence that the efficacy of policy interventions vary by race (Collins, Schmeiser, and Urban 2013)—explicitly designed for Latino and black homeowners. In a similar way, efforts to mollify the effects of concentrated foreclosure on local housing values, vacancy rates, crime, and population instability should be focused on neighborhoods of color—places at a vulnerable position in the geography of disadvantage even in the best economic times (Peterson and Krivo 2010; Sampson 2012). Finally, the extreme racial disparities documented here underscore the importance of identifying and redressing the underlying processes of stratification and discrimination—including those demonstrated by predatory lenders—that precipitated the crisis and its uneven impacts.

Notes

1. While Hispanics in the PSID may not be completely representative of the total U.S. Hispanic population, we retain them for our analysis here because of their growing contributions to U.S. demographic change and to their potentially heightened exposure to foreclosures.

2. The seventy-five missing metropolitan areas are mostly smaller metros but also include a few larger ones, such as Cincinnati and Dallas. Incomplete coverage in these areas reflects variation in the timing of RealtyTrac's efforts to establish data collection in all parts of these areas.

3. While the general pattern is for foreclosure rates to be markedly lower in white neighborhoods, regardless of location, some all-white neighborhoods had very high rates of foreclosure. The vast majority of these neighborhoods are located in the suburbs of metro areas hit hard by the crisis. Several of these neighborhoods appear to be parts of new housing developments or planned communities (e.g., near Sun City, AZ; Atlanta exurbs). Nevertheless, less than 5 percent of all neighborhoods with foreclosure rates above fifty were "all white," whereas all-white areas were about 40 percent of neighborhoods with no foreclosures.

References

Arcaya, Mariana, Maria Glymour, Prabal Chakrabarti, Nicholas Christakis, Ichiro Kawachi, and S. V. Subramanian. 2014. Effects of proximate foreclosed properties on individuals' systolic blood pressure in Massachusetts, 1987–2008. *American Journal of Public Health* 103:50–56.

Been, Vicki, Sewin Chan, Ingrid Gould Ellen, and Josiah R. Madar. 2011. Decoding the foreclosure crisis: Causes, responses, and consequences. *Journal of Policy Analysis and Management* 30:388–96.

Been, Vicki, Ingrid Gould Ellen, Amy Ellen Schwartz, Leanna Stiefel, and Meryle Weinstein. 2011. Does losing your home mean losing your school? Effects of foreclosures on the school mobility of children. *Regional Science and Urban Economics* 41:407–14.

Bennett, Gary, Melissa Scharoun-Lee, and Reginald Tucker-Seeley. 2009. Will the public's health fall victim to the home foreclosure epidemic? *PLoS Medicine* 6. doi:10.1371/journal.pmed.1000087

Bocian, Debbie Gruenstein, Wei Li, and Keith Ernst. 2010. *Foreclosures by race and ethnicity: The demographics of a crisis.* Durham, NC: Center for Responsible Lending.

Capone, Charles, and Albert Metz. 2003. *Mortgage default and default resolutions: Their impact on communities.* Chicago, IL: Federal Reserve Bank of Chicago.

Cheng, Ping, Zhenguo Lin, and Yingchun Liu. 2014. Racial discrepancy in mortgage interest rates. *Journal of Real Estate Finance and Economics*. doi:10.1007/s11146-014-9473-0.

Collins, J. Michael, Ken Lam, and Christopher Herbert. 2011. State mortgage foreclosure policies and lender interventions: Impacts on borrower behavior in default. *Journal of Policy Analysis and Management* 30:216–32.

Collins, J. Michael, and Maximilian Schmeiser. 2013. The effects of foreclosure counseling for distressed homeowners. *Journal of Policy Analysis and Management* 32:83–106.

Collins, J. Michael, Maximilian Schmeiser, and Carly Urban. 2013. Protecting minority homeowners: Race, foreclosure counseling and mortgage modifications. *Journal of Consumer Affairs* 47:289–310.

CoreLogic. 26 August 2010. *Real estate news and trends: CoreLogic data shows decline in negative equity.* CoreLogic Equity Report Q2 2010. Irvine, CA: CoreLogic.

Daneshvary, Nasser, and Terrence Clauretie. 2012. Toxic neighbors: Foreclosures and short-sales spillover effects from the current housing-market crash. *Economic Inquiry* 50:217–31.

Dwyer, Rachel E., and Lora A. Phillips Lassus. 2015. The great risk shift and precarity in the U.S. housing market. *The ANNALS of the American Academy of Political and Social Science* (this volume).

Economist. 24 May 2011. On a losing streak: The effects of America's worst property crash go very wide. *The Economist*.

Ellen, Ingrid Gould, Johanna Lacoe, and Claudia Ayanna Sharygin. 2013. Do foreclosures cause crime? *Journal of Urban Economics* 74:59–70.

Ellen, Ingrid Gould, Josiah Madar, and Mary Weselcouch. 2014. The foreclosure crisis and community development: Exploring REO dynamics in hard-hit neighborhoods. *Housing Studies*. doi:10.1080/02673037.2014.882496.

Engel, Kathleen, and Patricia McCoy. 2011. *The subprime virus: Reckless credit, regulatory failure, and next steps*. New York, NY: Oxford University Press

Faber, Jacob. 2013. Racial dynamics of subprime mortgage lending at the peak. *Housing Policy Debate* 23:328–49.

Farrell, Chad, and Barrett A. Lee. 2011. Racial diversity and change in metropolitan neighborhoods. *Social Science Research* 4 (4): 1108–23.

Gerardi, Kristopher, Stephen Ross, and Paul Willen. 2011. Understanding the foreclosure crisis. *Journal of Policy Analysis and Management* 30:382–88.

Hall, Matthew, Kyle Crowder, and Amy Spring. 2014. Neighborhood foreclosures, racial/ethnic transitions, and residential segregation. Cornell University Working Paper, Ithaca, NY.

Haughney, Christine, and Janet Roberts. 15 May 2009. Connecticut foreclosure crisis appears to be worsening. *New York Times*.

Houle, Jason. 2014. Mental health in the foreclosure crisis. *Social Science & Medicine* 118:1–8.

Houle, Jason, and Michael Light. 2014. The home foreclosure crisis and rising suicide rates, 2005 to 2010. *American Journal of Public Health* 104:1073–79.

Hoynes, Hilary, Douglas Miller, and Jessamyn Schaller. 2012. Who suffers during recessions? *Journal of Economic Perspectives* 26:27–48.

Hyra, Derek, Gregory Squires, Robert Renner, and David Kirk. 2013. Metropolitan segregation and the subprime lending crisis. *Housing Policy Debate* 23:177–98.

Immergluck, Dan. 2008. From the subprime to the exotic: Excessive mortgage market risk and foreclosures. *Journal of the American Planning Association* 74:59–76.

Immergluck, Dan. 2010. The accumulation of lender-owned homes during the U.S. mortgage crisis: Examining metropolitan REO inventories. *Housing Policy Debate* 20:619–45.

Immergluck, Dan. 2012. Distressed and dumped market dynamics of low-value, foreclosed properties during the advent of the federal neighborhood stabilization program. *Journal of Planning Education and Research* 32:48–61.

Kuehn, D. 2011. Metropolitan job growth patterns in the great recession. Working Paper 1-2011 Urban Institute. Washington, DC.

Mayer, Neil, Peter Tatian, Kenneth Temkin, and Charles Calhoun. 2012. *Has foreclosure counseling helped troubled homeowners? Evidence from the evaluation of the national foreclosure mitigation counseling program*. Washington, DC: Urban Institute.

Nettleton, Sarah, and Roger Burrows. 1998. Mortgage debt, insecure home ownership, and health: An exploratory analysis. *Sociology of Health & Illness* 20:731–53.

Osypuk, Theresa, Cleopatra Howard Caldwell, Robert Platt, and Dawn Misra. 2012. The consequences of foreclosure for depressive symptomatology. *Annals of Epidemiology* 22:379–87.

Peterson, Ruth, and Lauren Krivo. 2010. *Divergent social worlds: Neighborhood crime and the racial-spatial divide*. New York, NY: Russell Sage Foundation.

Pfeffer, Fabian T., Sheldon Danziger, and Robert F. Schoeni. 2013. Wealth disparities before and after the Great Recession. *The ANNALS of the American Academy of Political and Social Science* 650:98–123.

Pollack, Craig Evan, and Julia Lynch. 2009. Health status of people undergoing foreclosure in the Philadelphia region. *American Journal of Public Health* 99:1833–39.

RealtyTrac. 21 January 2011. Record 2.9 million U.S. properties receive foreclosure filing in 2010 despite 30-month low In December. Irvine, CA: RealtyTrac.

RealtyTrac. 14 January 2015. 1.1 million U.S. Properties with foreclosures filings in 2014, down 18 percent from 2013 to lowest level since 2006. Irvine, CA: RealtyTrac.

Rugh, Jacob S., and Douglas S. Massey. 2010. Racial segregation and the American foreclosure crisis. *American Sociological Review* 75:629–51.

Saegert, Susan, Desiree Fields, and Kimberly Libman. 2011. Mortgage foreclosure and health disparities: Serial displacement as asset extraction in African American populations. *Journal of Urban Health* 88 (3): 390–402.

Sampson, Robert. 2012. *Great American city: Chicago and the enduring neighborhood effect*. Chicago, IL: University of Chicago Press.

Stoll, Michael. 2014. Residential mobility in the United States and the Great Recession: A shift to local moves. In *Diversity and disparities: America enters a new century*, ed. John Logan, 139–80. New York, NY: Russell Sage Foundation.

U.S. Department of Housing and Urban Development (HUD) and U.S. Treasury Department. 2000. *Curbing predatory home mortgage lending*. Washington, DC: HUD and U.S. Treasury.

Wassmer, Robert. 2011. The recent pervasive external effects of residential home foreclosure. *Housing Policy Debate* 21:247–65.

Williams, Sonya, George Galster, and Nandita Verma. 2014. Home foreclosure and neighborhood crime dynamics. *Housing Studies* 29:380–406.

Zillow. 2011. Zillow real estate market report, September 2011 (Q3).

Understanding Residential Moves

A Comparison of Traditional and Discrete-Choice Approaches to the Analysis of Residential Mobility and Locational Attainment

By
LINCOLN QUILLIAN

This article contrasts traditional modeling approaches and discrete-choice models as methods to analyze locational attainment—how individual and household characteristics (such as race, socioeconomic status, age) influence the characteristics of neighborhoods of residence (such as racial composition and median income). Traditional models analyze attributes of a neighborhood as a function of the characteristics of the households within them; discrete-choice methods, on the other hand, are based on dyadic analysis of neighborhood attributes and household characteristics. I outline two problems with traditional approaches to residential mobility analysis that may be addressed through discrete-choice analysis. I also discuss disadvantages of the discrete-choice approach. Finally, I use data from the Panel Study of Income Dynamics to estimate residential mobility using traditional locational attainment and discrete-choice models; I show that these produce similar estimates but that the discrete-choice approach allows for estimates that examine how multiple place characteristics simultaneously guide migration. Substantively, these models reveal that the disproportionate migration of black households into lower-income tracts amounts to sorting of black households into black tracts, which on average are lower income.

Keywords: migration; residential mobility; locational attainment; segregation; discrete choice; neighborhood poverty; conditional logit

A handful of methods have dominated the analysis of data on migration and residential location in sociology. In most of these approaches, the outcome is an attribute of the place of destination, which is modeled as a function of individual and household character-

Lincoln Quillian is a professor of sociology and faculty fellow at the Institute for Policy Research at Northwestern University. His current work includes studies of the causes of racial and economic segregation in American cities and a comparative study of neighborhood economic segregation in the United States and France.

NOTE: My thanks to two anonymous reviewers for helpful comments.

DOI: 10.1177/0002716215577770

istics. Recently, however, a new approach has begun to see increasing use in sociology, one that employs models that have a dyadic structure, including discrete-choice models developed in the economics and transportation literatures.

The recent interest in the discrete-choice approach in sociology was sparked by Bruch and Mare (2012). Bruch and Mare's paper discusses the discrete-choice approach prominently but provided only a very brief discussion of traditional locational attainment approaches or how they differ from the discrete-choice approach. This article has two goals. First, it outlines the advantages and disadvantages of the discrete-choice approach to modeling residential mobility in contrast to more traditional approaches used in sociology and related disciplines. Second, it examines the additional substantive conclusions we reach when a discrete-choice model is applied to some of the most often used covariates in the locational attainment tradition.

Background

The neighborhoods that individuals live in have long been recognized as an important factor that influences both individuals' quality of life and their life chances (for a recent review see Sharkey and Faber 2014). Correspondingly, a long line of research in sociology has attempted to explain the formation of neighborhoods and the processes by which households are sorted into neighborhoods in ways that produce demographic and economic differences among neighborhoods. An apt turn of phrase to describe the process these studies consider has been "locational attainment" (Alba and Logan 1993; Logan and Alba 1993).

Studies of locational attainment most often start with a sample of households, including information on their neighborhood of residence. An attribute of the neighborhood of residence is the outcome, most commonly percentage white or median household income. The predictors are household characteristics, most importantly measures of household socioeconomic status and race or ethnicity. When panel data are used, predictors are often expanded to include characteristics of the tract of origin. The analysis of locational attainment generally then proceeds either by using cross-tabulations or, more commonly, some form of regression modeling.

Most cross-tabulation approaches calculate transition matrices (e.g., Massey, Gross, and Shibuya 1994; Quillian 2002). Origin and destination neighborhood types form the rows and columns of the tabulation. Rates of migration from origin types to destination types are calculated from panel or retrospective data and entered into the cells. Individual and household characteristics define groups for which separate transition matrices are calculated. While appealing in their simplicity, transition matrices require limited categorical representations of neighborhood types and household characteristics.

Another widely used approach is grounded in multiple regression, which allows for continuous variables and a richer treatment of individual and household predictors than the transition matrix approach. These regression models have often been called "locational attainment models" in sociology (following

Alba and Logan 1993; Logan and Alba 1993). Residential location is viewed as the outcome of an individual-level attainment process, similar to the way occupational status has been viewed in the status attainment tradition. The outcome is a measure of racial composition or affluence of the neighborhood of residence or destination, most commonly percentage white or median household income. The predictors are individual and household characteristics, such as race and ethnicity, income, and age.

While initially developed to analyze cross-sectional locational data using least-squares regression, locational attainment models have been widely applied with panel data and in studies of migration (e.g., South and Crowder 1998; Ellen 2000; Quillian 2002; Woldoff and Ovadia 2009; Crowder, Pais, and South 2012; Pais, South, and Crowder 2012). In migration studies, the dependent variable is often defined categorically as neighborhood type, similar to transition matrix studies, and modeled as a function of independent characteristics with multinomial or binary logistic or probit regression.

Locational attainment models provide estimates of neighborhood "payoffs" to household characteristics as the estimated slopes. Most often, studies focus on race and income or socioeconomic status measures, but studies have also focused on immigrant documentation status (Cort 2011), immigrant generation (Brown 2007), incarceration history (Massoglia, Firebaugh, and Warner 2012), and metropolitan-level characteristics such as rates of new housing construction (South and Crowder 1998; Crowder, Pais, and South 2012). Similar models—usually without the term locational attainment attached—have also been widely applied in other disciplines including urban planning (e.g., Freeman 2008), geography (e.g., Cadwallader 1992), and public policy (e.g., Ellen 2000). Whether called locational attainment or not, these methods have undeniably been productive, resulting in a number of worthwhile empirical results.

The representation of place in existing approaches

Nevertheless, traditional approaches have some important limitations in representing the substantive processes that underlie locational attainment. In locational attainment models, and most transition matrix studies, destinations are represented by a single attribute.[1] This attribute is usually a measure of either the economic or racial composition of the area, such as tract median income or percentage black.

While this allows for analysis of residential location using common statistical models such as regression, there are two ways in which these approaches poorly represent the substantive process of locational attainment. The first problem is places are actually bundles of multiple attributes, which matter simultaneously in how households choose destinations. The second problem is that the ecology of potential destinations influences the outcome of locational attainment processes but is not directly represented in most locational attainment models. I discuss each of these problems below.

The bundling problem

The bundling problem reflects the fact that neighborhoods, like households, are composed of multiple attributes. Moreover, specific neighborhoods are "bundled" combinations of these attributes. For instance, neighborhoods encompass attributes such as economic composition (e.g., median income, share poor, etc.), race and ethnic composition (e.g., share white, share black, share Hispanic), average housing price, crime rate, quality of the local schools, and distance from desirable amenities, among other characteristics.

In choosing a destination, households care about many of these dimensions simultaneously. Because an optimal choice on all dimensions is rarely available, choosing housing usually requires making tradeoffs across multiple dimensions to find a satisfactory destination.

In transition matrix and locational attainment modeling, one of these dimensions is taken as the outcome and is modeled as a function of household characteristics. Most often, the outcome is a measure of either economic or racial composition. While some studies consider two or more dimensions of place—with separate regressions for each dimension as an outcome—they do so by examining these features separately and one at a time rather than simultaneously. The locational attainment literature includes separate analyses of residential distribution or migration using percentage white (Alba and Logan 1993), median household income (Logan and Alba 1993), percentage black (in categories; South and Crowder 1998), or percentage poor (also in categories; South and Crowder 1997).

In practice, this has two implications. First, locational attainment models fail to fully represent the social process of locational attainment, because this process fundamentally involves consideration of multiple place attributes simultaneously. Second, this makes it impossible to know whether the association between independent variables and any outcome neighborhood attribute might not be capturing sorting based on other, correlated neighborhood attributes. In a traditional locational attainment model, if we find an individual characteristic to be a strong predictor of a neighborhood attribute, we cannot establish whether that association might actually reflect a link of the individual characteristic to another neighborhood attribute.

The ecological dependence problem

A feature of residential mobility is that households move to existing places. A household cannot move to a neighborhood that does not exist (or does not exist within the bounds set by the study, such as the neighborhoods in a metropolitan area from which a sample is drawn). In a metropolitan area with no majority Latino neighborhoods, for instance, a majority Latino neighborhood is not a possible within-metropolitan destination.

Moreover, the prevalence of potential destinations within the area defined by a study is likely to influence where households move. Households of all races that live in Los Angeles are probably more likely to move into Latino neighborhoods than those that live in Chicago, simply because the share of neighborhoods that are Latino is much higher in Los Angeles than in Chicago. Regarded in one way, this reflects a basic principle of statistical analysis: the marginal distribution of outcomes influences the probability of outcomes. In the case of residential mobility, the marginal distribution of potential outcomes (neighborhoods) is fixed before residential mobility occurs. Households confront the distribution of neighborhoods as a sort of list of possibilities and are constrained to select one destination. Given the fixed ecology of places households consider, we can improve models of residential mobility by including this as a factor guiding destinations in analyses of residential mobility.

Locational attainment models usually contain no direct representation of the influence of the ecological distribution—or the distribution of attributes of potential neighborhoods—on outcomes of migration or locational attainment. When linear regression is used, it is possible for a model to predict an outcome attribute that does not actually exist as a possible destination. Likewise, when models include data from multiple metropolitan areas, often they contain no representation of the fact that the distribution of destinations in the metropolitan area will strongly influence outcomes.

The mobility literature has taken some steps to incorporate availability in locational attainment models. In particular, several papers by South, Crowder, and Pais (e.g., South and Crowder 1997; Pais, South, and Crowder 2012) include measures of metropolitan characteristics as independent variables in their models of intrametropolitan migration. This clearly is an improvement over models without these predictors. In many cases, however, they enter metropolitan characteristics to capture substantive effects (for instance, how share black in a city is hypothesized to alter attitudes toward blacks among nonblacks) without directly representing the marginal distribution of neighborhood characteristics. Further, this approach is inherently limited in its potential to represent the marginal distribution, because neighborhood availability is represented using a single metropolitan summary statistic, when the full distribution of local available neighborhoods influences the outcome of locational attainment.

In summary, the most used statistical modeling frameworks employed by social scientists—linear and binary or multinomial logistic regression—are limited for representing the substantive process of locational attainment. This modeling problem, however, has had substantive and theoretical implications. As a result, mobility and attainment studies have focused on individual-level determinants of single place attributes. The role of neighborhood and metropolitan social and spatial structural factors that influence locational outcomes receive less direct focus in statistical models. To be sure, analysts have also tried to indirectly infer structural effects from results of traditional models. But this inference is more limited than if structural and place effects were better represented in models of migration and locational attainment.

Discrete choice and other dyadic approaches as alternatives

To address these problems requires a model that allows us to examine how multiple attributes of the potential neighborhoods of destination simultaneously influence destinations, and how family and individual characteristics combine with neighborhood attributes to make some neighborhoods attractive to people with particular characteristics. We would also like a model that recognizes that the distribution of neighborhoods available influences the outcome of locational processes.

Analyses based in dyads of movers and potential destinations are well suited to incorporate these qualities. Instead of modeling an attribute of the outcome, dyadic models define a universe of potential outcome neighborhoods for each household, called the choice set or destination set, and then include attributes of the potential destinations in predicting actual destinations. Among models with this property are p^* or exponential random graph models for social networks (Wasserman and Pattison 1996) as well as models of discrete choice (McFadden 1973).

In a discrete-choice model of location, destination is modeled as a function of multiple household characteristics and potential destination neighborhood attributes. The neighborhood attributes can include, for example, local race/ethnic composition, income level, housing costs, quality of local neighborhood schools, crime rates, and so on. The models can incorporate characteristics of families and individuals as they combine with attributes of neighborhoods to produce outcomes. This approach also allows for household-specific choice sets of neighborhood options that vary across households and represent the full distribution of possible outcomes.

Discrete-choice models may be understood as an analysis of dyadic data (see also Bader and Krysan, this volume, for a dyadic analysis). Imagine arranging the data so that the cases represent each possible household crossed by each destination that the household might move to. For instance, suppose each household in a survey of 1,000 households could move to 100 potential destination neighborhoods. In the discrete-choice analysis, each household would be represented by 100 cases (or 99 in an analysis that only looks at movers). The analysis then would include 100,000 cases representing all households by potential destination dyads (or 99,000 in an analysis that only looks at movers). This model can also be applied when the potential destinations differ for each household.

Households move to one of the destinations in the choice or destination set. An indicator variable, which is 1 for the tract the household resides in at destination and 0 for the origin tract, is the outcome. In the simplest discrete-choice model, conditional logit, the outcome is modeled with a type of logistic regression. Unlike typical logistic regression, the outcome is constrained so that the sum of probabilities of destinations for each household sum to 1 (since each household moves to one and only one destination place). This is formally equivalent to logistic regression models that include a household fixed effect; as with a fixed-effect model, characteristics of individuals on the outcome are not estimated, although these may be allowed to interact with characteristics of potential destinations.

The advantage of this data structure is that it can include characteristics of both the household moving and potential destinations in modeling the actual destination. In this representation, individual and household characteristics do not influence household destination choice alone, they do so relative to attributes of potential destination neighborhoods. Black households are likely to respond differently to majority black neighborhoods than nonblack households, for instance. Likewise, high-income households are likely to respond differently to expensive neighborhoods than are low-income households.

In economics, discrete-choice models were derived from random utility models of choice. In the underlying behavioral model, individuals choose the neighborhood that gives them greatest utility. Both observed and unobserved attributes of neighborhoods influence utility. Under some assumptions, coefficients in the model give estimates of preferences for neighborhood characteristics.

While preferences are coherent interpretations of the model parameters under some conditions, it is probably not a defensible interpretation in the case of residential mobility with observational data. The problem is that we cannot really separate household preferences from other factors that influence mobility, such as discrimination in housing markets, lack of information about some potential destinations, influence of an agent or broker on destination chosen, and so on. Fortunately, we can interpret the model without necessarily assuming that we are revealing preferences or that choice is the main factor driving the outcome. A more descriptive interpretation of the interaction between household and neighborhood characteristics is that the model captures *sorting* or *matching* between household and neighborhood attributes.

I now describe the simplest discrete-choice model—conditional logit—more formally. I denote households in the data with i; the potential neighborhoods they could move to with j; and time points (discrete) with t. Let p_{ijt} denote the probability that the ith household moves to the jth neighborhood at time t: this is the outcome. Note that p is the probability of moving to a single neighborhood (j) in the household's destination set, not the probability of moving to a type of neighborhood as has commonly been used in multinomial locational attainment models. For each household, summing probabilities over their destination set (summing over j) will sum to one because everyone moves to one of the possible destination neighborhoods.

Denote by U_{ijt} the attractiveness of the jth neighborhood to the ith household at time t. I define "attractiveness" as capturing whatever forces guide the ith household toward the jth destination, including preferences for neighborhood characteristics, but also information about certain types of neighborhoods, discriminatory processes in markets, specific types of destinations, and so on. The attractiveness of a neighborhood for a household depends on the attributes of the neighborhood and how their own characteristics shape responses to neighborhood attributes. We denote the mth characteristic for the ith household at time t with X_{mit} and the kth neighborhood attribute with Z_{kjt}. Let η_{jt} denote unobserved features of neighborhood j that affect attractiveness at time t. Then, the attractiveness of the jth neighborhood to the ith household is given by:

$$U_{ijt} = F(Z_{kit}, X_{mit}, \eta_{jt})$$

If F is a linear function, and limiting ourselves to two household characteristics (X_1, X_2) and two neighborhood attributes (Z_1, Z_2), then the basic model of attractiveness is:

$$U_{ijt} = \beta_1 Z_{1it} + \beta_2 Z_{2it} + \delta_{11} Z_{1it} X_{1it} + \delta_{21} Z_{2it} X_{1it} + \delta_{12} Z_{1it} X_{2it} + \delta_{22} Z_{2it} X_{2it} + \eta_{jt}$$

where β_k and δ_{km} are parameters to be estimated. Household characteristics (the Xs) do not influence attractiveness of destinations except as they interact with neighborhood characteristics. For this reason, main effects of household characteristics on destination choice are not estimated. This specification allows sorting by each household characteristic with each neighborhood characteristic; we may sometimes constrain some effects to not interact.

Given data on the characteristics of households and neighborhood attributes— and the neighborhoods moved to by households—and an assumed probability distribution of the unobserved characteristics of neighborhoods that influence desirability, one can estimate the model parameters. The nonrandom portion of attractiveness above is:

$$\hat{U}_{ijt} = \beta_1 Z_{1it} + \beta_2 Z_{2it} + \delta_{11} Z_{1it} X_{1it} + \delta_{21} Z_{2it} X_{1it} + \delta_{12} Z_{1it} X_{2it} + \delta_{22} Z_{2it} X_{2it}$$

If the errors follow an extreme value (Gumbel) distribution, we obtain the probabilities from the discrete-choice conditional logit model (see McFadden 1973; Train 2009, 34–75):

$$p_{ijt}(Z_{kjt}, X_{mit}, C_{(i)}) = \frac{\exp(\hat{U}_{ijt} - q_{ijt})}{\sum_{w=1}^{C_{(i)}} \exp(\hat{U}_{iwt} - q_{iwt})} \tag{1}$$

where $C_{(i)}$ denotes the set of neighborhoods available ("destination set") for the ith household, and w is an index used to sum over elements of this set for the ith household. In this model, β_k is an estimated coefficient indicating the attractiveness of the kth neighborhood attribute (Z_{kjt}). Positive coefficients indicate positive sorting (migration attraction) toward a neighborhood with that attribute; negative coefficients indicate negative sorting (migration aversion). The coefficient δ_{km} gives how the estimated effect of the kth neighborhood characteristic (Z_k) is altered by the mth household characteristic (X_m). The model shown in (1) allows for all household and neighborhood characteristics to interact. The term q_{ijt} represents an offset used to represent sampling from the destination set to reduce computation. The coefficients are estimated by maximum likelihood.

In this model, the sorting of a household characteristic with a neighborhood attribute is specified through a multiplicative interaction $(Z_{kwt} X_{mit})$. For instance, a dummy variable indicating the respondent is black and neighborhood share black might be interacted to allow for different responses to share black for black and nonblack respondents, capturing the likely positive race sorting (attraction) with share black for blacks and negative race sorting (aversion) with share black for whites. Although not shown in the equations above, sorting may also be specified through created variables that relate a household characteristic and a neighborhood attribute, such as a ratio or difference between them.

The plausible number of neighborhoods a household may be considering moving to may be large. For instance, in my application below I use all tracts in a household's metropolitan area, giving hundreds or thousands of potential destinations. To analyze the data efficiently, I randomly sample 100 percent of origin and destination tracts and 5 percent of other tracts as destinations. The offset term in equation (1), q, is the weight employed to represent this sampling.

The independence of irrelevant alternatives

The conditional logit model implies a type of proportional substitution across alternatives as the choice set is altered, an assumption often called the independence of irrelevant alternatives (or IIA; see Train 2009, 34–75). The assumption is that the odds ratios between any two destinations will be unchanged when a third alternative is removed as a possibility from the destination set.[2] This implies that alteration to the choice sets of destinations by dropping or changing alternatives should not alter model estimates, and correspondingly implies the model should generate valid predictions when the alternatives in the choice set are changed.

Several tests of the IIA property have been proposed, but unfortunately simulation studies suggest that these tests operate poorly. Train (2009) suggests that coefficients from a conditional logit model when IIA does not hold are still likely to be accurate estimates of average effects of characteristics on attractiveness (or average preferences in a preference interpretation). Further discussion of the IIA property is available in the online supplement.[3]

An Empirical Contrast of Results: Basic Locational Attainment and a Conditional Logit Model

To help understand the differences between traditional and discrete-choice models, I estimate some basic models using both methods and compare results. To perform this analysis, I use data on families from the Panel Study of Income Dynamics (PSID), a national longitudinal study of families. I use PSID from 1999 to 2009 to represent recent mobility. During this time, data on residence were available at two year intervals. Following standard practice, as proxies for neighborhood of residence I use data on census tracts. Tract data are taken from the 2000 and 2010 U.S. Censuses and the 2006 to 2010 American Community Survey. I use linear interpolation to fill in values for years between censuses.

Discrete-choice models require an explicit definition of the universe of potential destinations to which households may move. I focus on *intra*metropolitan area moves, defining the choice or destination set for each individual at each point in time to be all the census tracts in each respondent's metropolitan area. Many past studies have limited mobility to within-metropolitan moves (e.g., South and Crowder 1997, 1998). About 83 percent of moves in the PSID are within the same metropolitan area.

With this representation, the number of possible destinations for each household is large. A resident of Chicago would have roughly 3,700 census tract options in total, for instance. To allow for efficient estimation, I sample 100 percent of origin and destination dyads and 5 percent of all other dyads from the choice set and use the probability of selection (q_{ijt}) to weight the choice models as shown in equation (1).

I focus only on movers, using mover-year person years only (in which the household changed census tract from time $t-1$ to time t). It is possible, however, to build a model of the decision to move as well as destination, but this is not done here (see Bruch and Mare 2012).

Also following a number of past studies, I use only households with black and white household heads. The PSID has too few Latino households in this time period to produce precise estimates for Latinos.

I represent the data in person-year format, so households contribute up to seven mobility transitions each. To account for the clustering this induces, I use Huber-White cluster robust standard errors with households as the clusters.

Variables

In the locational attainment literature, most studies predict either the racial composition or the income level of tracts as an outcome. Since my goal is to compare differences in conclusions we might reach from a basic locational attainment and a similar discrete-choice analysis, I restrict the analysis to a set of simple, often-used covariates from past studies.

For individual characteristics (X_{it}), I use race coded as white or black and family income, based on an average of family income for up to five preceding years, using up to three years of income data because of the every-other-year sampling. Family income is in thousands of dollars, adjusted for inflation to constant year-2000 dollars, and is centered at the grand mean (grand mean family income = 0). In some models family income is recoded into quintiles.

In characterizing tracts, I use three tract attributes: tract racial composition, tract economic level, and population. For racial composition and income level, I use measures consistent with well-known prior studies such as South and Crowder (1997, 1998). In some models, I use percentage black as a continuous predictor. In others, I break race into categories of 0–20 percent black, 20–80 percent black, and 80+ percent black. For income, in some models I use median tract income. In other models, I characterize tracts as low-poverty or high-poverty. Low-poverty tracts have 20 percent or less of their residents in families with incomes below the poverty line, while high-poverty have 20 percent or more of residents with incomes above the poverty line. Log of tract population is entered to reflect the availability of housing for each neighborhood.

TABLE 1
Locational Attainment Models of Destination Tract Poverty and Percentage Black
Category

	Logistic Regression	Multinomial Logit (Ref.=0–20% black)	
	20%+ Poor	20–80% Black	80%+ Black
Household head is black (1 = yes)	1.898°°°	2.974°°°	5.053°°°
	(0.136)	(0.170)	(0.413)
Houehold income quintiles (Reference = 1st quintile)			
Household income 2nd quintile	−0.272	0.159	−1.169°
	(0.152)	(0.173)	(0.592)
Household income 3rd quintile	−0.650°°°	−0.234	−0.823
	(0.155)	(0.183)	(0.482)
Household income 4th quintile	−0.917°°°	−0.249	−1.819°°
	(0.162)	(0.181)	(0.641)
Household income 5th quintile	−1.257°°°	−0.538°°	−2.205°°
	(0.171)	(0.186)	(0.811)
Interactions of head race with household income			
Head black°	−0.100	−0.291	0.866
Household income 2nd quintile	(0.174)	(0.211)	(0.610)
Head black°	−0.227	−0.191	0.222
Household income 3rd quintile	(0.182)	(0.221)	(0.502)
Head black°	−0.171	−0.352	1.000
Household income 4th quintile	(0.196)	(0.226)	(0.659)
Head black°	−0.283	−0.280	0.667
Household income 5th quintile	(0.243)	(0.273)	(0.848)
Constant	−1.356°°°	−1.842°°°	−4.025°°°
	(0.122)	(0.148)	(0.400)
N (household-years)	9102	9102	
N (households)	5265	5265	

NOTE: Standard errors in parentheses are clustered by household.
$°p < .05.$ $°°p < .01.$ $°°°p < .001.$

Results

The first column of Table 1 shows a logistic regression model predicting residence in a poor (> 20 percent poor) neighborhood.

Consistent with prior research, I find that black households are much more
likely than white households to move into poor neighborhoods (with income
quintile controlled), and that higher income households are less likely to move
into poor neighborhoods. I also find the effects of income on reducing the odds
of entering a poor neighborhood are not much different for black than for white

households. Notable in these results is how large the race effect is: the effect of black race on the odds of moving to a poor tract is larger than the effect of a household switching from the lowest to the highest income quintile.

In the two rightmost columns of Table 1, coefficients of a locational attainment model are shown with categories of tract percentage black as the outcome. Percentage black has the categories of 0–20 percent black, 20–80 percent black, and more than 80 percent black. The model is multinomial logistic regression with 0–20 percent black as the base category.

The results show that head race is the strongest predictor, as we would expect, with black headed households much more likely to enter a mixed or predominately black neighborhood. Higher-income households of both races are also less likely to move into a mixed or (especially) predominately black neighborhood.

Table 2 shows the same variables analyzed with conditional logit models. The outcome in the model is the destination from tracts in the metropolitan area, and tract characteristics are used to predict this outcome in combination with household characteristics.

The first model in Table 2 is similar in specification to model 1 in Table 1: destination is predicted by whether a neighborhood is 20 percent or more poor, and the interaction of whether a neighborhood is 20 percent or more poor with several individual characteristics: black race, household income quintile, and black race interacted with household income quintile. In the discrete-choice model, the neighborhood attributes that were previously outcomes become predictors of destination. The results are also similar to the first model of Table 1: the coefficient of the "black × neighborhood 20 percent or more poor" variable is large and significant, indicating much higher entry rates to poor neighborhoods for black households than for white households. Higher income households have lower rates of entry into poor neighborhoods, although as I found previously the effect of a black household head has a greater effect on the odds of entering a poor neighborhood than the difference between the lowest and highest income quintile households. The general magnitudes of the coefficients in the discrete-choice and traditional models are also similar.

In Table 2, model 2 shows estimates of a model that is similar to the multinomial logit model in Table 1. The model shows that black households are much more likely to enter black neighborhoods and high-income households are significantly less likely to enter black neighborhoods. Both the magnitude and the significance of results are similar between the conditional logit and the similar multinomial logit models.

Model 3 in Table 2 shows estimates of a conditional logit model that combines variables from both model 1 and model 2. Racial composition and the tract poverty rate are simultaneously allowed to determine destination. This is a model with no general equivalent using standard locational attainment approaches.[4]

When both race and income sorting are controlled together in model 3 of Table 2, there are two notable changes. First, the coefficient of "black head × neighborhood 20 percent or more poor" becomes much smaller than in model 1. Second, many of the interactions of the tract percentage black categories and the household income categories from model 2 become smaller and nonsignificant.

TABLE 2
Conditional Logit Models of Destination Tract, Categorical Variables Specification

Variable	Model 1	Model 2	Model 3
Neighborhood > 20% poor (1 = yes)	0.149		0.314
	(0.172)		(0.185)
Percentage black 20–80% (1 = yes)		0.034	−0.089
		(0.201)	(0.210)
Percentage black 80%+ (1 = yes)		−0.929°	−1.153°
		(0.448)	(0.459)
Interaction of head race and 20% poor tract			
Black head × neighborhood > 20% poor	2.125°°°		0.758°°°
	(0.189)		(0.204)
Interactions of head race and tract racial composition			
Head black × Tract percentage black 20–80%		2.657°°°	2.392°°°
		(0.233)	(0.242)
Head black × Tract percentage black 80%+		5.007°°°	4.452°°°
		(0.473)	(0.486)
Interactions of household income and 20% poor tracts			
Household income 2nd quintile ×	−0.348		−0.403
Neighborhood > 20% poor	(0.210)		(0.230)
Household income 3rd quintile ×	−0.713°°°		−0.593°
Neighborhood > 20% poor	(0.214)		(0.232)
Household income 4th quintile ×	−1.167°°°		−1.059°°°
Neighborhood > 20% poor	(0.227)		(0.244)
Household income 5th quintile ×	−1.416°°°		−1.219°°°
Neighborhood > 20% poor	(0.231)		(0.245)
Interactions of head race with household income and 20% poor tracts			
Black × Household income 2nd quintile ×	−0.161		0.034
Neighborhood > 20% poor	(0.236)		(0.260)
Black × Household income 3rd quintile ×	−0.327		−0.347
Neighborhood > 20% poor	(0.244)		(0.266)
Black × Household income 4th quintile ×	−0.092		−0.080
Neighborhood > 20% poor	(0.264)		(0.289)
Black × Household income 5th quintile ×	−0.352		−0.130
Neighborhood > 20% poor	(0.310)		(0.349)
Interactions of tract 20%–80% black and tract income quintile			
Tract 20%–80% black × Household income		0.176	0.333
2nd quintile		(0.230)	(0.242)
Tract 20%–80% black × Household income		−0.521°	−0.308
3rd quintile		(0.239)	(0.249)
Tract 20%–80% black × Household income		−0.550°	−0.218
4th quintile		(0.238)	(0.250)
Tract 20%–80% black × Household income		−0.934°°°	−0.572°
5th quintile		(0.240)	(0.250)

(continued)

TABLE 2 (CONTINUED)

Variable	Model 1	Model 2	Model 3
Interactions of tract 80% + black and tract income quintile			
Tract 80%+ black × Household income 2nd quintile		-1.280°	-0.982
		(0.646)	(0.668)
Tract 80%+ black × Household income 3rd quintile		-1.011	-0.583
		(0.524)	(0.543)
Tract 80%+ black × Household income 4th quintile		-2.472°°	-1.706°
		(0.808)	(0.780)
Tract 80%+ black × Household income 5th quintile		-2.742°°	-1.956°
		(0.880)	(0.866)
Interactions of head black, tract 20%–80% black, and tract income quintile			
Head black × Tract 20%–80% black × Household income 2nd quintile		-0.534	-0.530
		(0.281)	(0.292)
Head black × Tract 20%–80% black × Household income 3rd quintile		-0.199	-0.059
		(0.293)	(0.302)
Head black × Tract 20%–80% black × Household income 4th quintile		-0.409	-0.331
		(0.295)	(0.307)
Head black × Tract 20%–80% black × Household income 5th quintile		-0.211	-0.104
		(0.342)	(0.362)
Interactions of head black, tract 80%+ black, and tract income quintile			
Head black × Tract 80%+ black × Household income 2nd quintile		0.583	0.598
		(0.675)	(0.700)
Head black × Tract 80%+ black × Household income 3rd quintile		-0.035	0.243
		(0.559)	(0.583)
Head black × Tract 80%+ black × Household income 4th quintile		1.100	1.170
		(0.833)	(0.811)
Head black × Tract 80%+ black × Household income 5th quintile		0.479	0.651
		(0.929)	(0.933)
Log (tract population)	0.764°°°	0.809°°°	0.859°°°
	(0.033)	(0.035)	(0.036)
N (household-years)	371300	371300	371300
N (households)	5201	5201	5201

NOTE: Standard errors in parentheses and are clustered by household.
$°p < .05.$ $°°p < .01.$ $°°°p < .001.$

What do these changes indicate? Sorting of black heads toward poor neighborhoods was very strong in model 1, indicating odds 8.4 (exp(2.125)) times as great of entering a poor tract for a black compared to a white household. When race sorting is added to the model (model 3), the odds drop to 2.13 times (exp(.758)) as great. Most of the apparent "attraction" to poor neighborhoods reflects sorting into black neighborhoods, which on average have higher poverty rates.

TABLE 3
Locational Attainment Models, Continuous Variables

	Tract Outcome	
Variable	Median Income	Percentage Black
Household head black (1 = yes)	−13.407°°°	43.233°°°
	(0.645)	(0.857)
Family income (1000s $)	0.111°°°	−0.014°°°
	(0.014)	(0.003)
Head black × Family income (1000s $)	0.073°°°	−0.160°°°
	(0.019)	(0.021)
Constant	−3.965°°°	−11.004°°°
	(0.397)	(0.258)
N (household-years)	9102	9102
N (households)	5265	5265

NOTE: Standard errors in parentheses and are clustered by household. Models estimated with OLS regression.
°°°$p < .001$.

The interactions of income and tract race type become weaker in the final model, although the change is less stark than for the black × poor neighborhood interaction. These results indicate that some of the apparent sorting of higher income households away from black neighborhoods is because higher income households are less likely to enter poor neighborhoods, and poorer neighborhoods are more often black neighborhoods (consistent with population models of segregation and poverty concentration; see Massey and Denton 1993; Quillian 2012).

Tables 3 and 4 estimate similar models but use continuous variable specifications for tract racial composition, tract income, and household income. The models in Table 3 are linear regressions, with outcomes of median income of tract of destination and percentage black of tract of destination. The predictor variables are race of household head, family income, and head race.

The results in Table 3 show results similar to those with the discrete variable locational attainment models in Table 1, except for a stronger interaction of race and income in predicting neighborhood percentage black in Table 3. Family income is centered at the mean, so the household head black coefficient shows that at average income black households on average live in tracts that are 43 percent higher in percentage black and with $13,400 lower median income.

In Table 4, the first two columns show estimates of discrete-choice models similar to Table 3. Income sorting is accounted for by creating a variable that is the absolute difference between household and tract median income.[5] The discrete-choice models indicate many similar patterns to the linear regressions. In model 1 the higher the median income of a tract, the more likely a white

TABLE 4
Conditional Logit Models of Destination, Continuous Variables

Variable	Model 1	Model 2	Model 3
Tract percentage black		-0.017***	-0.017***
		(0.001)	(0.001)
Head is black × Tract percentage black		0.049***	0.048***
		(0.002)	(0.002)
Tract percentage black × Family income		-0.019***	-0.005
(in 1000s $), Coef × 100		(0.004)	(0.003)
Head black × Tract percentage black ×		-0.003	0.000
Family income (in 1000s $), Coef × 100		(0.004)	(0.004)
Tract median income (in 1000s $)	0.004***		-0.001
	(0.001)		(0.001)
Black × Tract median income (in 1000s $)	-0.047***		-0.009***
	(0.002)		(0.002)
Absolute difference	-0.016***		-0.015***
Family and tract median income	(0.001)		(0.001)
Head black × Absolute difference	-0.007***		-0.006**
Family and tract median income	(0.002)		(0.002)
Log of tract population	0.842***	0.895***	0.945***
	(0.035)	(0.036)	(0.037)
N (household-years)	371300	371300	371300
N (households)	5201	5201	5201

NOTE: Standard errors in parentheses and are clustered by household.
**$p < .01$. ***$p < .001$.

household is to move into that tract, and the less likely a black household is to move in. In model 2, the higher the percentage black of a tract, the more likely a black household is to move in, and the less likely a white household is to move in.

When race sorting is added in Table 4, model 3, however, the relationship between tract income and the chance a white or black household will enter becomes nonsignificant. Again, race differences in the income level of migration destinations are mostly accounted for by race segregation. Also repeated from the discrete variables specification, when we add income and race sorting together the tendency of higher income families to avoid black neighborhoods is weakened and becomes nonsignificant.

The main substantive point gained from the conditional logit analyses of neighborhood percentage black and income segregation is that the huge gap in the odds of moving into a poor neighborhood between whites and blacks mostly reflects the fact that whites move into white neighborhoods and blacks move into black neighborhoods, and there is a large average difference in poverty rates and income between white and black neighborhoods. This is not a point that was

clear from locational attainment models, which left it plausible that blacks might
have trouble attaining residence in low-poverty neighborhoods for other reasons,
for instance because of reduced ability to translate income into more affluent
neighborhoods regardless of racial composition, or perhaps because racial wealth
gaps made it difficult for black households to afford housing in nonpoor areas.
We see it is actually pure race sorting combined with the strong negative association of neighborhood percentage black and neighborhood average income that
account for most of the racial gap in the chance of entering a poor or low-income
neighborhood.

Neighborhood attributes beyond racial composition and economic level

Beyond neighborhood income level, many neighborhood attributes are potentially important bases of household sorting that produce segregation. Some attributes that are likely to be important are neighborhood percentage Hispanic, quality
of local public schools, distance from work, and distance from closest kin. A major
strength of the discrete-choice approach is that given available data, these can be
incorporated into the model in a straightforward fashion by adding them as predictors and allowing their interactions with household characteristics.

To incorporate these factors, we need measures of how each potential destination would score on that factor. For percentage Hispanic, this would be simply
percentage Hispanic in the tract; for quality of the local public school, we need a
measure of school quality for the public school that a resident of each potential
tract of destination would attend; for distance from work, this could be a measure
of distance for each tract from the workplace of the household head; for distance
from closest kin, this would be distance of each potential destination from residence of the respondent's closest kin. Coefficients of these variables provide estimates of the importance of each of these factors in determining mobility
destinations. Coefficients of interactions of household characteristics and neighborhood attributes provide estimates of how response to the neighborhood attribute is changed by household characteristics. See Quillian (2014) for an example
that adds multiple measures of neighborhood class and spatial distance.

Ecological dependence in traditional and discrete-choice models

Ecological dependence refers to the fact that the outcomes of locational attainment processes will be influenced by the distribution of neighborhood attributes
in the choice or destination sets. When not all cases in the data have the same
choice or destination sets, we want to incorporate the ecological structure in
predicting outcomes.

In the discrete-choice approach, the marginal distribution is represented
through the data structure, since neighborhoods in the destination set are
reflected as cases. A variable with the population of each tract is used to represent a baseline probability of entry into each tract.

To illustrate, I calculate for each metropolitan area with PSID respondents in
it: (1) the actually observed proportion of within-metropolitan moves by PSID

FIGURE 1
Rates of Entry to Poor Neighborhoods

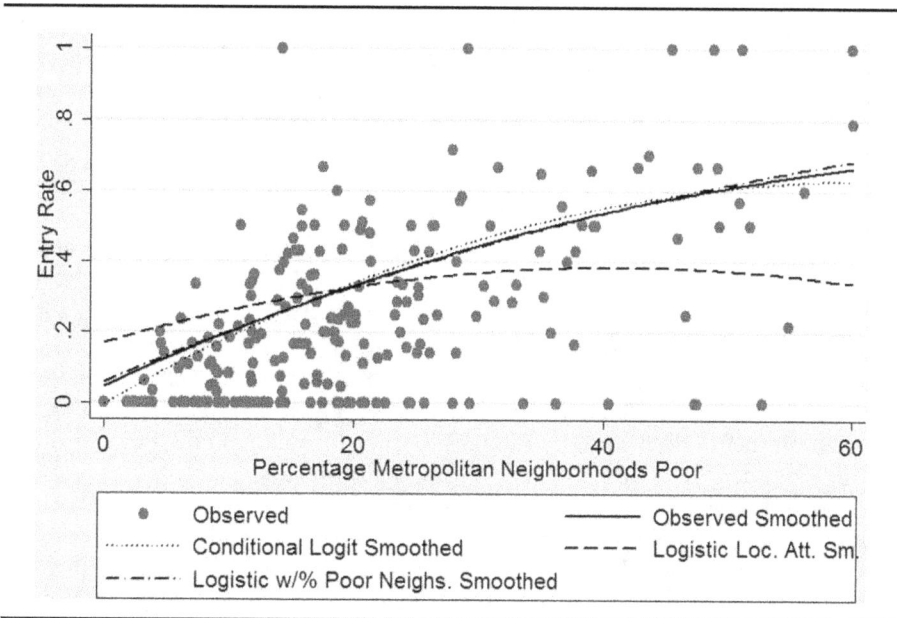

• Observed	——— Observed Smoothed
············ Conditional Logit Smoothed	– – – – Logistic Loc. Att. Sm.
– · – · –· Logistic w/% Poor Neighs. Smoothed	

respondents that end in poor neighborhoods; (2) the prediction of the proportion of entries to poor neighborhoods from the conditional logit model; and (3) the prediction of the proportion of entries to poor neighborhoods from the locational attainment model. I use the poor neighborhood (> 20+ percent poor) outcome model, model 1 of Table 2, for the discrete-choice model, and model 1 of Table 1 for the locational attainment model.

Figure 1 shows the actual and predicted rates of entry into poor neighborhoods graphed against the percentage of neighborhoods poor in the metropolitan area. There is significant jitter in the actual dots, shown for observed entries only, so the figure shows a quadratic smoothed line through the points for each model. The observed proportion of entries into poor neighborhoods increases nearly linearly with the share poor neighborhoods in the metropolitan area; the discrete-choice prediction increases similarly.[6] The traditional locational attainment model (logit), however, shows no similar increase. The traditional model thus fails to represent ecological dependence.

We may, however, build into traditional locational attainment models the role of the marginal distribution of neighborhoods through adding parameters. For instance, we can add a measure of share of population living in poor neighborhoods in a metropolitan area as a predictor. When this is done in a locational attainment model, this also provides a satisfactory representation of the effects of ecological context, as shown in Figure 1.[7]

Conclusion

This article compares discrete-choice and traditional locational attainment approaches to modeling locational attainment and residential mobility. First, I sought to clarify the advantages and disadvantages of discrete-choice approaches to analyzing locational attainment as contrasted with traditional locational attainment models. Second, I considered how a discrete-choice approach can enrich our understanding of "classic" locational attainment models by employing the most-used covariates from past research.

The major advantage of the discrete-choice approach is its ability to represent place outcomes as multidimensional, or as composed of a bundle of attributes that simultaneously influence locational attainment. Discrete choice then allows us to address the question of the relative importance of each neighborhood attribute in determining destinations, and of how the effects of locational attributes vary with household characteristics. Incorporating place attributes as simultaneous predictors is a major advantage because fundamentally neighborhood sorting is guided by multiple attributes of neighborhoods.

A second advantage is that discrete-choice models, estimated with specific neighborhoods as choices, incorporate outcomes that depend in part on the destinations that are available. Predictions from the model only produce predictions that exist, as neighborhoods and changes in frequencies of neighborhoods available will automatically tend to influence model predictions.

A potential disadvantage of the discrete-choice approach, however, is the independence of irrelevant alternative assumptions of the simplest discrete-choice model, or the difficulties in estimation of discrete-choice models that do not make the IIA assumption. However, traditional locational attainment models based on regression also fail to have coefficients that are invariant to changing the potential neighborhood outcomes. Some experts suggest that coefficients of conditional logit represent approximate average attractiveness if IIA does not hold.

Another downside of discrete choice is simply the greater complexity of the models and their interpretation. To a large extent, however, this complexity reflects real complexity in migration: migration is guided by multiple place attributes and multiple individual or household attributes. Accounting for both together involves a more complex model.

How would major substantive findings from locational attainment models be different under a discrete-choice approach? The focus of a great deal of the work using locational attainment models has been on the effects of income and race on neighborhood income level and racial composition. But these analyses treat neighborhood racial composition and income level separately, when they both simultaneously guide migration. When neighborhood race and income are considered together, most of what appears to be race sorting into neighborhoods of different income levels is actually race sorting by racial composition. This sorting then produces racial inequality in neighborhood outcomes due to the fact that black neighborhoods are poorer, on average.

This result suggests that if our goal is to address the very large gap in the income level of neighborhood environments between black and white households, a primary way to do this should be by either reducing racial segregation or increasing the affluence level of residents of black neighborhoods. Race sorting and the low income levels of black neighborhoods, rather than difficulty among black households in moving into higher income neighborhoods per se, are the main causes of racial gaps in destination neighborhood income levels.

In situations where our main interest is describing the association between multiple individual or household characteristics and a single neighborhood attribute, traditional methods would still be the right tool for the job. The real power of the discrete-choice approach and other dyadic approaches is their ability to incorporate a much wider set of place characteristics as simultaneous influences on destination, such as variables representing housing prices, school quality, neighborhood spatial location, and so on. Because many attributes of places are important simultaneously for locational attainment, we need to include them to build better models of locational attainment as a process. Despite some drawbacks resulting from IIA issues and complexity, discrete-choice and related dyadic models represent the clearest ways to move forward to more realistic and informative models of the locational attainment process.

Notes

1. A few studies using cross tabulation or multinomial logit have used categories representing multiple dimensions of place simultaneously. Massey, Gross, and Shibuya (1994), for instance, create nine categories combining three categories of percentage black and three categories of percentage poor. Yet this approach can only accommodate a few dimensions of place represented categorically before the number of categories becomes unmanageable.

2. The widely used multinomial logit models also make the IIA assumption, although relatively few applications of multinomial logit discuss testing it.

3. See http://ann.sagepub.com/supplemental.

4. In this case, it would be possible to create six cross-classified types of neighborhood poverty by neighborhood racial composition and analyze the results with multinomial logit. This solution is not available, however, for continuous tract attributes or when cross tabulation of categorical attributes would produce more than a small number of categories. Discrete choice, on the other hand, can accommodate many neighborhood attributes and continuous attributes.

5. Other approaches could include taking ratios or log ratios of household and tract median income. I elected for this specification because most of the locational attainment uses unlogged income and for simplicity of interpretation.

6. If the discrete-choice model diverges sharply from the observed marginal rate of entry, this would suggest IIA violation. In some cases certain types of IIA violation may be corrected by adding parameters representing the marginal distribution to the conditional logit model.

7. A table of estimated coefficients of this model is in the online supplement: http://ann.sagepub.com/supplemental.

References

Alba, Richard D., and John R. Logan. 1993. Minority proximity to whites in suburbs: An individual level analysis of segregation. *American Journal of Sociology* 98:1388–1427.

Bader, Michael D. M., and Maria Krysan. 2015. Community attraction and avoidance in Chicago: What's race got to do with it? *The ANNALS of the American Academy of Political and Social Science* (this volume).

Brown, Susan K. 2007. Delayed spatial assimilation: Multigenerational incorporation of the Mexican-origin population in Los Angeles. *City & Community* 6 (3): 193–209.

Bruch, Elizabeth E., and Robert D. Mare. 2012. Methodological issues in the analysis of residential preferences, residential mobility, and neighborhood change. *Sociological Methodology* 42 (1): 103–54.

Cadwallader, Martin. 1992. *Migration and residential mobility: Macro and micro approaches.* Madison, WI: University of Wisconsin Press.

Cort, David A. 2011. Reexamining the ethnic hierarchy of locational attainment: Evidence from Los Angeles. *Social Science Research* 40:1521–33.

Crowder, Kyle, Jeremy Pais, and Scott J. South. 2012. Neighborhood diversity, metropolitan constraints, and household migration. *American Sociological Review* 77 (3): 325–53.

Ellen, Ingrid Gould. 2000. *Sharing America's neighborhoods: The prospects for stable racial integration.* Cambridge, MA: Harvard University Press.

Freeman, Lance. 2008. Is class becoming a more important determinant of neighborhood attainment for African-Americans? *Urban Affairs Review* 44:3–26.

Logan, John R., and Richard D. Alba. 1993. Locational returns to human capital: Minority access to suburban community resources. *Demography* 30:243–68.

Massey, Douglas S., and Nancy Denton. 1993. *American apartheid: Segregation and the making of the underclass.* Cambridge, MA: Harvard University Press.

Massey, Douglas, Andrew Gross, and Kumiko Shibuya. 1994. Migration, segregation, and the geographic concentration of poverty. *American Sociological Review* 59 (3): 425–45.

Massoglia, Michael, Glenn Firebaugh, and Cody Warner. 2012. Racial variation in the effect of incarceration on neighborhood attainment. *American Sociological Review* 78 (1): 142–65.

McFadden, Daniel. 1973. Conditional logit analyses of qualitative choice behavior. In *Frontiers in economics*, ed. Paul Zarembka, 105–35. New York, NY: Wiley.

Pais, Jeremy, Scott South, and Kyle Crowder. 2012. Metropolitan heterogeneity and minority neighborhood attainment: Spatial assimilation or place stratification? *Social Problems* 59 (2): 258–81.

Quillian, Lincoln. 2002. Why is black-white segregation so persistent? Evidence on three theories from migration data. *Social Science Research* 31:197–229.

Quillian, Lincoln. 2012. Segregation and poverty concentration: The role of three segregations. *American Sociological Review* 77:354–79.

Quillian, Lincoln. 2014. Race, class, and location in neighborhood migration: A multidimensional analysis of locational attainment. Unpublished manuscript, Northwestern University, Department of Sociology.

Sharkey, Patrick, and Jacob W. Faber. 2014. Where, when, why, and for whom do residential contexts matter? Moving away from the dichotomous understanding of neighborhood effects. *Annual Review of Sociology* 40 (1): 559–79.

South, Scott J., and Kyle D. Crowder. 1997. Escaping distressed neighborhoods: Individual, community, and metropolitan influences. *American Journal of Sociology* 102 (4): 1040–84.

South, Scott J., and Kyle D. Crowder. 1998. Leaving the 'hood: Residential mobility between black, white, and integrated neighborhoods. *American Sociological Review* 63:17–26.

Train, Kenneth. 2009. *Discrete choice methods with simulation.* 2nd ed. New York, NY: Cambridge University Press.

Wasserman, Stanley, and Philippa Pattison. 1996. Logit models and logistic regression for social networks: I. An introduction to Markov graphs and p^*. *Psychometrica* 61 (3): 401–25.

Woldoff, Rachael A., and Seth Ovadia. 2009. Not getting their money's worth: African-American disadvantages in converting income, wealth, and education into residential quality. *Urban Affairs Review* 45:66–91.

Community Attraction and Avoidance in Chicago: What's Race Got to Do with It?

By
MICHAEL D. M. BADER
and
MARIA KRYSAN

We argue that the relative persistence of racial segregation is due, at least in part, to the process of residential search and the perceptions upon which those searches are based—a critical but often-ignored component of the residential sorting process. We examine where Chicago-area residents would "seriously consider" and "never consider" living, finding that community attraction and avoidance are highly racialized. Race most clearly shapes the residential perceptions and preferences of whites, and matters the least to blacks. Latinos would seriously consider moving to numerous neighborhoods, but controls for demographics and distance from the respondents' home make Latino preferences much like those of whites. Critically, the geography of existing segregation begets further segregation: distance from current community significantly affects perceptions of the communities into which respondents might move. While neighborhood perception may cause persistent segregation, it may also offer hope for integration with appropriate policy interventions.

Keywords:　racial residential segregation; housing search; race/ethnicity; community perceptions; Chicago

Chicago is a city of neighborhoods. Chicago's neighborhoods cast into relief the city's identity of world-class ambitions with small-town feel. Politicians like to extol the virtues of neighborhoods providing unique identities to the places where Chicagoans live their

Michael Bader is an assistant professor of sociology at American University in Washington, DC. He studies how patterns of neighborhood change have evolved since the civil rights movement, processes that perpetuate spatial inequality, and measurement of neighborhood environments.

Maria Krysan is a professor in the Department of Sociology and at the Institute of Government and Public Affairs at the University of Illinois at Chicago. She studies residential segregation, racial attitudes, and survey methods. Her current work examines how community perceptions, knowledge, and experiences influence housing searches and perpetuate segregation.

DOI: 10.1177/0002716215577615

everyday lives (e.g., Daley 2010). While the mayor and city boosters extol the virtues of the diversity of experiences across its neighborhoods, what they do not highlight is that Chicago continues to be one of the most segregated cities within one of the most segregated metropolitan areas in the country (Logan and Stults 2011). While one might experience substantial diversity traveling across the neighborhoods and communities in the Chicago area, there is little diversity within each neighborhood.

There is little controversy about the fact that segregation is associated with inequality. To take a few examples, most suburban schools outperform urban ones; everyday amenities are more readily available in white neighborhoods; and community hospitals face chronic funding and staff shortages not experienced by their private or nonprofit counterparts (e.g., Logan, Oakley and Stowell 2008; Immergluck 2002; Small and McDermott 2006). The negative consequences of racially segregated neighborhoods are so consequential that segregation could be considered the structural linchpin of persistent racial inequality in the United States (Bobo 1989; Sharkey 2013). The unequal distribution of resources across metropolitan communities hampers the ability of racial and ethnic minorities in particular to succeed. Therefore, there is a compelling need to understand how people come to live in the nondiverse (and ultimately unequal) neighborhoods that make up this unequal landscape.

To date, racial residential segregation is usually explained in one of three ways (Charles 2003; Quillian 2002): the reduced purchasing power of minorities relative to whites due to economic inequality (e.g., Alba and Logan 1993; Iceland and Wilkes 2006); racially biased residential preferences (e.g., Clark 2009; Krysan et al. 2009; Lewis, Emerson, and Klineberg 2011); and discrimination by housing market actors, such as real estate agents and loan officers (e.g., Massey and Denton 1993; Roscigno, Karafin, and Tester 2009; Ross and Turner 2005; Squires 2007). All three are important factors, but they fall short of describing the forces that perpetuate racial segregation. Economic inequality has declined, racial attitudes have become more tolerant, and illegal discrimination has declined; yet high levels of segregation persist. In this article, we follow a new line of inquiry to examine the perceptions that residents hold about actual communities in the Chicago metropolitan area and the potential consequences of those perceptions on racial residential segregation. We view these perceptions as a critical but often-ignored component of the residential sorting process that translates into persistent segregation or offers the potential for racial integration (see Krysan, Crowder, and Bader 2014 for a more thorough discussion).

We examine two questions in this article. First, where would residents of the Chicago metropolitan area seriously consider searching for housing? Second, where would those same residents never consider searching for a house or apartment? The first provides a sense of where residents would initially channel their

NOTE: The authors gratefully acknowledge the support of the National Science Foundation (SES-0317740), the Ford Foundation, and the University of Illinois at Chicago, which funded the data collection reported in this article. The authors wish to acknowledge their conversations and collaboration with Kyle Crowder, which contributed to the ideas in this article.

resources in a housing search. The second identifies where residents would impose limits on their housing search if it expanded beyond their initial search destinations—limits that would make moves to those neighborhoods very unlikely. To answer both questions, we rely on responses that residents gave about real communities in the Chicago metropolitan area, not ideal or hypothetical neighborhoods. By studying real neighborhoods, we can estimate the factors that influence search patterns, especially in our case the importance of a community's racial composition on the willingness to consider or likelihood of avoiding particular communities.

Background

The long history of racial segregation scholarship has focused on the three reasons cited above: interracial economic inequality, racialized residential preferences, and racial discrimination. We have argued elsewhere (Krysan, Crowder, and Bader 2014) that the primacy of these arguments has neglected a focus on the *process* of residential selection. This means that we have little evidence about how people come to know about and perceive communities in their metropolitan area. We also fail, then, to know how residents' knowledge and perceptions influence their housing searches that, in turn, structure patterns of racial segregation. Asking what those perceptions are and the influence that race exerts on those perceptions can help to uncover systematic processes that perpetuate racial segregation. To be clear, though, these neighborhood perceptions no doubt interact with, affect, and result from interracial economic inequality, racialized residential preferences, and racial discrimination (see Krysan, Crowder and Bader 2014 for a detailed discussion of this argument).

There is a small body of related research that focuses on the idea of community perceptions and the influence of racial composition on it. Some (Quillian and Pager 2001; Sampson and Raudenbush 2004) focus on perceptions of one's existing neighborhood rather than where one might move. Other research does examine perceptions of other communities (Charles 2000; Krysan 2002) and finds that racial composition influences neighborhood perceptions, but this research has examined only a handful of neighborhoods. Finally, Krysan and Bader (2007) most closely matches our current effort, but did not control for important community covariates and examined perceptions of blacks and whites only.

This article contributes to the literature in three ways. First, and most importantly, our theoretical framework is specifically situated within questions about the *processes* of moving that might maintain (or, ideally, abate) racially segregated neighborhoods. We investigate in which communities Chicago-area residents would seriously consider searching for a home and communities they would never consider; the latter is a part of the search process that the existing literature generally neglects.

Our emphasis on both attraction and avoidance comes from a need to focus on how residential preferences translate into searches. As economic geographers note, housing searches are expensive in terms of both time and resources (Brown and Moore 1970), and knowledge gleaned from one search is rarely applicable in the next search since people search for houses infrequently (MacLennan 1982). Information constraints place bounds on how well people can optimize their preferences and, as a result, residents quickly reduce their choice-set to a manageable size (MacLennan 1982). To reduce the size of the choice-set, residents likely eliminate communities from consideration (and would thus "never consider" them) while prioritizing relatively few places to initially search (and thus "seriously consider" them). Though we ask about hypothetical future moves, the responses are given about communities in a real metropolitan area that have real identities. This means that our work is situated in new theoretical frameworks that focus on how individuals think about options in metropolitan areas (also see Quillian, this volume).

Second, this article extends previous research by including important community characteristics as control variables. Some evidence suggests that social class characteristics that are correlated with racial composition explain racialized residential preferences (Harris 1999, 2001). Our previous study of Detroit (Krysan and Bader 2007) isolated the association between racial composition and community perceptions among thirty-three communities. That article, however, did not measure the influence of two of the most important contextual variables associated with race and thought to explain perceptions: school quality and crime (Emerson, Chai, and Yancey 2001; Goyette, Farrie, and Freely 2012). Even for residents without students, schools influence home prices sufficiently to affect where residents without children would consider living, and evidence suggests that racial composition of school districts affects metropolitan-level segregation (Logan, Oakley, and Stowell 2008).

Third, our previous article studied Detroit, a traditionally black and white segregated city. In this study, we study the multiethnic metropolis of Chicago and offer for the first time a close look at the community perceptions of Latinos.

Data and Methods

Our analysis is based on data from the 2004–5 Chicago Area Study (CAS), a face-to-face multistage area probability sample of adults 21 years and older living in households in Cook County, Illinois. Cook County (which includes the city of Chicago) was first stratified by racial/ethnic composition based on counts from tracts in the 2000 U.S. Census, and oversamples were drawn of African Americans, Latinos, and those living in racially mixed neighborhoods. A total of 789 interviews were completed in the CAS, with a 45 percent overall response rate (based on RR4 in American Association of Public Opinion Researchers [2008]). Interviews were conducted in either English or Spanish from August 2004 through August 2005. All analyses are weighted for probability of selection and adjusted for nonresponse.

For one module of the interview, respondents were given a booklet of maps showing major roads and forty-one communities (see Figure 1 for the map used in data collection). Next to each of the forty-one areas labeled on the map were checkboxes, which allowed respondents to mark any community in which they would seriously consider looking for a house or apartment. On a separate map they were asked to mark any community in which they would never consider searching for housing. Chicago neighborhoods were defined using Community Areas (though we named South Lawndale "Pilsen/Little Village" and combined Oakland and Kenwood into "Bronzeville" on the map), and suburban communities were defined using Census Designated Places. The responses to each of these map questions constitute the dependent variables in this analysis.

We limited the map to forty-one communities to reduce respondent burden since more than 300 neighborhoods and communities exist in the Chicago metropolitan area. We purposively selected the forty-one areas to reflect a variety of communities including places inside and outside the city with a range of housing prices and a variety of racial compositions. To orient readers to the racial composition of these communities within the context of other Chicago communities, we plot the proportion of white, black, and Latino residents in Chicago Community Areas and Census Designated Places in Figure 2. For a complete overview of the comparison of neighborhoods identified on the maps and those not, see Krysan and Bader (2009). Relative to the distribution of Chicago neighborhoods, the fifteen Chicago neighborhoods labeled on the map underrepresent "all black" community areas and overrepresent racially mixed communities. The twenty-six suburban communities labeled on the map underrepresent "all-white" communities and overrepresent mostly black and racially mixed communities compared with the region's suburbs.

Analytic approach

Our initial descriptive analysis examines the proportion of respondents of each racial group that indicates they would seriously or never consider communities. We supplement our general overview presented here with detailed online materials.[1] To summarize similarities and differences across racial groups, we rank-ordered the list of communities by the proportion of respondents endorsing the community for each racial group and calculated Spearman's rank-order correlations on those rank-ordered lists.

For each dimension—seriously consider and never consider—we examined the influence of respondent race on endorsing each community by conducting a series of forty-one logistic regressions. We first conducted baseline regressions with respondent race as the only predictor before adding controls for individual-level attributes. We controlled for demographic characteristics, income, education, and the distance from the respondent to the community.

Finally, we used multilevel logistic regression models to assess the influence of race among the entire sample of communities simultaneously. Conceptually, these multilevel models combine the forty-one separate regressions into a single model to estimate the overall effect of race on endorsing communities.

FIGURE 1
Map Used to Measure Community Attraction and Avoidance, 2004–5 Chicago Area Study

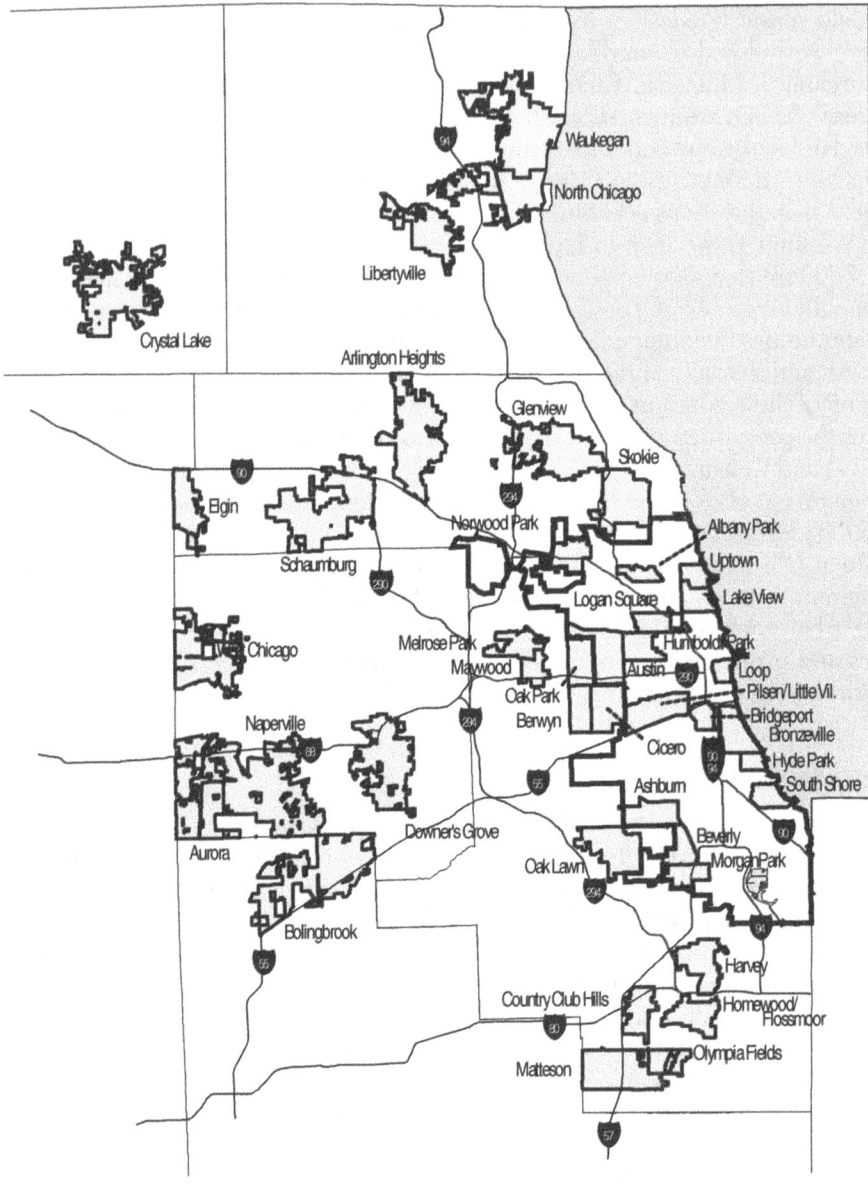

4. Where would you seriously consider looking for a house or apartment?

Conducting this analysis allowed us to model whether responses to each dimension were a function of individual characteristics of respondents, characteristics of the community being evaluated, or an interaction between the two.

FIGURE 2

Racial Composition of Chicago Area Census Designated Places and Chicago Community Areas, 2000

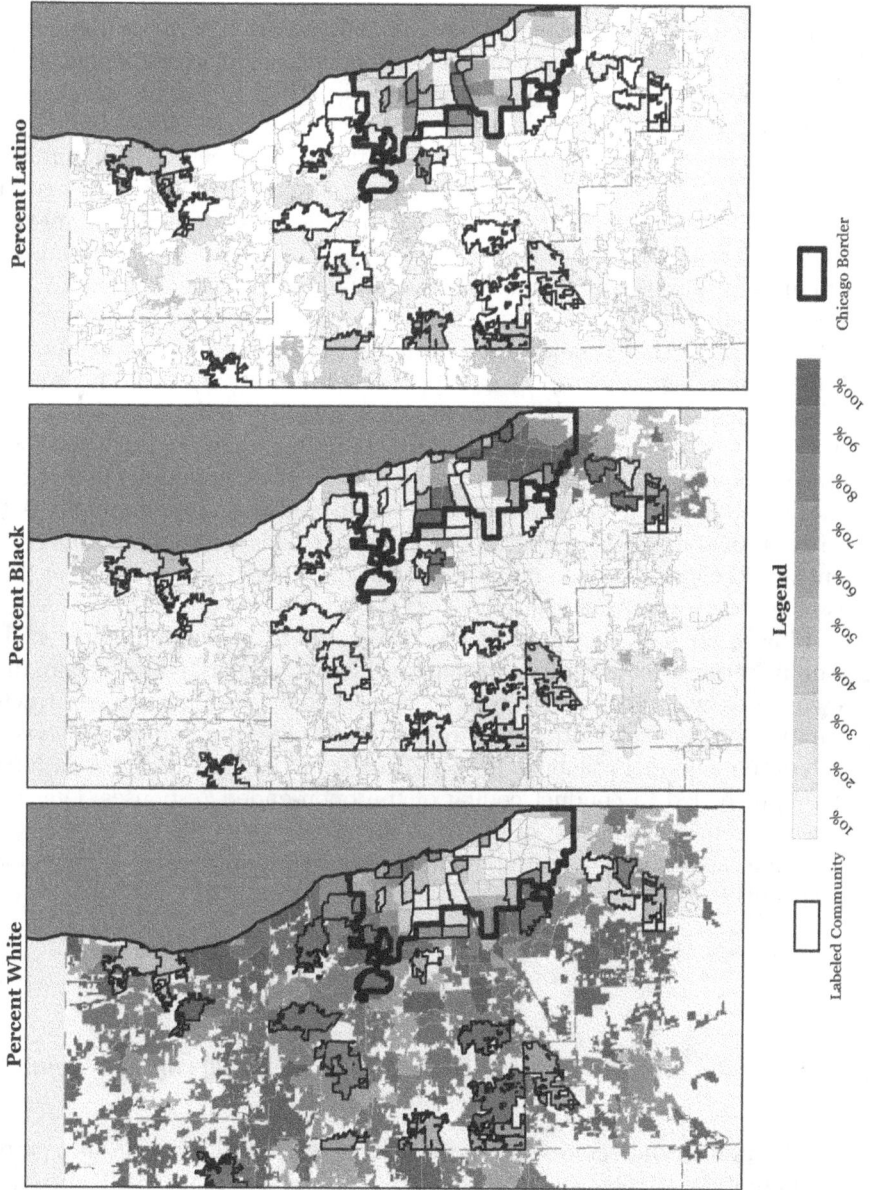

Percent White

Percent Black

Percent Latino

Legend

Labeled Community

10% 20% 30% 40% 50% 60% 70% 80% 90% 100%

Chicago Border

$$\ln\left(\frac{P_{ci}}{1-P_{ci}}\right) = \gamma_{00} + \sum_{}^{R}\gamma_{0r}X_{ri} + \sum_{}^{Q}\gamma_{q0}W_{qci} + \sum_{}^{R,Q}\gamma_{qr}X_{ri}W_{qci} + u_{0i} \qquad (1)$$

Equation 1 shows the model used to estimate each of these three types of associations. P_{ci} is the probability that respondent i endorses community c for the particular dimension (attraction or avoidance). The coefficients γ_{0r} measure the influence of each of R respondent-level variables, X_{ri}, on endorsement for individual i. The coefficients γ_{q0} measure the influence of each of Q community-level variables, W_{qci}, on endorsement for each community c that individual i evaluates. The coefficients γ_{qr} measure the relative difference of community-level variable W_{qci} for a respondent with individual-level variable X_{ri} on the endorsement of a community. This is how we measure the influence of respondent race (X_{RACEi}) on the evaluation of communities based on the racial composition of communities ($W_{RACECOMPci}$). The cross-level interactions, γ_{qr}, change the interpretation of the "main effects," γ_{0r} and γ_{q0}: γ_{0r} is the predicted level of X_r at the intercept (i.e., when W_q equals zero) and γ_{q0} becomes the level of W_q at the intercept (i.e., when X_r equals zero). For example, if X_r were an indicator where 1 = black and 0 = white and W_q equals the percent African American in the community, then γ_{0r} equals the log-odds that a black respondent would endorse the community with no African Americans, and γ_{q0} becomes the change in log-odds that a white respondent would endorse a community for each percentage point increase in the percentage of African Americans.

Finally, for a small subset of the communities, respondents who said they would "seriously consider" or "never consider" it were asked to explain, in their own words, why they felt that way. A complex coding scheme was constructed to capture the themes mentioned by respondents. Once our two research assistants achieved intercoder reliability in excess of 80 percent agreement for each theme, they proceeded to conduct coding of the open-ended responses. Our analysis includes a brief discussion of the main themes that emerged in response to the question about why selected communities were attractive or unattractive, respectively.[2]

Results

Community attraction: Where would people "seriously consider" searching for housing?

The percentages of residents endorsing the idea of living in any particular community were relatively low. They ranged from 4 percent to 20 percent. Only thirteen communities were endorsed by at least 10 percent of respondents. Whites were the choosiest racial group: only ten communities were endorsed by 10 percent or more of whites while at least 10 percent of Latinos endorsed sixteen communities and at least 10 percent of blacks endorsed eighteen

FIGURE 3

Statistically Distinguishable Racial Differences in Community Attraction With and Without Controls for Socioeconomic Status and Distance

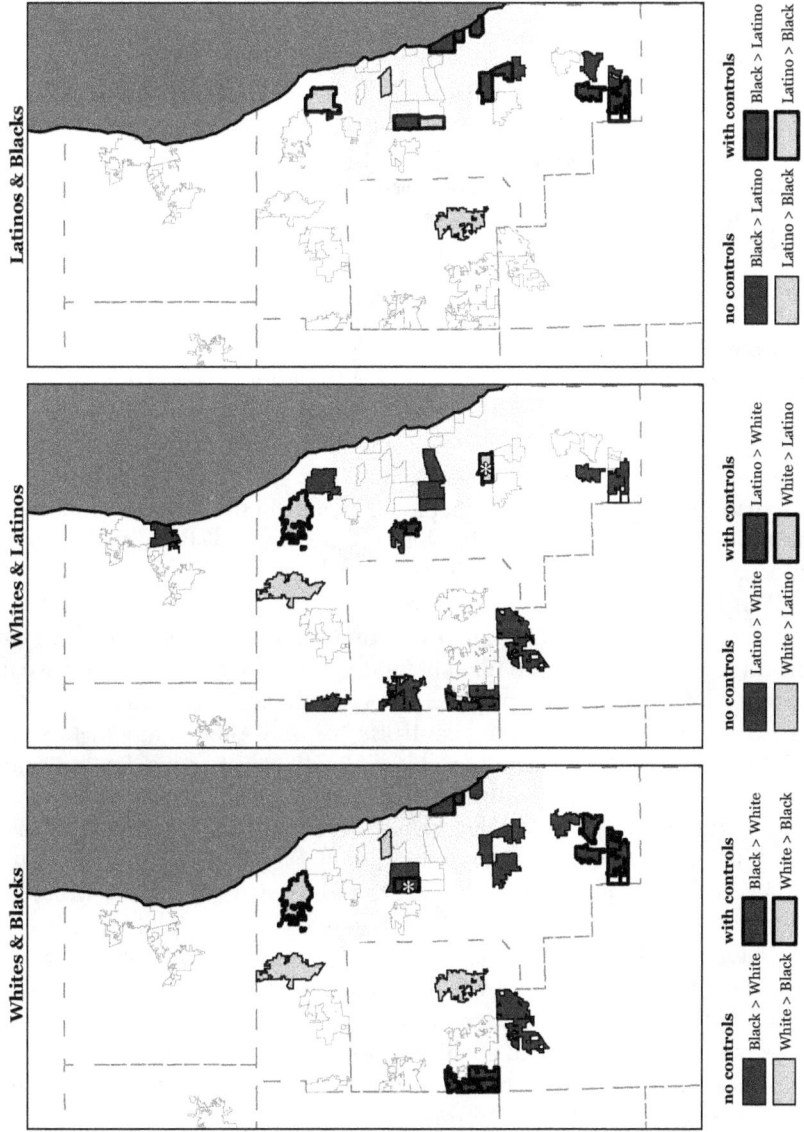

Whites & Blacks

no controls
- Black > White
- White > Black

with controls
- Black > White
- White > Black

Whites & Latinos

no controls
- Latino > White
- White > Latino

with controls
- Latino > White
- White > Latino

Latinos & Blacks

no controls
- Black > Latino
- Latino > Black

with controls
- Black > Latino
- Latino > Black

Communities with asterisks (*) indicate that there was no statistically distinguishable racial difference without controls, but a distinguishable racial difference with controls communities. The list of ten communities endorsed by at least 10 percent of whites was, with only a few exceptions, made up of predominantly white suburbs. The list of communities endorsed by at least 10 percent of Latinos and blacks was more diverse and included racially mixed and predominantly white communities in addition to communities where their own group had a substantial presence.

The Spearman correlation coefficient, which focused on the rankings of most-to least-endorsed across racial groups, showed that white and Latino endorsements were positively correlated ($\rho = .60$, $p < .001$). The ranking of communities by blacks was, however, uncorrelated with both white ($\rho = .02$, n.s.) and Latino ($\rho = -.07$, n.s.) rankings. These initial findings show that whites are more selective in where they would seriously consider living, but that there are similarities in the communities that whites and Latinos find attractive.

What happens when we control for demographics and distance?

Figure 3 graphically presents the results of the series of forty-one logistic regressions. Each of the maps compares the odds of selecting each community across two racial groups. Communities where the two groups were statistically different in their likelihood of choosing it are filled with a color based on the group more likely to consider the community. Communities outlined with a thin line had a statistically distinguishable difference in the race-only model, and communities outlined in a thick line had a statistically distinguishable racial difference after the full set of controls was added.

The left panel of Figure 3 shows racial differences in community attraction between whites and blacks. The four communities whites were more likely than blacks to find attractive are filled in light gray; three communities were predominantly white suburbs and one was the gentrifying, mostly Latino neighborhood of Logan Square. After adding controls, it is only Glenview, which is 82 percent white, where whites remain more likely than blacks to consider, indicated by the light-gray fill and heavy outline.

The fifteen communities that blacks were more likely to consider than whites are filled with dark gray in the left panel of Figure 3. Seven were in the city of Chicago: three overwhelmingly black neighborhoods, two black-white integrated neighborhoods, and two multiethnically integrated neighborhoods. The other eight were suburban communities. Among these eight are all five southern suburbs labeled on the map, of which two are predominantly black, two are black-white integrated communities with black majorities, and one is black-white integrated community with a white majority. The remaining three suburbs were two multiethnically integrated western suburbs and one nearly all-white inner-ring suburb. After adding controls, blacks were more likely than whites to consider two Chicago neighborhoods (overwhelmingly black Bronzeville and multiethnic Hyde Park), all but one southern suburb, and one multiethnic western suburb (shown with gray fill and heavy outlines). Blacks were more likely to consider integrated Oak Park (shown on the map with an asterisk) after adding controls, even though there was no distinguishable racial difference without controls. After controlling for individual attributes and geographic distance, whites

FIGURE 4

Statistically Distinguishable Racial Differences in Community Avoidance With and Without Controls for Socioeconomic Status and Distance

271

were more likely to consider one overwhelmingly white community whereas blacks were more willing to consider eight communities, including four with majority-white populations.

The middle panel of Figure 3 shows the comparison between whites and Latinos. Whites were more likely to consider two predominantly white northern suburbs, both filled with light gray, and whites remained more likely to consider one (shown with a heavy outline) after adding controls. Adding controls revealed that whites were more likely than Latinos to consider the multiethnic South Side neighborhood of Ashburn (shown with an asterisk). Latinos were more likely than whites to consider thirteen of the communities in the model without controls (communities filled with dark gray). One community remained more likely to be considered by Latinos after adding controls: the predominantly black suburb of Maywood (11 percent Latino, 84 percent black), shown with dark gray fill and a heavy outline.

The right panel in Figure 3 compares the communities that blacks and Latinos found attractive. Blacks were more likely to consider eleven communities (filled with dark gray). Six of the eleven were in the city of Chicago, all of which were on the South Side. Three were predominantly black neighborhoods, one black-white integrated community, and two three-group racially integrated communities. Four of the five suburban communities with these differences were southern suburbs. The remaining suburban community was the black-white integrated community of Oak Park. The racial differences in all but three communities were robust to controls (those with significant differences after adding controls are outlined with a heavy outline).

Community avoidance: Where would Chicago-area residents "never consider" looking for a place to live?

The percentages of residents saying they would avoid any particular community were high. Out of the forty-one communities on the map, at least 40 percent of respondents overall would never consider thirty-three, and more than 60 percent would never consider twenty-three communities. The higher prevalence of "never considering" communities is consistent with the idea that people limit their choice sets. But whites were, on average, much more likely than blacks or Latinos to dismiss all kinds of communities. Latinos were the least likely and blacks were in between. Forty percent or more of whites would avoid all but three communities. But there were only eighteen communities avoided by at least 40 percent of blacks, and only four avoided by at least 40 percent of Latinos. While 60 percent of whites avoided a majority of communities (twenty-three), there are no (0) communities avoided by 60 percent of blacks or Latinos. Although whites were more likely to never consider communities, the rank-order of communities endorsed as places whites would never consider was similar to Latinos, as indicated by the high Spearman's rank-order correlation ($\rho = .71$, $p < .001$). There is no correlation between the ordering of communities by whites and blacks ($\rho = -.13$, n.s.) and a weak correlation in Latino and black ordering ($\rho = .27$, $p < .10$).

What happens when we control for demographics and distance?

The community-by-community assessments of racial differences in community avoidance are presented in the maps in Figure 4. In the left panel, the results show that whites were more likely than blacks to avoid thirty-one of the forty-one communities (shown with light gray fill). The remaining ten communities that whites were equally likely to avoid are all northern and western suburbs, except for the North Side (predominately white) Chicago neighborhood of Norwood Park. After adding controls, statistically distinguishable results disappear in ten of the communities (those filled with light gray and outlined with a thin border): three diverse but majority white far western suburbs and a mixed white-Latino western suburb; two multiethnically diverse far northern suburbs; a majority-Latino near-in suburb; and two Chicago neighborhoods. Adding controls also revealed that whites were more likely than blacks to avoid the overwhelmingly white North Side neighborhood of Lakeview. There are no communities that blacks avoid more than whites.

The middle panel of Figure 4 reports the comparison between whites and Latinos. Whites were more likely than Latinos to never consider forty of the forty-one communities (filled with light gray on the map). The sole exception is Arlington Heights, a community whites and Latinos were equally likely to never consider. After adding demographic and geographic controls, however, whites were more likely to not consider only nine of these forty communities (those communities with a heavy outline filled with light gray). The nine communities include one racially mixed and five predominantly black Chicago neighborhoods. The remaining three were a predominantly Latino inner-ring suburb and two predominantly black southern suburbs.

The comparison of neighborhoods blacks and Latinos would avoid is in the right panel of Figure 4. Blacks were more likely than Latinos to indicate that they would never consider thirteen communities (shown with dark gray fill on the map). These include six predominantly white suburbs and three more integrated suburbs. All of these differences disappear after introducing demographic and distance controls. Introducing controls, however, revealed three communities that Latinos were more likely to avoid than African Americans (shown in light gray fill with heavy outlines on the map).

Is it just a racial proxy?

The logistic regressions suggested that racial composition influences how different racial groups evaluated communities. To answer this question systematically, we tested whether a community's racial composition had an influence on community evaluations and whether this influence persisted after controlling for levels of crime, school test scores, and other social class characteristics of the community.

Seriously consider. Column 1 of Table 1 shows the results of the models estimating individual- and community-level influences on where residents would

TABLE 1

Generalized Hierarchical Linear Model of Respondents Seriously Considering and Never
Consider Communities by Community and Individual Characteristics

	Seriously Consider	Never Consider
Individual Characteristics		
Race (reference = Non-Hispanic white)		
Non-Hispanic black	0.019	−0.359
Latino/a	−0.191	−0.801
Income (reference = $20,000–$40,000)		
Less than $20,000	−0.171	−0.019
$40,000–$70,000	−0.435	1.060°°
$80,000+	−0.491	1.204°°
Education (reference = high school degree)		
Less than *high school* degree	0.304	0.005
Some college, less than a bachelor's degree	0.454	−0.008
Bachelor's degree or higher	0.832°°°	0.177
Home owner	−0.185	−0.201
Age	−0.027°°°	0.032°
Female	−0.175	0.405
At least one child under 18 present	0.052	−0.079
Currently married	−0.281	0.098
Years lived in the Chicago metro area	0.014	0.014
Chicago resident	0.044	0.172
Community Characteristics		
Percent African American in 2000	−0.012°°	0.017°°°
Black × Percent African American in 2000	0.023°°°	−0.020°°°
Latino × Percent African American in 2000	0.003	−0.005
Percent Hispanic in 2000	−0.011°	0.018°°°
Black × Percent Hispanic in 2000	−0.009	−0.004
Latino × Percent Hispanic in 2000	0.011°	0.010
Median home value in $1,000s	0.000	0.001
Less than $20,00 × Median home value	−0.002	0.001
$40,000–$70,000 × Median home value	0.002	0.000
$80,000+ × Median home value	0.007°°°	−0.006°°°
Total population in 1,000s	0.008°°°	−0.002°
Percentage owner-occupied in 2000	−0.008°	0.010°°°
Distance to downtown Chicago (in km)	0.022°°°	−0.015°°°
Distance to community (in km)	−0.114°°°	0.083°°°
Distance to community squared (in km-sq.)	0.001°°°	−0.001°°°
School achievement	0.033°°°	−0.010°°
Avg. UCR violent crime rate per 100,000; 2004 (logged)	−0.078	−0.016
Avg. UCR property crime rate per 100,000; 2004 (logged)	0.552°°°	−0.083
Intercept	−2.101°°°	−3.460°°°

NOTE: All community-level variables except percent black and percent Latino centered at their
grand mean.
°*p* < .05. °°*p* < .01. °°°*p* < .001.

seriously consider living. The individual-level variables in the top panel of Table 1 show that residents with a BA degree or more education tended to select more communities overall, and older respondents tended to select fewer communities to seriously consider. Consistent with geographic search models, communities with larger population sizes were more likely to be "seriously considered," likely because they were better known. The farther away the community was from the respondent, the less likely a respondent was to seriously consider that community, an effect that declined as distance increases.

The cross-level interactions reveal that race influenced where people would consider moving. Whites' willingness to seriously consider neighborhoods declined rapidly as the percent black and Latino increased (reported as the "main effect" of racial composition). A 10-percent increase in the percentage of blacks or Latinos reduced the odds that a white respondent would consider the community by 11 percent and 10 percent, respectively. Blacks were much more likely than whites to consider neighborhoods as the percentage of blacks in communities increased (reported as "Black × Percent African American"), but were not statistically different from whites as the percentage of Latinos increased. Latinos were more likely than whites to consider neighborhoods with larger percentages of Latinos. The interaction ("Latino × Percent Hispanic") was the same magnitude in the opposite direction as the main effect (both have an absolute value of 0.011), which means that Latinos were largely unaffected by the percentage of co-ethnics present in communities. Latinos were not statistically different from whites in their odds of endorsing communities as the percentage of blacks increased.

These racial influences existed even in the presence of controls that are thought to explain differences in racial preferences. These include home values, school scores, and crime rates. Following economic arguments, residents with the highest incomes were more likely than those with modest incomes to consider moving to communities with higher home values. Higher school test scores were positively correlated with respondents seriously considering the community. Counter-intuitively, however, higher property crime rates were also associated with higher desirability. This association was due to the Loop, which was a very desired community with a high property crime rate. Removing the Loop eliminated this association.

Never consider. Column 2 of Table 1 reports the results for residential avoidance. A positive coefficient indicates that a respondent was more likely to "never consider" the community, meaning positive coefficients indicate increased odds of avoidance. Whites were more likely to avoid communities as both the percentage of African Americans and Latinos increased. A 10-percent increase in either increased the odds that whites avoided the community by just under 20 percent. Both blacks and Latinos were less likely than whites to avoid communities as the proportion of black residents increased (though the influence is only significant among African Americans). The statistically insignificant coefficients on both interactions with percent Latino mean that blacks, whites, and Latinos were all about equally likely to avoid a neighborhood as the proportion of Latino residents increased, even when controls for demographics, distance, and the school quality and crime characteristics of the neighborhood were added. By contrast, as the percent of African Americans increased, blacks were less likely to avoid the community relative to whites. Blacks and Latinos were as likely as whites to avoid nearly all-white communities (neighborhoods without blacks or Latinos); if anything, blacks and Latinos were less likely to avoid white communities.

Again these racial differences persisted after controlling for neighborhood characteristics. The wealthy were more likely to avoid neighborhoods with lower home values. Communities with higher school test scores were less likely to be avoided. Crime, however, was not associated with community avoidance as neither violent nor property crime rates had independent effects on whether a community would be avoided.

Consequences of the context of perceptions of place

We conclude with a final analysis that confirms the importance of perceptions of places and illustrates the context in which people develop their perceptions of places. This analysis reports on the results from open-ended follow-up questions asking respondents to explain why they would (or would not) consider two communities: Schaumburg and Homewood/Flossmoor. The responses illustrate the subtle manner in which attraction and avoidance were racialized even when the basis of responses was similar across racial groups.

In particular, the responses to Schaumburg and Homewood/Flossmoor reveal the importance of geography. A substantial number of whites said they would seriously consider Schaumburg and those who would seriously consider it explained that they found the proximity to family and friends attractive, in addition to its amenities and the quality of homes. By contrast, Schaumburg was *unlikely* to be seriously considered by blacks because they felt that it was isolated, too far from school and jobs, and that it is away from the city. The inverse was true of the racially diverse southern suburb of Homewood/Flossmoor. Blacks favored Homewood/Flossmoor because of its community environment, quality of housing, and good schools, whereas whites and Latinos avoided the community because they felt that it was too isolated, far from school and jobs, and they were unfamiliar with it. Because geography is relative to the person rating, being "proximate" or "too far" is a function of where one lives. Due to the ongoing racial segregation of Chicago, what is proximate or too far is racialized. While whites and blacks give the same reasons for their perceptions, they are literally coming at each of these communities from different places.

Discussion and Conclusion

A more thorough understanding of racial influences on the housing search process is necessary to understand continued racial segregation. We focused on the influence of race on two search-related dimensions of community perceptions:

where people would seriously consider and where they would never consider living. We find that race structures the communities that respondents find attractive and the communities that they would avoid.

Race most clearly shapes the residential perceptions of whites. Our results show that they were most likely to start their searches in white communities. Our results further suggest that failing to find a place to live in one of those communities, whites were unlikely to expand their search into communities with more than a token percentage of blacks or Latinos. This penchant for self-segregation cannot be explained by nonracial community factors such as schools or crime, as previous research has suggested (e.g., Harris 1999, 2001). Whites, we found, were likely to maintain segregation at both the initial and subsequent phases of the residential search.

Racial composition influences blacks the least. They would seriously consider a larger number of places than whites and consider a more diverse mix of communities, from nearly all-white to nearly all-black and almost everything between. Blacks also crossed off many fewer communities than did whites. Our final models showed that they were as likely as whites to consider all-white communities and not any more likely than whites to avoid all-white communities. Blacks' more expansive and inclusive choice sets contradict evidence and presuppositions that minorities would have more geographically constrained choice sets (e.g., Huff 1982). These results also undermine the idea that black self-segregation is responsible for metropolitan patterns of segregation.

Latinos are the most complicated group. Chicago-area Latinos would seriously consider quite a few places and exclude just a small set of communities. But, once we controlled for demographics and distance from the respondents' home to the community, Latinos' preferences became much like whites' preferences. This convergence might suggest that Latinos were following the classic spatial attainment model; at the same time, Latino perceptions differ enough from whites that the convergence expected by the spatial attainment model might not hold. The differences are also complicated by the fact that many whites found the largely Latino gentrifying neighborhoods in Chicago attractive.

One implication of our study is that we need more research on the housing search process. Research tends to focus on studies of residential preferences or revealed preference studies. The former are usually unmoored from reality while the latter, by showing only the final disposition, fail to reveal how inequality seeps into the process. The result is that we know very little about how searching and moving perpetuates inequality or how to intervene in the process. In the absence of empirical data on the search process, scholars fall back on the assumption of an economically rational model of housing search where people attempt to optimize their preferences.

Here, we find the major difference between the two dimensions—attraction and avoidance—instructive. Respondents were far more likely to avoid communities than seriously consider living in them. This difference is consistent with consumer choice theory that people attempt to create manageable choice sets to guide their decision-making (Schwartz 2000; Iyengar and Lepper 2000). It suggests an important way that the rational behavior of people searching for homes is bounded by existing information. It also provides a possible reason for the slippage between stated preferences and revealed outcomes that moves beyond existing explanations that focus on housing market discrimination (e.g., through racial steering) or institutions (e.g., through interracial economic inequality).

A second implication of our study is the dynamic role of geography. We find that the geography of existing segregation begets further segregation as distance affects respondents' evaluations of communities and influences how they think about different communities. Respondents found nearby communities more attractive than those farther away, just as they were more likely to avoid distant communities. This follows closely with economic search theory positing that the probability of searching in a place will be inversely proportional to the community's distance from the person and that searchers are more likely to already know about nearby communities (Clark and Smith 1979; Krysan and Bader 2009). The legacy of past segregation increases the chances that communities near respondents are likely to be ones composed largely of their own racial or ethnic group.

What this implies is that any influence of racial composition on community perceptions after controlling for geographic distance could, therefore, be considered a contemporary effect of race on perceptions of place. The legacy of past segregation structures residents' knowledge of and perceptions about communities. Part of that likely happens through social ties. Having friends or family in communities influences how people perceive those communities and makes them more or less desirable. Racial homophily in social networks might exacerbate these racialized perceptions of place and perpetuate racial segregation (see Krysan, Crowder, and Bader 2014 for more detailed argument).

We believe our results imply that the analysis of racial segregation needs to happen on two levels. On one, we need to investigate how people decide where to move. Community attraction and avoidance influence different parts of the search process and, thus, affect racial segregation differently. On the other, our findings highlight the way in which places have reputations (Semyonov and Kraus 1982).

Perception of place and the perils of policy

This research provides a warning of sorts for policy-makers. American housing policy has moved away from place-based, supply-side solutions to one in which individuals in need of housing assistance are provided vouchers so that they can "choose" where they want to live (e.g., Pattillo, Delale-O'Connor, and Butts 2014; Sharkey 2013; Goering and Feins 2003). In this context, understanding how people decide where to live becomes increasingly important. Though there is a growing body of rich research into this process among the very poor, rounding out our understanding to include the spectrum of income levels is important to understanding the complete landscape. This is a critical step if we are interested in understanding the mechanisms through which inequality is perpetuated so that we can shape policies that might help to alleviate it.

Failing to account for the perceptions of place might perpetuate inequality and concentrate poverty and affluence. If the aggregate result of individual moves reinforces existing segregation because of racialized perceptions of place and the dynamic role of geography, housing policies that encourage individual moves might exacerbate existing inequality. This is especially true because whites seek to live among whites and avoid living among minorities. Our results provide solid evidence for this while also contradicting the assertion that the particularly high levels of segregation of African Americans are due to "self-segregating" preferences (e.g., Thernstrom and Thernstrom 1997; Patterson 1997). In our study, we found at least some white communities that would be seriously considered by all three groups. But we did not find a single all/mostly white community that was excluded from consideration by a substantial number of any racial/ethnic group.

The policy inaction that stems from this assumption—that is, if people are choosing to self-segregate, then policy has no role—is equally inappropriate. Instead, our results show that African Americans would seriously consider (and in other research we show, would also actually search in) communities with a number of different types of racial compositions. Thus, racially biased proclivity among African Americans cannot explain findings demonstrating the greater likelihood of blacks to move to predominantly black neighborhoods (e.g., Crowder 2001; South and Crowder 1998, Sampson and Sharkey 2008), at least in initial stages of the search process.

Comprehending the ongoing experience of racial residential segregation as we approach the fiftieth anniversary of the Fair Housing Act requires insights not found in the traditional theoretical molds. Our study points research on racial segregation in new directions. We emphasize the need to understand the search process and provide evidence for a plausible model of search behavior. We also show the need to understand how perceptions structure place reputations. New data are needed to investigate how these processes combine to perpetuate racial segregation and potentially influence the success of American housing policy.

Notes

1. See the online supplement: http://ann.sagepub.com/supplemental.

2. Up to 30 percent of some racial groups were not asked about particular communities due to interviewer error. The randomness of the interviewer error, however, did not likely introduce a systematic error that would undermine this exploratory analysis.

References

Alba, Richard D., and John R. Logan. 1993. Minority proximity to whites in suburbs: an individual-level analysis of segregation. *American Journal of Sociology* 98 (6): 1388–427.

American Association of Public Opinion Researchers (AAPOR). 2008. *Standard definitions: Final dispositions of case codes and outcome rates for surveys.* 5th ed. Lenexa, KS: AAPOR.

Bobo, Lawrence D. 1989. Keeping the linchpin in place: Testing the multiple sources of opposition to residential integration. *International Review of Social Psychology* 2 (3): 305–23.

Brown, Lawrence A., and Eric G. Moore. 1970. The intra-urban migration process: A perspective. *Geografiska Analer. Series B, Human Geography* 52 (1): 1–13.

Charles, Camille Zubrinsky. 2000. Residential segregation in Los Angeles. In *Prismatic metropolis: Inequality in Los Angeles*, eds. Lawrence D. Bobo, Melvin L. Oliver, James H. Johnson Jr., and Abel Valenzuela Jr., 167–219. New York, NY: Russell Sage Foundation.

Charles, Camille Zubrinsky. 2003. The dynamics of racial residential segregation. *Annual Review of Sociology* 29:167–207.

Clark, William A. V. 2009. Changing residential preferences across income, education, and age: Findings from the multi-city study of urban inequality. *Urban Affairs Review* 44 (3): 334–55.

Clark, William A.V., and T. R. Smith. 1979. Modelling information use in a spatial context. *Annals of the Association of American Geographers* 69:575–88.

Crowder, Kyle D. 2001. Racial stratification in the actuation of mobility expectations: Micro-level impacts of racially restrictive housing markets. *Social Forces* 79 (4): 1377–96.

Daley, Richard M. 2010. Introduction. In *Explore Chicago: Eat, play, love our neighborhoods*. Chicago, IL: Chicago Department of Tourism.

Emerson, Michael O., Karen J. Chai, and George Yancey. 2001. Does race matter in residential segregation? Exploring the preferences of white Americans. *American Sociological Review* 66 (6): 922–35.

Goering, John M., and Judith D. Feins. 2003. *Choosing a better life? Evaluating the moving to opportunity social experiment*. Washington, DC: Urban Institute Press.

Goyette, Kimberly A., Danielle Farrie, and Joshua Freely. 2012. This school's gone downhill: Racial change and perceived school quality among whites. *Social Problems* 59:155–76.

Harris, David R. 1999. "Property values drop when blacks move in, because…": Racial and socioeconomic determinants of neighborhood desirability. *American Sociological Review* 64 (3): 461–79.

Harris, David R. 2001. Why are white and blacks averse to black neighbors? *Social Science Research* 30 (1): 100–16.

Huff, James O. 1982. Spatial aspects of residential search. In *Modelling housing market search*, ed. William A. V. Clark, 106–29. London: Croom Helm.

Iceland, John, and Rima Wilkes. 2006. Does socioeconomic status matter? Race, class, and residential segregation. *Social Problems* 53 (2): 248–73.

Immergluck, Daniel. 2002. Redlining redux: Black neighborhoods, black-owned firms, and the regulatory cold shoulder. *Urban Affairs Review* 38:22–41.

Iyengar, Sheena S., and Mark R. Lepper. 2000. When choice is demotivating: Can one desire too much of a good thing? *Journal of Personality and Social Psychology* 79 (6): 995–1006.

Krysan, Maria. 2002. Community undesirability in black and white: Examining racial residential preferences through community perceptions. *Social Problems* 49 (4): 521–43.

Krysan, Maria, and Michael D. M. Bader. 2007. Perceiving the metropolis: Seeing the city through a prism of race. *Social Forces* 86 (2): 699–733.

Krysan, Maria, and Michael D. M. Bader. 2009. Racial blind spots: Black-white-Latino differences in community knowledge. *Social Problems* 56 (4): 677–701.

Krysan, Maria, Kyle Crowder, and Michael D. M. Bader. 2014. Pathways to residential segregation. In *Choosing homes, choosing schools*, eds. Annette Lareau and Kimberly A. Goyette, 27–63. New York, NY: Russell Sage Foundation.

Krysan, Maria, Mick P. Couper, Reynolds Farley, and Tyrone Forman. 2009. Does race matter in neighborhood preferences? Results from a video experiment. *American Journal of Sociology* 115 (2): 527–59.

Lewis, Valerie A., Michael O. Emerson, and Stephen Klineberg. 2011. Residential segregation and neighborhood racial composition preferences of whites, blacks, and Latinos. *Social Forces* 89 (4): 1386–407.

Logan, John, Deirdre Oakley, and Jacob Stowell. 2008. School segregation in metropolitan regions, 1970–2000: The impacts of policy choices on public education. *American Journal of Sociology* 113 (6): 1611–44.

Logan, John R., and Brian Stults. 2011. The persistence of segregation in the metropolis: New findings from the 2010 census. Census Brief prepared for Project US2010. Available from http://www.s4.brown .edu/us2010.

MacLennan, Duncan. 1982. *Housing economics: An applied approach*. New York, NY: Longman.

Massey, Douglas S., and Nancy A. Denton. 1993. *American apartheid: Segregation and the making of the underclass*. Cambridge, MA: Harvard University Press.

Patterson, Orlando. 1997. *The ordeal of integration: Progress and resentment in America's "racial" crisis*. New York, NY: Civitas/Counterpoint.

Pattillo, Mary, Lori Delale-O'Connor, and Felicia Butts. 2014. High stakes choosing. In *Choosing homes, choosing schools,* eds. Annette Lareau and Kimberly A. Goyette, 237–67. New York, NY: Russell Sage Foundation.

Quillian, Lincoln. 2002. Why is black-white residential segregation so persistent? Evidence on three theories from migration data. *Social Science Research* 31:197–229.

Quillian, Lincoln. 2015. A comparison of traditional and discrete-choice approaches to the analysis of residential mobility and locational attainment. *The ANNALS of the American Academy of Political and Social Science* (this volume).

Quillian, Lincoln, and Devah Pager. 2001. Black neighbors, higher crime? The role of racial stereotypes in evaluations of neighborhood crime. *The American Journal of Sociology* 107:717–67.

Roscigno, Vincent J., Diana L. Karafin, and Griff Tester. 2009. The complexities and processes of racial housing discrimination. *Social Problems* 56 (1): 49–69.

Ross, Stephen L., and Margery Austin Turner. 2005. Housing discrimination in metropolitan America: Explaining changes between 1989 and 2000. *Social Problems* 52 (2): 152–80.

Sampson, Robert J., and Stephen W. Raudenbush. 2004. Seeing disorder: Neighborhood stigma and the social construction of "broken windows." *Social Psychology Quarterly* 67:319–42.

Sampson, Robert J., and Patrick Sharkey. 2008. Neighborhood selection and the social reproduction of concentrated racial inequality. *Demography* 45 (1): 1–29.

Schwartz, Barry. 2000. Self-determination: The tyranny of freedom. *American Psychologist* 55 (1): 79–88.

Semyonov, Moshe, and Vered Kraus. 1982. The social hierarchies of communities and neighborhoods. *Social Science Quarterly* 63 (4): 780–89.

Sharkey, Patrick. 2013. *Stuck in place: Urban neighborhoods and the end of progress toward racial equality*. Chicago, IL: University of Chicago Press.

Small, Mario, and Monica McDermott. 2006. The presence of organizational resources in poor urban neighborhoods: An analysis of average and contextual effects. *Social Forces* 84:1697–1724.

South, Scott J., and Kyle D. Crowder. 1998. Leaving the 'hood: Residential mobility between black, white, and integrated neighborhoods. *American Sociological Review* 63 (1): 17–26.

Squires, Gregory D. 2007. Demobilization of the individualistic bias: Housing market discrimination as a contributor to labor market and economic inequality. *The ANNALS of the American Academy of Political and Social Science* 609 (1): 200–14.

Thernstrom, Stephan, and Abigail Thernstrom. 1997. *America in black and white: One nation, indivisible*. New York, NY: Simon & Schuster.

Arab American Housing Discrimination, Ethnic Competition, and the Contact Hypothesis

By
S. MICHAEL GADDIS
and
RAJ GHOSHAL

This study uses a field experiment to study bias against living with Arab American women, a group whose position in the U.S. race system remains uncertain. We developed fictitious female white and Arab American identities and used the audit method to respond to 560 roommate-wanted advertisements in four metro areas: Los Angeles, New York, Detroit, and Houston. To focus on social—rather than purely economic—biases, all responses identified the sender as college-educated and employed and were written in grammatically correct English. We compare the number of replies received, finding that Arab-origin names receive about 40 percent fewer replies. We then model variation in discrimination rates by proximity to mosques, geographic concentration of mosques, and the percentage of Arabs living in a census tract so as to test ethnic competition theory and the contact hypothesis. In Los Angeles and New York, greater discrimination occurred in neighborhoods with the highest concentration of mosques.

Keywords: Arab Americans; audit; bias; discrimination; housing; Muslims; race

According to the U.S. census, Arab Americans are white. But theories of racial formation posit that regardless of official classification, ethnic or cultural groups can undergo a process of racialization in the wake of political, economic, or cultural changes (Omi and Winant 1994) and become marked as threatening or inferior in the popular imagination of

S. Michael Gaddis is a Robert Wood Johnson Scholar in Health Policy Research and an assistant professor of sociology at the Pennsylvania State University. His research interests include the sociology of education, education policy, racial inequality, and mental health.

Raj Ghoshal is an assistant professor of sociology at Goucher College in Baltimore, MD. He studies the role of historic U.S. racial violence in contemporary culture, racial discrimination, social movements, and U.S. politics and culture. Prior publications address social movement incubation, same-sex marriage politics, and implicit bias.

DOI: 10.1177/0002716215580095

other groups. In the wake of September 11, 2001, Arab Americans, Muslim Americans, and Americans of Middle Eastern descent were conflated into one category, and many people came to see this group as fundamentalist or violent (Jamal and Naber 2008; Panagopoulos 2006). While the key event that sparked increased negativity toward Arabs and Muslims[1] is a decade and a half in the past, negative perceptions of these groups have lingered, raising the question of to what degree the official classification of Arab Americans as white matches the social boundaries that people draw in everyday behavior.

In this article, we test to what extent Arab Americans are treated as white, as opposed to being marked as racial others, in one consequential realm. Whether people seeking roommates for already-rented apartments ignore inquiries from individuals with Arab names is significant because it reveals the extent to which Arab Americans are racialized, shapes the potential for cooperative social contact that might reduce divisions, and affects the neighborhoods that groups have access to. Because most prior research on Arab Americans' racialization has highlighted the experiences of men, and because racial biases are frequently gendered (Robnett and Feliciano 2011), we focus on bias against women. We develop fictitious racially coded names and email addresses for four white and four Arab American female identities and use the audit method to respond to several hundred roommate-wanted advertisements in the Detroit, New York, Houston, and Los Angeles areas. All our emails identify the sender as college-educated and employed, and are written in grammatically correct English. We compare the number of replies received by the two groups, examining discrimination rates in the aggregate and differences across metro areas. We find substantial bias against Arab-origin names, with the lowest levels of discrimination occurring in Los Angeles. We then use data on the location of mosques and on individuals with Arab ancestry to examine whether neighborhood proximity to Arab Americans and Muslim religious life affects Arab Americans' treatment, testing claims from ethnic competition theory and Allport's (1954) contact hypothesis. In Los Angeles and New York we find some evidence of greater discrimination in neighborhoods with the greatest concentration of mosques, a finding in line with ethnic competition theory. We close by considering why proximity to mosques may exacerbate bias.

Literature Review

Audit studies have found discrimination against various nonwhite groups in hiring and in renting apartments. For example, identical job applications with an identifiably white name rather than an identifiably black name received

NOTE: The authors are equal contributors to this article. An earlier version of this article was presented at Penn State's 2014 Stratification Conference. Chris Gaddis provided invaluable assistance with the implementation of the data collection methods. Larry D. Schoen provided access to birth record data from New York. We thank conference organizers, participants, and two anonymous reviewers for their extremely helpful comments.

50 percent more callbacks (Bertrand and Mullainathan 2004), and rental housing inquiries with African American names received 60 percent fewer responses than those with white names in one city (Carpusor and Loges 2006). There is also evidence of similar, albeit smaller, bias against Latinos (U.S. Department of Housing and Urban Development 2013). These studies are consistent with broader evidence that African Americans and Latinos face substantial economic disadvantages, such as lower incomes and wealth relative to whites, in the modern United States (Lowrey 2013). But the status of Arab Americans in the U.S. racial system is less clear: Arab Americans are officially classified as white, and Arabs' average incomes and homeownership rates are similar to whites (Asi and Beaulieu 2013). Yet they are also deeply marked as cultural "others," especially in the post–9/11 years (Bayoumi 2009). Could this group, well-integrated by most economic measures, face substantial bias in economic and social interactions?

Several prior studies have tested for bias against Arabs or Muslims in primarily *economic* relationships. In the realm of job discrimination, Widner and Chicoine (2011) found that resumes with an Arab male name received only half as many replies as identical resumes with a white male name. This meshes with evidence that Arab-Muslim men's economic security dropped after September 11, 2001, possibly due to increased reluctance by employers to engage with those of Middle Eastern origin (Kaushal, Kaestner, and Reimers 2007; Davilla and Mora 2005). Studies of bias in renting apartments have yielded similar results. Three studies found that landlords in Los Angeles County, Sweden, and London replied at lower rates to inquiries from Arab-sounding male names than to white male names (Carpusor and Loges 2006; Ahmed and Hammarstedt 2008; Carlsson and Eriksson 2014), while bias against Arabs was moderate for men and low for women seeking housing in Toronto (Hogan and Berry 2011).

Moving beyond principally economic interactions, a few studies have examined anti-Arab or anti-Muslim biases in intimate relationships. For example, Middle Eastern and South Asian men received slightly fewer replies to their messages than do white men on an online dating site, even after accounting for differences in personal characteristics (such as religiosity, gender attitudes, and so on), while Middle Eastern and South Asian women received slightly more replies than white women.[2] Robnett and Feliciano (2011) found that about two-thirds of women using Yahoo's dating site overtly excluded South Asian and Middle Eastern men as potential romantic matches, while about half of men excluded South Asian and Middle Eastern women. Robnett and Feliciano argue that negativity toward Arabs and Muslims (a largely economically successful group in the United States) challenges purely economic models of racial incorporation, instead suggesting that America's racial system is based on two separate hierarchies: one for economically valued traits and success, and another for non-"foreignness" or cultural assimilation (Xu and Lee 2013). The different treatment of Middle Eastern men and women raises the question of whether the gendering of anti-Arab biases may hold true in other types of relationships.

While these lines of research suggest biases against those seen as Arab, Muslim, or Middle Eastern in economic and intimate realms, little attention has been paid to another important arena: the formation of roommate relationships, which combine economic cooperation with close social contact. Whether individuals seeking roommates online are racially biased is an important question for three reasons. First, especially for those moving to a new city or neighborhood, the formation of roommate ties can pull outsiders into new social networks, analogous to a new friendship or romantic relationship between previously unacquainted individuals. Racial patterning in roommate searches can prefigure more integrated social networks, whose impacts can endure for decades (Granovetter 1973), just as interracial friendships can prefigure interracial romance and marriages (Shiao 2013). While most useful to the individuals directly affected, this process could also reduce social distance between Arab Americans and others at the macro level (Bogardus 1925; Parrillo and Donoghue 2005), shaping broader patterns of racialization and race relations. Second, bias against Arabs in roommate searches could directly influence inequality by affecting the type of neighborhoods they have access to. Since few young people can afford to live on their own in parts of many cities, sharing housing is a prerequisite for living in neighborhoods that promote opportunities for, and the well-being of, residents. Neighborhood characteristics are tied to access to good jobs and good schools, crime victimization and involvement, availability of health care and social services, civic engagement, mental health, and more (Massey and Denton 1993; Wilson 1987; Yinger 1995; Sampson 2012; Sharkey 2013). Third, sharing housing is common. Tens of millions of Americans, especially young Americans who might be expected to be at the vanguard of changes in race relations, live with roommates. While prior research has addressed bias against various racial groups by landlords and property managers, we are aware of no published research that considers the impact of biases of the individuals who live with those they select (or reject) as roommates.

Given the importance of roommate relationships for race relations and racial inequality, our study examines whether those who post roommate-wanted ads online are biased against replying to Arab Americans. Nearly all prior studies of anti-Arab biases focus on men, on the grounds that negative images depicting Middle Easterners as violent usually highlight men. Given this gender imbalance, one might reasonably wonder whether Arab American women encounter discrimination in social encounters to the degree that men do or are instead effectively treated more like whites. We therefore test the existence and scope of biases against female Arab names in roommate searches. Our first research question asks: *Will replies to roommate-wanted ads sent from Arab-sounding female names receive replies at a lower rate than those sent from white-sounding female names?*

Whether Arab Americans face discrimination in the aggregate when seeking roommates might depend on demographic contexts that shape others' perceptions. Two sets of theories yield different predictions about how social contact between a racialized minority group and members of the majority (or other groups) will unfold. First, ethnic competition theory (Coser 1956; Olzak 1992; Savelkoul et al. 2011) posits that when a group has been racialized, a greater presence can spur a heightened sense of threat by majority group members. For example, whites in states with large African American populations vote against policies to benefit African Americans more than whites in states with low black populations. While ethnic competition theory has been deployed in the U.S. context most extensively to explain white-black race relations, its applicability to Arab Americans is uncertain. A "high" concentration of Arabs or Muslims in a U.S. metro area would still be below 5 percent, far different than the figures for African Americans in the U.S. South. At the same time, media depictions of Muslims and Arabs as threatening were omnipresent in the wake of 9/11, raising the question of whether a sense of group threat has been decoupled from group numbers. Most of the seven American states that have banned Sharia law (Brown 2013) have low to very low Arab populations, raising the possibility that negativity may be higher in areas with fewer Muslims and Arabs.

Allport's contact hypothesis (1954) posits a somewhat different connection between a group's numbers and its social reception than does competition theory. The contact hypothesis contends that social contact erodes perceived distinctions, leading to the reduction of negative stereotypes and the softening of group boundaries. Studies have supported the contact hypothesis for white-black relations (Sigelman and Welch 1993; Pettigrew and Tropp 2006), heterosexuals' views of homosexuals (Herek and Capitanio 1996), and others. Since having more group members living in an area affords more opportunities for social contact, other things being equal, it would seem that majority group members in neighborhoods with a larger minority presence would become less prejudiced and more accepting of that minority group.

However, there are important qualifications to the contact hypothesis. As originally developed, contact must take place on largely equal terms, in service of a common goal, with cooperation between members of the two groups (but see Pettigrew and Tropp 2006, who argue that these conditions are helpful but not necessary). If whites and Arab Americans share a neighborhood but remain segregated in friendship networks, religious institutions, businesses patronized, and so on, indifference or even hostility between the groups might continue to dominate. Further, negative contact can worsen stereotypes more than positive contact can improve them (Barlow et al. 2012; Paolini, Harwood, and Rubin 2010). Research testing attitudes in the Netherlands toward Muslims has found some support for both ethnic competition and contact models (Savelkoul et al. 2011), as neighborhood copresence predicts negative attitudes toward Muslims, but this effect is reduced once Muslim populations are large enough that social contact becomes commonplace.

The racial systems of the United States and the Netherlands are quite different, and attitudes as measured in surveys are of only moderate use in predicting behavior, especially around race (Bonilla-Silva 2006). We therefore investigate how claims from the contact hypothesis and ethnic competition theory play out in actual behavior toward Arab Americans in the United States. Our second research question asks: *How will the concentration and proximity of Arabs or Muslim religious life in a neighborhood affect the treatment of inquiries from Arab names by roommate-wanted ad-posters?*

Data and Methods

Between May and August 2014, we conducted a computerized audit study to examine the different experiences of white and Arab American individuals in seeking a roommate online. We responded to 560 roommate-wanted ads on Craigslist, each with two different emails, for a total of 1,120 data points. Below we detail the three main steps to implement this field experiment: (1) creating profiles, (2) selecting a sample, and (3) conducting the audit.

Creating profiles

We created four white and four Arab American profiles, each with a racial/ethnic name and Gmail account to respond to roommate-wanted ads on Craigslist. In preliminary searches, we found that roommate-wanted ads are often restricted by gender: women frequently look only for other women as roommates (excluding men), but men often look for either gender in their search. Because most prior research on discrimination against Arabs and Muslims has used male identities, and to ensure comparability across our cases, we created only female profiles but emailed both men and women who posted ads.

We represented race through names and used multiple data sources to select theses names. First, we used birth records obtained from the New York State Department of Health to select white first names (names of children born to predominantly white mothers) and data from the U.S. Census Bureau to select white last names (U.S. Census Bureau 2008). The four white names we used were Brenda Olson, Heidi Wood, Joan Peterson, and Melany McGrath.[3] The data from New York do not record race/ethnicity beyond white, black, Hispanic, and Asian. Thus, for our four Arab American names, we used lists of common Arab and Muslim female names from the websites BabyNames.com and BehindTheName.com to generate identities. Our names were Fatima Al-Jabiri, Basimah Hadad, Iman Farooq, and Maryam Qasim. While we cannot know what share of these names are Arab or Muslim, we believe that most people seeing these names interpret them as Arab or Muslim in origin.

Selecting a sample

For many major urban areas in the United States, hundreds of new roommate-wanted ads are posted on Craigslist every day. One of our goals in this experiment was to select areas that had a sufficient number of roommate-wanted ads along with large Arab American populations and a number of mosques to explore differential response rates across the metro area. Thus, we selected four U.S. metropolitan statistical areas (MSAs) in the top ten largest Arab American populations: Detroit, Houston, Los Angeles, and New York City (U.S. Census Bureau 2014).[4]

Our sampling frame began with all ads posted in the last three days in each metro area. We then implemented some limitations. As a proxy for likely seriousness of the ads and to realistically model what a serious roommate-seeker might reply to, we (a) screened out all ads without pictures of the dwelling, (b) only responded to ads with rent above $200 per month, and (c) excluded postings by landlords instead of roommates. From the remaining ads, we selected 155 ads each at random from the Los Angeles and New York metro areas. Because fewer qualifying ads were available in the Detroit and Houston regions, we selected 65 ads each from those areas and two weeks later selected another 60 ads each, for a total of 125 ads from those areas. Ads that fit our criteria were most abundant in the New York metro area, followed by the Los Angeles (~70 percent as abundant as NYC), Houston (~25 percent), and Detroit (~15 percent) metro areas. We screened to verify that we had no duplicate ads before sending any emails.

Conducting the audit

To implement the field experiment, we randomized the order of ads for each area and assigned a unique ID to each ad. We then assigned an order for each pair (white and Arab American), so that each profile sent an email first half of the time and second the other half of the time. We assigned one of three textually different but substantively identical messages to each sender, so each email profile sent each message text one-third of the time. Each email text attempted to remove economic considerations from the response decision by including information about education (each sender reported that they were a recent college graduate) and employment (each sender reported that they had a full-time job in the area). All the email texts were written in grammatically correct English and mention being female, ensuring that any difference in assumed cultural integration or communication ability is due only to the senders' names. We waited four to six hours between sending the first and second email of each pair. Full texts of the emails are available in the online supplement.[5]

After sending each email we recorded all responses and denoted positive or potentially positive (e.g., "Yes, we can discuss this further on the phone or meet in person.") versus negative (e.g., "I am sorry, I have already found a roommate") responses, along with nonresponses. For this article, we focus on only positive or potentially positive responses.[6]

Descriptive results

Table 1 shows relevant information about roommate listings and responses in each of the four metro areas. Overall, we received at least one reply from 55.4 percent of ads, with a low of 50.4 percent in Detroit and a high of 62.6 percent in New York City. Each ad in our sample included information on price, and the vast majority of ads included geographic information, either in the form of the nearest cross-streets or other clues that helped us to pinpoint the location of the residence.[7] The average listed rental price varied considerably from $485 in Detroit to $1,131 in New York City. Full descriptive statistics comparing census

TABLE 1
Descriptive Statistics by Metro Area

Variable	Detroit	Houston	Los Angeles	New York	Total (Average)
Number of ads responded to	125	125	155	155	560
Total data points	250	250	310	310	1120
Received any response	63	65	85	97	310
	50.4%	52.0%	54.8%	62.6%	(55.4%)
Average listed rental price	$485	$606	$820	$1,131	($786)
Complete geographic data	110	110	135	151	506
	88.0%	88.0%	87.1%	97.4%	(90.4%)
Census tract Arab population	2.8%	0.6%	1.3%	1.0%	(1.3%)
Nearest mosque (in miles)	2.9	2.2	2.2	0.6	(1.9)
Total mosques within 5 miles	4.4	6.4	5.4	56.7	(20.7)
Total mosques within 3 miles	1.8	2.6	2.1	22.4	(8.2)

tracts with ads to all census tracts for each metro area are available in the online supplement.

We converted each address or cross-street location from the ads to latitude and longitude coordinates using the Geocode3 package in Stata (Bernhard 2013). We then compiled lists of mosques in each MSA using two sources cross-referenced through Google searches (Burke 2010).[8] With latitude and longitude coordinates for residences and mosques, we used the geoNear package in Stata to calculate the distance between each residence and the nearest mosque in miles and the total number of mosques within a set distance (Picard 2010). Mosque proximity measures are shown in the bottom half of Table 1. These descriptive results suggest that the geographic density in New York leads residences to be in close proximity to mosques (average distance to nearest mosque is 0.6 miles) while residencies in the other metro areas are not as near to mosques (2.9 miles in Detroit and 2.2 miles in Houston and Los Angeles).

Methods of analysis

We conduct logistic regression analyses predicting odds-ratios for our main results. These regressions control for all observed characteristics, return estimates that are weighted based on the sample size difference across MSAs, and allow for cluster-corrected standard errors at the roommate advertisement level:

$$\ln\left(\hat{p}/\left[1 - \hat{p}\right]\right) = \alpha_i + \beta_1 R_i + \beta_2 MSA_i + \beta_3 RP_i + \beta_4 AP_i + \beta_5 SO_i + \beta_6 E_i \quad (1)$$

In the equation above, α_i is the individual-level intercept, the β coefficients 1–6 represent the coefficients for race, MSA, rental price, Arab population, submission order, and email version, respectively. We include controls for Arab population (using the 2012 American Community Survey five-year estimates)

FIGURE 1
Roommate Response Rates by Racial/Ethnic Name and Location

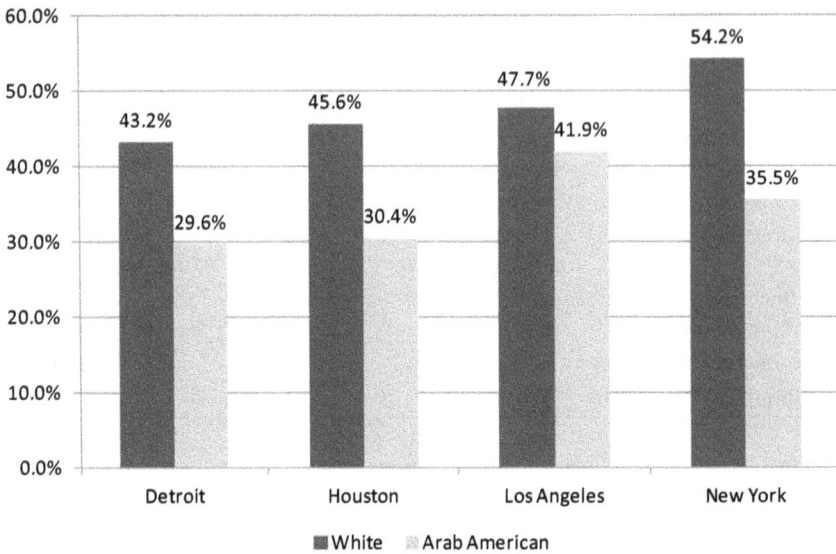

and rental price that are divided into quartiles by each MSA. Additional equations include interaction terms when appropriate.

Results

Figure 1 shows the bivariate results of responses by race. In the Detroit area, 43 percent of whites' inquiries received responses, whereas Arab Americans' response rate was 30 percent (a statistically significant difference at $p < .01$). The results are somewhat similar for Houston (46 percent to 30 percent, respectively). In Los Angeles, the response rates were much closer to parity; whites received responses at a rate of 48 percent, compared with 42 percent for Arab Americans ($p < .10$). Finally, the largest absolute gap occurred in the New York MSA: whites received responses at a rate of 54 percent, compared with only 36 percent for Arab Americans ($p < .01$).

Table 2 shows results from our logistic regression analysis. Overall, Arab Americans are only 57 percent as likely as whites to receive a response to a roommate-request email. We also include an interaction term between MSA and race, to test for differences in discrimination levels across metro areas. These results (shown in Table 3) suggest that overall response rates are higher in New York City, while Arab Americans face the least bias in Los Angeles (both marginally significant at $p < .10$). The sum total of these results so far suggests that individuals looking for a roommate on Craigslist are less likely to respond to emails from Arab Americans than whites, despite our email text controlling for

TABLE 2
Logistic Regressions Predicting Roommate Response

Variable	Odds Ratio (se)
Arab American (ref: White)	0.573***
	(0.049)
MSA: Houston (ref: Detroit)	1.061
	(0.251)
MSA: Los Angeles	1.410
	(0.316)
MSA: New York	1.410
	(0.303)
Rental price: Lowest quartile (ref: middle two quartiles)	1.103
	(0.220)
Rental price: Highest quartile	1.630**
	(0.297)
Arab population: Top quartile	0.994
	(0.160)
Submission order: second	0.896
	(0.077)
Email version: 2	0.963
	(0.122)
Email version: 3	0.873
	(0.118)
Constant	0.739
N	1120

NOTE: All cases are included. Standard errors are cluster-corrected at the roommate advertisement level.
**$p < .01$. ***$p < .001$.

potential economic and cultural integration concerns. Additionally, the level of discrimination varies somewhat by MSA, with considerably less discrimination occurring in Los Angeles.

To test predictions from ethnic competition theory and the contact hypothesis, we conducted additional logistic regression analyses that include interaction terms between Arab American respondents and (1) distance to mosque, (2) Arab population within census tract, and (3) mosque concentration. Distance to mosque captures the distance to the nearest mosque in miles from each residence, Arab population within census tract uses the percent Arab ancestry variable from the ACS, and mosque concentration calculates the number of mosques within three or five miles of the residence. We standardize each of these variables by dividing them into quartiles by MSA.

We find no significant interaction results for distance to mosque at the aggregate. Additional individual MSA-level models show a negative and significant

TABLE 3
Logistic Regressions Predicting Roommate Response with MSA × Race Interactions

Variable	Odds Ratio (se)
Arab American (ref: White)	0.549°°°
	(0.097)
MSA: Houston (ref: Detroit)	1.090
	(0.282)
MSA: Los Angeles	1.181
	(0.2891)
MSA: New York	1.536⁺
	(0.377)
Arab American × Houston	0.942
	(0.233)
Arab American × Los Angeles	1.441⁺
	(0.322)
Arab American × New York	0.836
	(0.216)
Constant	0.753
N	1120

NOTE: All completed cases are included. Models also control for rental price, Arab population within census tract, submission order, and email version. Standard errors are cluster-corrected at the roommate advertisement level.
$^{+}p < .10.$ $^{°°°}p < .001.$

interaction for proximity to mosques and responses to Arab Americans in New York: neighborhoods closer to mosques display more bias. We also find no significant aggregate interaction results for Arab population within census tract, but we find interesting trivariate patterns by MSA (shown in Figure A1 in the appendix). Arab Americans in Detroit are less likely to receive a response in high-percent Arab census tracts compared with low/mid-percent Arab census tracts (21 percent vs. 34 percent; $p < .10$). The pattern reverses in Los Angeles; Arab Americans are more likely to receive a response in census tracts with more Arab-origin residents (38 percent vs. 48 percent; $p < .10$).

Finally, we find marginally significant interaction effects for mosque concentration and Arab American respondents. Table 4 shows that Arab Americans are even less likely to receive responses from residences in the highest mosque concentration areas. Appendix Figure A2 shows that this is driven by results in two MSAs; Arab Americans in Los Angeles and New York City are less likely to receive a response in high mosque concentration areas compared with low/mid mosque concentration areas (32 percent vs. 44 percent, $p < .10$; 21 percent vs. 41 percent, $p < .05$, respectively).

These results are somewhat tentative due to differences in significant effects across the types of variables (mosque proximity, mosque concentration, and

TABLE 4

Logistic Regressions Predicting Roommate Response with Mosque Concentration × Race
Interactions

Variable	5 miles (se)	3 miles (se)
Arab American (ref: White)	0.608***	0.611***
	(0.057)	(0.061)
MSA: Houston (ref: Detroit)	0.940	0.948
	(0.240)	(0.241)
MSA: Los Angeles	1.296	1.297
	(0.310)	(0.309)
MSA: New York	1.359	1.382
	(0.303)	(0.308)
Mosque concentration: Top quartile (ref: bottom three quartiles)	1.489	0.988
	(0.378)	(0.218)
Arab American × Mosque concentration	0.598+	0.661+
	(0.181)	(0.159)
Constant	0.732	0.784
N	1012	1012

NOTE: All completed cases are included. Odds ratios shown. Models also control for rental price, Arab population within census tract, submission order, and email version. Standard errors are cluster-corrected at the roommate advertisement level.
+$p < .10$. ***$p < .001$.

percent Arab ancestry).[9] The findings suggest that ethnic competition may be at work in both Detroit and New York City since higher Arab population and higher mosque concentration, respectively, result in more Arab American discrimination. The mixed findings for Los Angeles paint an unclear picture; Arab American discrimination is lower overall, nonexistent in high-percent Arab census tracts, and higher in areas with high mosque concentration. We suggest there is more overall support for ethnic competition models than the contact hypothesis; in addition, perhaps collective memory processes are at work in New York City. We consider further the interpretation of these findings below.

Discussion

Previous research suggests that racial discrimination occurs in a variety of economic settings (Pager 2003; Bertrand and Mullainathan 2004; Gaddis, forthcoming) and that racial bias factors in romantic relationship formation (Robnett and Feliciano 2011). However few studies have systematically tested for potential discrimination when economic considerations are minimized and in nonintimate relations, as in the primarily social arena of roommate selection (although see Ghoshal and Gaddis 2014). Additionally, few studies have examined racial discrimination among other groups beyond whites, blacks, and occasionally Hispanics. In this research we set out to explore discrimination against Arab Americans by responding to roommate-wanted ads on Craigslist. We found that individuals posting ads in four major urban areas discriminate against Arab Americans when selecting potential roommates. This discrimination is lower in Los Angeles overall, but higher in Los Angeles and New York City in areas of high mosque concentration. Further, the bias is large: the Arab name penalty of about 40 percent is close in size to the penalty for African Americans seeking housing and employment found in previous research (e.g. Bertrand and Mullainathan 2004). Because we carefully controlled for our applicants' educational backgrounds, economic position, and command of English, we can be confident that different response rates are indeed due to the names' racial/ethnic connotations, a finding consistent with Ahmed, Andersson, and Hammarstedt's (2010) contention that including additional information on housing applications does little to reduce discrimination.

Our second goal in this article was to test competing predictions from the contact hypothesis and ethnic competition theory by examining differences in bias against Arab Americans based on the percentage of Arabs within a census tract, mosque proximity, and mosque concentration. Although we find no anti-Arab discrimination in the highest percentage Arab census tracts in Los Angeles, there is no additional evidence that high Arab population share, low mosque proximity, or high mosque concentration reduces discrimination. Instead, in Detroit, Los Angeles, and New York City, at least one of these characteristics seems to exacerbate discrimination.

New York City gives the clearest signal since the findings suggest that both low mosque proximity and high mosque concentration increase Arab American discrimination. Because the September 11 attacks targeted New York City, and because collective memory of major historic events is place-specific (Griffin 2004), the presence of Muslims in a New York City neighborhood may induce a sense of threat to a greater degree than elsewhere. If the September 11 attacks left a particular mnemonic mark on the New York region or on New Yorkers living near mosques, this could interfere with the positive conditions (such as groups coming together on equal terms with a sense of shared goals) necessary for contact to weaken boundaries, and instead intensify ethnic competition. This could be especially true if mosques draw many of their members from outside the neighborhood. If non-Muslim residents live in the presence of a symbol of Islam and are around Muslims before and after religious services but do not have other meaningful social contact with Muslims, they would have little opportunity to develop an understanding of Arabs or Muslims as more than members of a religion that is viewed negatively.

The finding in Detroit that discrimination is higher in census tracts with the highest percentages of Arabs may be a product of the high relative and absolute representation of Arabs in the MSA. Detroit ranks second in absolute and first in relative Arab population in the United States. None of the other MSAs rank in the top ten in relative Arab population. Recall that the percent Arab in census tract measure is divided into quartiles by MSA so that it is standardized within

MSA. Thus, it may be that the difference in Arab visibility across these neighborhoods is most noticeable in Detroit due to the MSA's overall larger relative Arab population. Future research should attempt to further test these and other measures of Arab and Muslim life as moderators of discrimination against Arab Americans.

There are a few potential shortcomings with this work so far. First is the generalizability of this field experiment to the MSAs as a whole. We suggest that this study closely matches the real-world experience of young whites and Arab Americans moving to a new urban area. Most new college graduates cannot buy a house, and many cannot afford to rent on their own, particularly in middle-class areas and in Los Angeles and New York City. These individuals likely would be forced to find roommates either through their own social networks or on a website such as Craigslist. Our field experiment attempts to replicate the experience of searching for a roommate.[10] As appendix Table A1 shows (see online supplement), the census tracts from which these ads are drawn are slightly more advantaged in a number of ways than the MSA as a whole. However, it is likely that Craigslist roommate-wanted ads rarely come from those census tracts not found in our sample. The small differences between our sample and the MSA therefore do not threaten generalizability, because potential Craigslist roommate searchers would not find housing in those omitted census tracts anyway. Additional analysis of a larger sampling frame over a longer period of time is necessary to fully adjudicate this point.

Additionally, we measured discrimination at only one stage of the roommate-seeking process—the stage of initial email contact. In the real world, roommate-seekers would usually then go on to meet one or more prospective co-renters in person, as well as perhaps talk over the telephone, check references, and more. Interactions in these later stages could trigger additional differences in outcomes. For example, if prospective Arab American renters dress differently than whites, speak with a more noticeable accent, or have references with unusual sounding names, these could trigger additional favoritism toward whites. These differences would not be solely about race "itself," but rather cultural and network attributes correlated with race. They would still, however, affect renters' access to a desired living space. Future research might consider the cumulative effects of racial and cultural favoritism at all stages of the roommate-selection process.

The widespread reluctance by many to live with Arab Americans reduces Arabs' housing opportunities and social integration. Thus, systemic harm results from the accumulation of thousands of seemingly individual, personal decisions about whom to live with. This research describes and analyzes the experiences of fictional Arab American young people in searching for housing, but what policies might help to curb the discrimination we find? Mandating race-blindness in roommate selection seems problematic: the United States has strong norms of personal choice in roommate selection, and the U.S. 9th Circuit Court of Appeals ruled in 2012 that the Fair Housing Act does not apply to roommate searches (Williams 2012). Further, enforcing nondiscrimination in roommate arrangements would be logistically near-impossible, as nonresponse is the most common strategy to screen out applicants, and it is difficult to prove racial motivation in any particular individual's selection. We therefore suggest two strategies that could reduce anti-Arab biases in roommate selection, without unduly restricting choice.

First, websites for roommates could implement a filter that restricts individuals from being able to see the names of those with whom they correspond, until they have used the site to make plans to meet (or moved beyond the stage of initial contact, in some meaningful way). Individuals who repeatedly broke plans to meet could then be penalized by the site; for example, someone who screens out many potential roommates, based on their names, could be barred from using the site for a period of time. In this way, roommate-seekers would be forced to look deeper into messages' contents, rather than immediately jumping to a name heuristic. Highly motivated discriminators might figure out how to circumvent this process, but many individuals both desire to see themselves as good people and believe that outright bias is unacceptable (Kleinman and Copp 2009). Confronting these individuals with a system that they must deliberately trick if they want to privilege their biases, rather than one in which they can immediately and easily make race-driven decisions based on names, would change their path of least resistance (Johnson 2001), potentially reducing discrimination. Recent court activity suggests that the Fair Housing Act applies to websites that facilitate roommate matching, so this type of solution could be implemented by websites such as Craigslist or Roommates.com as legal protection and compliance (Sachson 2010).

Second, while our analysis has focused on individuals within the private housing market, colleges and universities could help to facilitate roommate matches for recent graduates and older alumni to minimize racial discrimination. The bond of being from the same institution might provide just enough information to reduce the importance of stereotypes about an individual from a different race/ethnicity and increase the likelihood that the two individuals might set up an initial meeting. Some institutions already provide similar informal services through message boards, but a formal and more high profile service would likely garner more use. Roommate-matching websites or colleges could also do more to increase the salience of other kinds of nonracial information. For example, roommate search sites could ask users to complete a standard short set of questions on their interests and social habits, generate percent matching scores, and send these scores to users, reducing the power of race as a heuristic.

We close with three final suggestions for researchers interested in measuring social biases against Arab and Muslim Americans. First, because the categories "Arab" and "Muslim" are somewhat conflated in the popular imagination, our study cannot say whether the bias found against the names we used is more about perceived ethnicity or perceived religion. Research could examine what happens when candidates with Arab-sounding names include information identifying themselves as something other than Muslims, or when candidates with white-sounding names represent themselves as Muslims. This would provide more information about the extent to which information-challenging rapid categorizations can change biased behavior (Ahmed, Andersson, and Hammarstedt 2010). Second, while we studied female names in four areas with high Arab populations, research could consider whether bias is indeed worse against male names in this realm, and whether bias varies in locations with fewer Arab Americans. Finally, we used both first and last names to signal racial/ethnic identity, but some Arab-origin individuals adopt "Americanized" first names as a way of signaling cultural integration. Whether this strategy reduces others' perceptions of Arabs' cultural difference has not yet been tested. Would Brenda al-Jabiri be treated like Fatima al-Jabiri, or like Brenda Olson? Examining this question would provide additional insight into how perceived racial/ethnic origin interacts with perceived cultural integration, with implications for understanding how racialization works in the twenty-first century.

Notes

1. While the categories Arabs, Muslims, and those of Middle Eastern origin are not identical, at times we refer to members of these groups collectively as "Arab Americans," "Arabs," or "Arabs and Muslims." We do this to reference that these categories are blurred in popular understandings and for brevity. We also refer to Arab Americans as a separate group than whites, as racialization theory highlights the importance of popular understandings of race over official categories.

2. Rudder, Christian. 2014. How your race affects the messages you get. *OK trends: Dating research from OK Cupid*. Available from http://blog.okcupid.com (accessed August 2014).

3. For more information on selecting racialized names and measuring race and class origins of names, see Gaddis (forthcoming, 2014) and Ghoshal and Gaddis (2014). For more information on differences in results by name, see appendix Figure A3 in the online supplement.

4. According to responses on the ancestry question the three-year ACS estimates rank the MSAs in absolute terms of Arab population as: New York #1, Detroit #2, Los Angeles #3, and Houston #10; and in relative terms of Arab population as: Detroit #1, New York #11, Los Angeles #19, and Houston #35.

5. See http://ann.sagepub.com/supplemental.

6. Less than 5 percent of all responses were negative.

7. We were able to increase the percentage of the sample with geographic information by sending generic emails from a separate email account asking about the nearest cross-streets. Some respondents from the audit included additional geographic information in their email response as well.

8. Wikipedia. List of mosques in the United States. en.wikipedia.org/wiki/List_of_mosques_in_the_United_States (accessed August 2014).

9. Space limitations prevent us from showing all aggregate and individual MSA models. These are available from the authors upon request.

10. Although other paid websites such as Roommates.com exist, postings across the Internet on news websites, message boards, and review websites suggest that Craigslist processes the vast majority of roommate searches conducted online due to no costs and ease of use. Additionally, we can find no evidence of Arab American–specific roommate locator or matching services online.

References

Ahmed, Ali M., Lina Andersson, and Mats Hammarstedt. 2010. Can discrimination in the housing market be reduced by increasing the information about the applicants? *Land Economics* 86 (1): 79–90.

Ahmed, Ali M., and Mats Hammarstedt. 2008. Discrimination in the rental housing market: A field experiment on the Internet. *Journal of Urban Economics* 64 (2): 362–72.

Allport, Gordon. 1954. *The nature of prejudice*. Cambridge, MA: Perseus Books.

Asi, Maryam, and Daniel Beaulieu. 2013. Arab households in the United States: 2006–2010. *American Community Survey Briefs*. Available from www.census.gov.

Barlow, Fiona Kate, Stefania Paolini, Anne Pedersen, Matthew J. Hornsey, Helena R. M. Radke, Jake Hardwood, Mark Rubin, and Chris G. Sibley. 2012. The contact caveat: Negative contact predicts increased prejudice more than positive contact predicts reduced prejudice. *Personality and Social Psychology Bulletin* 38 (12): 1629–43.

Bayoumi, Moustafa. 2009. *How does it feel to be a problem? Being young and Arab in America*. New York, NY: Penguin.

Bernhard, Stefan. 2013. GEOCODE3: Stata module to retrieve coordinates or addresses from Google geocoding API version 3. Statistical Software Components S457650. Boston, MA: Boston College Department of Economics, revised 25 July 2013.

Bertrand, Marianne, and Sendhil Mullainathan. 2004. Are Emily and Greg more employable than Lakisha and Jamal? A field experiment on labor market discrimination. *American Economic Review* 94 (4): 991–1013.

Bogardus, Emory S. 1925. Social distance and its origins. *Journal of Applied Sociology* 9:216–26.

Bonilla-Silva, Eduardo. 2006. *Racism without racists: Color-blind racism and the persistence of racial inequality in the United States*. 2nd ed. Lanham, MD: Rowman & Littlefield.

Brown, Matthew. 28 August 2013. North Carolina becomes 7th state to ban Muslim Sharia law. *Deseret News*. Available from deseretnews.com.

Burke, Kevin. 2010. Mosques, USA & Canada 8.2010. Dataset. Available from http://cloud.geoiq.com/overlays/1264 (accessed 3 July 2014).

Carlsson, Magnus, and Stefan Eriksson. 2014. Discrimination in the rental market for apartments. *Journal of Housing Economics* 23:41–54.

Carpusor, Adrian G., and William E. Loges. 2006. Rental discrimination and ethnicity in names. *Journal of Applied Social Psychology* 36 (4): 934–52.

Coser, Lewis. 1956. *The functions of social conflict*. Glencoe, IL: Free Press.

Davilla, Alberto, and Marie T. Mora. 2005. Changes in the earnings of Arab men in the U.S. between 2000 and 2002. *Journal of Population Economics* 18:587–601.

Gaddis, S. Michael. Forthcoming. Discrimination in the credential society: An audit study of race and college selectivity in the labor market. *Social Forces*.

Gaddis, S. Michael. 2014. By any other name: Two experiments to examine perceptions of race and social class from names. Working Paper.

Ghoshal, Raj Andrew, and S. Michael Gaddis. 2014. Finding a roommate on Craigslist: Racial discrimination and residential segregation in urban areas. Working Paper.

Granovetter, Mark. 1973. The strength of weak ties. *American Journal of Sociology* 78 (6): 1360–80.

Griffin, Larry J. 2004. "Generations and collective memory" revisited: Race, region, and memory of civil rights. *American Sociological Review* 69 (4): 544–57.

Herek, Gregory, and John Capitanio. 1996. "Some of my best friends": Intergroup contact, concealable stigma, and heterosexuals' attitudes toward gay men and lesbians. *Personality and Social Psychology Bulletin* 22 (4): 412–24.

Hogan, Bernie, and Brent Berry. 2011. Racial and ethnic biases in rental housing: An audit study of online apartment listings. *City & Community* 10 (4): 351–72.

Jamal, Amaney, and Nadine Naber. 2008. *Race and Arab Americans before and after 9/11: From invisible citizens to visible subjects*. Syracuse, NY: Syracuse University Press.

Johnson, Allan G. 2001. *Privilege, power, and difference*. Boston, MA: McGraw-Hill.

Kaushal, Neeraj, Robert Kaestner, and Cordelia Reimers. 2007. Labor market effects of September 11th on Arab and Muslim residents of the United States. *Journal of Human Resources* 42:275–308.

Kleinman, Sherryl, and Martha Copp. 2009. Denying social harm: Students' resistance to lessons about inequality. *Teaching Sociology* 37 (3): 283–93.

Lowrey, Annie. 28 April 2013. Wealth gap among races has widened since recession. *New York Times*.

Massey, Douglas S., and Nancy A. Denton. 1993. *American apartheid: Segregation and the making of the underclass*. Cambridge, MA: Harvard University Press.

Olzak, Susan. 1992. *The dynamics of ethnic competition and conflict*. Stanford, CA: Stanford University Press.

Omi, Michael, and Howard Winant. 1994. *Racial formation in the United States: From the 1960s to the 1990s*. 2nd ed. New York, NY: Routledge.

Pager, Devah. 2003. The mark of a criminal record. *American Journal of Sociology* 108 (5): 937–75.

Panagopoulos, Costas. 2006. The polls-trends: Arab and Muslim Americans and Islam in the aftermath of 9/11. *Public Opinion Quarterly* 70 (4): 608–24.

Paolini, Stefania, Jake Harwood, and Mark Rubin. 2010. Negative intergroup contact makes group memberships salient: Explaining why intergroup conflict endures. *Personality and Social Psychology Bulletin* 36 (12): 1723–38.

Parrillo, Vincent N., and Christopher Donoghue. 2005. Updating the Bogardus social distance studies: A new national survey. *Social Science Journal* 42 (2): 257–71.

Pettigrew, Thomas F., and Linda R. Tropp. 2006. A meta-analytic test of intergroup contact theory. *Journal of Personality and Social Psychology* 90 (5): 751–83.

Picard, Robert. 2010. GEONEAR: Stata module to find nearest neighbors using geodetic distances. Statistical Software Components S457146. Boston, MA: Boston College Department of Economics, revised 22 February 2012.

Robnett, Belinda, and Cynthia Feliciano. 2011. Patterns of racial-ethnic exclusion by Internet daters. *Social Forces* 89 (3): 807–28.

Sachson, Molly. 2010. The big bad Internet: Service provider immunity under Sec. 230 to protect the private individual from unrestrained Internet communication. *Journal of Civil Rights and Economic Development* 25:353–78.

Sampson, Robert J. 2012. *Great American city: Chicago and the enduring neighborhood effect.* Chicago, IL: University of Chicago Press.

Savelkoul, Michael, Peer Scheepers, Jochem Tolsma, and Louk Hagendoorn. 2011. Anti-Muslim attitudes in the Netherlands: Tests of contradictory hypotheses derived from ethnic competition theory and intergroup contact theory. *European Sociological Review* 27 (6): 741–58.

Sharkey, Patrick. 2013. *Stuck in place: Urban neighborhoods and the end of progress toward racial equality.* Chicago, IL: University of Chicago Press.

Shiao, Jiannbin. 2013. The influence of interracial friendships on the likelihood of interracial intimacy. Paper presented at the annual meeting of the American Sociological Association. August 10–13. New York, NY.

Sigelman, Lee, and Susan Welch. 1993. The contact hypothesis revisited: Black-white interaction and positive racial attitudes. *Social Forces* 71 (3): 781–95.

U.S. Census Bureau. 2008. Census report data file A: Top 1000 names. Available from http://www.census .gov.

U.S. Census Bureau. 2014. American fact finder. Available from http://factfinder2.census.gov.

U.S. Department of Housing and Urban Development. 2013. Housing discrimination against racial and ethnic minorities 2012. Available from www.hud.gov.

Widner, Daniel, and Stephen Chicoine. 2011. It's all in the name: Employment discrimination against Arab Americans. *Sociological Forum* 26 (4): 806–23.

Williams, Carol J. 3 February 2012. Roommate-finder doesn't facilitate discrimination, court rules. *Los Angeles Times.*

Wilson, William Julius. 1987. *The truly disadvantaged.* Chicago, IL: University of Chicago Press.

Xu, Jun, and Jennifer C. Lee. 2013. The marginalized "model" minority: An empirical examination of the racial triangulation of Asian Americans. *Social Forces* 91 (4): 1363–97.

Yinger, John 1995. *Closed doors, opportunities lost: The continuing costs of housing discrimination.* New York, NY: Russell Sage Foundation.

Neighborhood Change

White Entry into Black Neighborhoods: Advent of a New Era?

By
LANCE FREEMAN
and
TIANCHENG CAI

This article considers whites' entry into black neighborhoods. The historical review in the first part of the article shows such entry to have been exceedingly rare during the twentieth century. Our analysis of trends in white entry into black neighborhoods for the period 1980–2010 documents a substantial increase in white entry for the 2000–10 decade. We speculate that the increase in white entry into black neighborhoods was due to declining racism among whites and dramatically declining crime rates in the 1990s. We also use multivariate regression to explain which black neighborhoods were most likely to experience an influx of whites. Factors associated with gentrification appear to offer the most promising explanations. We discuss the implications of these findings in the conclusion.

Keywords: segregation; gentrification; neighborhood change

Since blacks' arrival in significant numbers in cities at the beginning of the twentieth century, they have been and continue to be among the most segregated ethnic/racial groups in the developed world (Musterd 2005). While there is considerable debate about the causes of the spatial isolation of blacks, on one reason there is near unanimity—whites' avoidance of black neighborhoods. Neither the discrimination faced by blacks attempting to move into white neighborhoods nor white flight from neighborhoods into which blacks move will necessarily result in the apartheid-like landscape that

Lance Freeman is a professor and director of the Urban Planning program at Columbia University in New York City. He is also the author of There Goes the Hood: Views of Gentrification from the Ground Up *(Temple University Press 2006).*

Tiancheng Cai holds an MS in Urban Planning from Columbia University. His major research interests include land use planning and regulation, and housing and social justice.

DOI: 10.1177/0002716215578425

characterizes much of urban America without whites concomitantly avoiding black neighborhoods.

The purpose of this article is threefold. We first place white avoidance and its converse, white invasion, into historical context. Our second aim is to describe trends in white entry into black neighborhoods for the period 1980–2010. Finally, we explain the spatial variation in white entry into black neighborhoods in the 2000–10 decade. We focus on the last decade of our study period because this was a time when white entry into black neighborhoods achieved unprecedented levels. We begin by reviewing historical patterns of whites' entry into black neighborhoods. After documenting the patterns of white entry over recent decades, we turn to explaining both macrolevel secular trends as well as neighborhood-level factors that predict white entry into black neighborhoods. We conclude with a discussion of the implications of our findings for the longstanding racial divide in American cities.

Historical Trends

The Chicago School model of urban ecology predicts that social relations between groups will be reflected in spatial relations. Segregation between blacks and whites is therefore a consequence of the two groups' social distance. Urban ecology also emphasizes invasion and succession—manifesting in white flight— as the dynamic responsible for neighborhood racial turnover and persistent segregation. A critique of this perspective, the place stratification model, emphasizes discrimination by whites in maintaining segregation (Logan and Alba 1993). Yet white avoidance, in addition to the oft-written-about mechanisms of discrimination and white flight, would seem to also be a necessity for whites to maintain their spatial distance from blacks.

For much of America's urban history, avoidance was indeed a tool for maintaining segregation between whites and blacks. The racism that motivated discrimination and white flight also kept whites from invading black spaces. Late nineteenth century observers of housing conditions noted whites' studious avoidance of buildings or blocks inhabited by blacks (Riis 1890/1971). The available evidence suggests that the stigmatization of black spaces and whites' subsequent avoidance of these spaces continued into the early twentieth century, coinciding with the Great Migration. The Chicago Commission on Race Relations (1922), in an exhaustive study of the causes of the Chicago riot of July 1919, wrote that "A symptom of the general prejudice is the very prevalent belief that if Negroes have once occupied property its value is thereby 'destroyed' for white persons" (p. 205). A subsequent study of black housing during the Great Depression echoed this sentiment, reporting the feeling by white realtors that "Where Negro property of any sort is concerned, it is usually felt that the buying market is restricted to Negroes" (President's Conference on Home Building and Home Ownership 1931, 93).

The first studies that were able to systematically document white invasion found it to be exceedingly rare. Duncan and Duncan (1957) in their study of residential segregation in mid-twentieth-century Chicago found only one tract with a significant black presence where the number of whites increased. Only in a handful of southern cities was there what Taeuber and Taeuber (1965) called pure or growing displacement, where the white population actually increased in tracts with a significant nonwhite presence.

Defining the displacement of blacks as a decline in the black population of at least 5 percent, Wood and Lee (1991) found that 7.3 percent of tracts in the 1940s, 8.8 percent of tracts in the 1950s, and 10.9 percent of tracts in the 1960s experienced such displacement. Overall the trend was upward, suggesting an increased willingness of whites to move into black neighborhoods. But we should be careful not to overstate the importance of this trend. Wood and Lee (1991) included all tracts that were between 10 percent and 90 percent black.

We can also infer the rarity of whites avoiding black spaces from the lack of attention to this scenario by mid-twentieth-century integrationists. Such integrationists thought of integration in terms of creating integrated spaces anew or opening up white areas to blacks. Black spaces were assumed to be destined to remain black. Weaver (1948) advocated the development of new interracial housing developments that would not be burdened by the legacy of an already segregated identity and that would serve as models of interracial harmony. Describing the actions of the grassroots groups, two housing reformers (Levenson and Fisher 1962, 28) wrote:

> These indigenous groups of "citizens for integrated living" are working to make the goal a reality in their own communities. If they live in an "all white" section, they want it to become integrated. If they live in an interracial area, they want to keep it that way.

Notably absent from the description of the fair housing groups or Weaver's prescription was an indication that the integration of existing ghettos should be considered. Given when these items were written, this is not surprising. Blacks were for the most part hemmed into some of the most overcrowded and dilapidated housing in many of the nation's cities (Weaver 1948).

Other studies of neighborhood racial transition in other settings during the turbulent 1960s show a continued pattern of relatively little white influx into black neighborhoods. Studying racial transition in Los Angeles and San Francisco in the 1960s, Massey and Mullan (1984) found 8.3 percent and 0.9 percent, respectively, of black tracts experiencing both white gain and black loss. Even these relatively low numbers have to be discounted by their generous definition of a black tract—one that had at least 250 black residents.

From the late nineteenth century on, the empirical evidence, anecdotal reports, and the foci of the studies themselves all point to the conclusion that white entry into black spaces was exceedingly rare in the pre–civil rights era.

Invading Black Spaces in the Wake of the Civil Rights Era

With the passage of the 1968 Civil Rights Act, citizens were hopeful that the successes of the civil rights movement might begin to crack the walls of the ghetto.

Urban ecology would predict that to the extent that the civil rights reforms reflected decreasing social distance between blacks and whites, segregation would decline as well. This shift in race relations between whites and blacks thus requires a subtle shift in theorizing about the causes of white avoidance. Prior to the civil rights era, whites' animus toward blacks was sufficient to explain white avoidance. With racism in decline, however, conditions in black neighborhoods both perceived and real took on additional importance. Whites might avoid black neighborhoods because conditions were objectively worse there in many instances or because they assumed conditions were worse (Krysan et al. 2009).

Beyond improved race relations, several macrolevel forces appeared to be converging to spur what was then called "the back to the city" movement. On the demand or consumer side were shifting demographics. While the suburban amenity package of single-family homes with backyards, good schools, and shopping malls held great allure for young families with children, this package was less attractive to the growing number of childless adults. Furthermore, the counterculture movement influenced the residential tastes of many baby boomers, who looked for "authenticity" in all avenues of life including their choice of homestead. For many, sterile suburbs were the antithesis of authenticity. Instead, the older central city neighborhoods with distinctive architecture and ethnic culture were held up as real neighborhoods (Ley 1996). On the supply side of the equation, the decline and disinvestment prevalent in many inner-city neighborhoods left an abundance of affordable residential spaces that attracted professionals who often were poor (e.g., artists, writers). This back to the city movement came to be known as gentrification.

However, early studies of black to white succession in ten cities—Atlanta, Baltimore, Columbus, Dallas, New Orleans, Philadelphia, St. Paul, San Francisco, Seattle, and Washington, DC—found significant white displacement of blacks only in San Francisco and Washington, DC. In the other cities the gentrifying areas actually saw increases in their black population (Lee, Spain, and Umberson 1985).

Despite the fanfare that accompanied gentrification, systematic studies of neighborhood racial change during the 1970s tell a story of relatively infrequent white invasion into black space. Lee (1985) found that 13 percent of black neighborhoods experienced a decline in their proportion black by at least 5 percentage points between 1970 and 1980, and 72 of 3,303 neighborhoods experienced nonblack increase and black decrease between 1970 and 1980. In a second study with another colleague, Lee found that 15 percent of the neighborhoods they studied experienced a decrease in their black population (Lee and Wood 1991). Keep in mind, however, that the decreases in the black population measured by Lee and his colleagues were not necessarily accompanied by an increase in whites.

A few studies of neighborhood change in the 1980s and 1990s, while not focusing on white entry into black neighborhoods, nevertheless documented transitions from predominantly black status to a mixed black-white status. A study of 24 racially diverse metropolitan areas found that only 17 out of 556 majority black tracts in 1980 had a significant white presence (defined as at least 50 percent of the white population in the wider metropolitan area) in 2000 (Logan and Zhang 2010). Farrell and Lee (2011) in a study of tracts in the largest 100 metropolitan areas found that only 1.6 percent of majority black tracts (defined as 90 percent black) in 1990 transitioned to having a significant white presence by 2000.

Thus, from the late nineteenth century until 2000, white entry into black neighborhoods was the exception. With the advent of gentrification there was a noticeable increase in whites moving into black neighborhoods, yet overall white invasion remained exceedingly rare. While discrimination and white flight, the other instruments of segregation, fell out of favor or at least declined, white avoidance appears to have been a durable mechanism through which segregation has persisted in urban America. In the ensuing decades since scholars' work on gentrification and white invasion, there has been relatively little in the way of systematic analyses of white invasion of black neighborhoods. What is the overall trajectory of white invasion in recent decades? The most recent studies end in 2000 and none consider trends in patterns across several decades. To the extent that white invasion does occur, what are the precipitating factors? Patterns of white invasion into black neighborhoods at the end of the twentieth and beginning of the twenty-first centuries have yet to be studied—a gap this article aims to fill.

Recent Trends in White Invasion

To describe white entry into black neighborhoods during 1980–2010, we make use of the Longitudinal Tract Database. We start our analysis in 1980 because the definition of non-Hispanic whites is consistent across decades starting in 1980. As with previous studies of the late twentieth century, neighborhoods are defined as tracts. The Longitudinal Tract Database has the advantage of linking tracts in a way that accounts for tract boundary changes, resulting in comparable areal units for the entire 1980–2010 span (Logan, Xu, and Stults 2012). Black neighborhoods are defined as either 50 percent non-Hispanic black or 90 percent non-Hispanic black, respectively. Both thresholds are admittedly arbitrary, but both would certainly conform to most colloquial definitions of a black neighborhood. Moreover, the 50 percent and 90 percent thresholds have been used previously to define neighborhoods racially (Ellen 2000).

We define white invasion as an increase in the white population that represents at least 5 percent of the total population at the beginning of the decade. For example, in a tract with 1,000 persons in 2000, the white population would have had to increase by at least 50 people by 2010 to register as white invasion. We use this definition rather than the percentage increase relative to the white population because neighborhoods with very small white populations might register large percentage increases that were actually small in absolute terms.

Figure 1 graphically illustrates the fluctuations in the proportion and number of black neighborhoods experiencing white invasion between 1980 and 2010. Most notable is the surge in white invasion in the last decade. When compared with the 1980s or 1990s the first decade of the twenty-first century witnessed a

FIGURE 1
Proportion of Black Neighborhoods Experiencing White Invasion

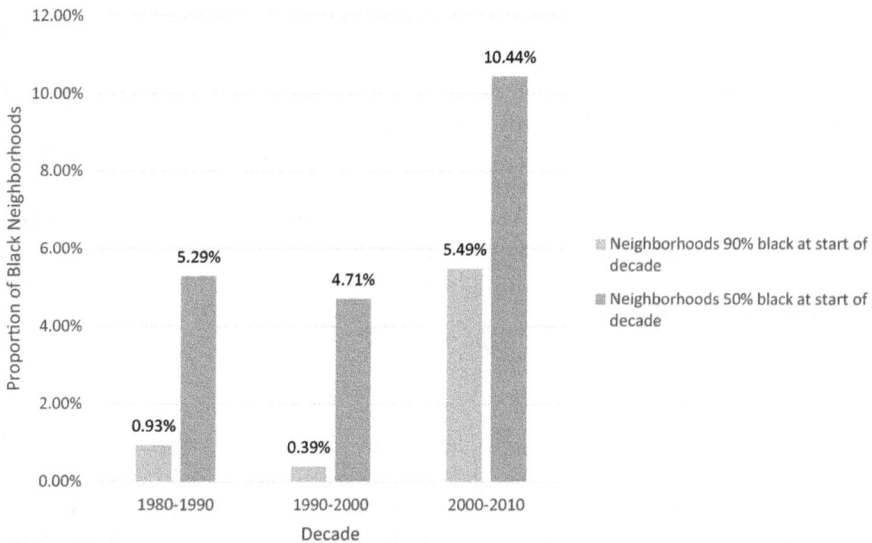

much higher proportion of black neighborhoods experiencing white invasion. This is true regardless of whether we define black neighborhoods as at least 50 percent black or at least 90 percent black. The proportion of black neighborhoods gaining a significant number of whites is more than twice as great in the last decade as it was in any previous decade for which we have data.

The dramatic increase in white entry into black neighborhoods in the 2000–10 decade raises a host of interesting questions. We consider two here. First, why was there an uptick in white invasion in this decade? Second, what explains the spatial variation in white invasion during this same period?

Why the surge in white invasion?

Why did we witness historic increases in white invasion of black spaces in the first decade of the twenty-first century? If we reverse the logic and consider why in the past whites avoided black spaces, we may gain a better understanding. Broadly speaking, whites avoided black space for two reasons: racism and the perception of inferior quality of many black neighborhoods. If whites' racist attitudes toward living among blacks changed, this could lead to more whites being willing to move into black neighborhoods. In addition, if conditions in black neighborhoods improved, whites might exhibit an increased willingness to move into these neighborhoods.

Is there any reason to believe whites' tolerance of integration increased between 2000 and 2010? Analyzing whites' responses to survey questions on integration from 1942 to 1983, Bobo, Schuman, and Steeh (1986, 163) concluded

that "the change over the past four decades has been away from both the principle and, to an extent, the practice of absolute segregation—and in this sense it
has been a genuine and large change." Evidence from the Detroit Area studies,
which asked whites how comfortable they would be living in hypothetical neighborhoods with varying representation of blacks, also shows a modest increase in
whites' willingness to live among blacks between 1976 and 1992 (Farley et al.
1993). The General Social Survey shows the proportion of white respondents
favoring laws outlawing housing discrimination rising from 37 percent to 69 percent between 1972 and 2008 (Smith, Hout, and Marsden 2013). While being
against discrimination is not the same as being for integration, Bobo, Schuman,
and Steeh (1986) have shown that these two attitudes are related. Quite possibly
this continued openness toward integration contributed to whites' willingness to
move into black neighborhoods.

Did black neighborhoods improve?

It seems plausible that the quality of life in black neighborhoods also would
have some bearing on whites' willingness to move into and remain in these neighborhoods. Some of whites' avoidance is due to negative stereotypes that do not
accurately reflect conditions in black neighborhoods (Krysan et al. 2009). Some
of that same avoidance might be due to anxieties about how residing in a black
neighborhood might affect their own status (Greenberg 1996). At least some of
whites' reticence, however, is due to the fact that on many indicators average
black neighborhoods are worse when compared with majority white neighborhoods. Black neighborhoods have been found to have higher crime rates, inferior
schools, less proximity to jobs, and, perhaps most importantly, more poverty
(Massey, Condran, and Denton 1987).

If we look at the most commonly used measure of neighborhood quality, poverty concentration, there is little evidence to suggest that black neighborhoods
are getting better. Blacks' representation in neighborhoods that were at least 40
percent poor hardly budged between the 2000 census and the 2005–9 American
Community Survey (Kneebone, Nadeau, and Berube 2011). Other research suggests that for black neighborhoods the first decade of the twenty-first century
represented a continuation of the twentieth century. Black neighborhoods continued to be marginalized and were the targets of new forms of spatial discrimination. Subprime lending, for example, has often targeted black neighborhoods
(Immergluck 2009).

There was one social shift in particular, however, that may have had major
implications for whites' invading black neighborhoods—the dramatic decline in
crime. For reasons that are not completely understood, America experienced the
greatest decline in crime since reliable records began in the twentieth century
(Zimring 2007; also see Friedson and Sharkey, this volume).

The dramatic decline in crime may be relevant here because black spaces have
been associated with criminality since blacks began migrating North in large
numbers at the end of the nineteenth century. According to Muhammad (2010),
marking blacks as inherently criminal was motivated by beliefs about black

FIGURE 2

Proportion of Black Tracts Experiencing White Invasion Based on Neighborhoods That Were at Least 50 Percent Black in 2000

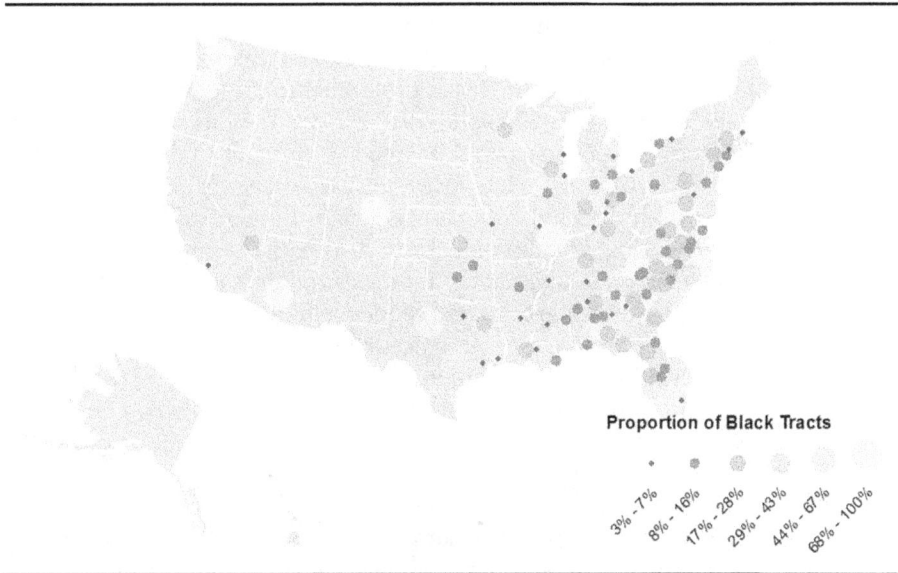

Proportion of Black Tracts

3% - 7%   8% - 16%   17% - 28%   29% - 43%   44% - 67%   68% - 100%

inferiority, and by white immigrants' criminality being explained away by the context of their circumstances. For our purposes, what is of import is the persistence of the belief that black neighborhoods are dangerous and crime ridden (Krysan et al. 2009).

Consider the well-known black neighborhood of Bedford-Stuyvesant in Brooklyn, New York. Known colloquially as "Do-or-Die Bed-Stuy" during its rougher days, this neighborhood witnessed between 1990 and 2013 a decline in homicides from 120 to 19—an 84 percent decline (Gregor 2014). Bedford-Stuyvesant is also a neighborhood that, perhaps not coincidentally, has recently experienced an influx of whites (Roberts 2011). The decline in crime in many urban centers represents a change in the quality of black neighborhoods. Many black neighborhoods were now viewed as safer, even if not as safe as white neighborhoods. In an absolute sense, black neighborhoods improved—they were safer.

Thus two macrolevel trends that may have contributed to the increase of white entry into black neighborhoods are the dramatic decline in crime coupled with decreasing antipathy toward black spaces among whites. These factors were national trends that probably influenced patterns of white invasion across the entire country. But most (89 percent or 95 percent depending on how black neighborhoods are defined) black neighborhoods did not experience white invasion. As can be seen in the map in Figure 2, white invasion was also more likely to happen in some metropolitan areas than in others. In the next section we attempt to shed light on the factors associated with white invasion into black neighborhoods.

Explaining Spatial Variation in White Invasion

In this section we attempt to model the factors likely to be associated with white invasion. We used year 2000 values for our independent variables in an attempt to predict which predominantly black neighborhoods experienced white invasion over the 2000–10 period. This approach is preferable to using the change in the independent variables between 2000 and 2010 for two reasons. First, in several instances (e.g., tract-level poverty rates) it would be difficult to discern whether neighborhood traits were driving white invasion or vice versa. Moreover, identifying exogenous factors that might explain changes in the independent variables but not white invasion would be difficult. Second, as both the dependent variable and several of the independent variables are measured only in 2000 and 2010, it would be impossible to know the temporal sequence of the observed change. For example, did the change in poverty rates occur in the beginning of the decade followed by white invasion, or did white invasion in the early part of the decade precede changes in the poverty rate? With the available data it would be impossible to infer which, if either, of these scenarios was correct.

The early urban ecology explanations of segregation and invasion and succession stressed the importance of social distance between groups. Class-based explanations of housing segregation make a similar argument, positing that the status differences between blacks and whites lead to spatial segregation between them (Bobo and Zubrinsky 1996). Because blacks are stereotyped as, and indeed are on average, poorer than whites, whites avoid black neighborhoods to avoid mixing with the lower class. At the neighborhood level this suggests that black neighborhoods of higher socioeconomic status would be more attractive to whites than other black neighborhoods. To capture neighborhood socioeconomic status we include three measures: the homeownership rate, the poverty rate, and the median household income of the black population.

Where whites harbor less racial animus, their attitudes toward integration are likely more positive and we might expect whites to be more willing to invade black space. Most residential moves are local and hence most likely to originate from the surrounding metropolitan area. Although we lack metropolitan-level measures of racial attitudes, we can use the dissimilarity index as a proxy for these attitudes. The dissimilarity index measures segregation, specifically unevenness in the spatial distribution between blacks and whites. As such, it does to some extent reflect racial attitudes in the local area.

A second theoretical perspective that can inform our explanation of white invasion comes from demand-side explanations of gentrification. As noted above, changing demographic patterns and tastes made central city living more popular among the "new middle class" starting in the late 1950s and 1960s (Spain and Laska 1980). These are not new factors and cannot by themselves explain why there was an increase in white invasion in the 2000s. But because of dramatically declining crime rates along with the slow cessation of white racism, the literature on gentrification suggests that certain neighborhoods may have been ripe for white invasion. We describe those neighborhoods below.

A key tenet of the gentrification literature is the preference for older central city neighborhoods. These neighborhoods' chief asset is proximity to a broad array of urban amenities. Operationalizing this concept with data that span the entire United States is a challenge, as the spatial configuration of American cities varies a great deal. We use the distance of the centroid of the census tract to the centroid of the central business district (CBD) for that metropolitan area to proxy for proximity to central city amenities.[1]

The long and enduring pattern of white avoidance, described above, led to blacks' housing typically being cheaper than similar housing occupied by whites by the end of the twentieth century (Cutler, Glaeser, and Vigdor 1999). We might anticipate that the "bargain" represented by housing in black neighborhoods would be especially attractive where housing prices in general are expensive relative to income. To test this hypothesis, we include the ratio of median household income to median rents in the metropolitan area as a predictor of white invasion.

Gentrification, and consequently white invasion, may have also gotten a boost from the supply side in the form of HOPE VI redevelopment schemes. Initiated in 1993, HOPE VI prompted the demolition of distressed public housing developments and often replaced this housing with mixed income, including market rate, developments (see Chaskin and Joseph, this volume). Although HOPE VI aimed to promote income—and not necessarily racial—integration, distressed developments were disproportionately black. At least some of the redeveloped sites attracted white residents (Holin et al. 2003). HOPE VI began in 1993, but most units did not come online until after 2000. We therefore test the hypothesis that a tract that was the site of a HOPE VI redevelopment scheme would have been more likely to experience white invasion.

Finally, we include controls for proximity to a majority white neighborhood, the presence of Asians and Latinos in the neighborhood, and the crime rate. We reason that whites are more likely to move into a nearby black neighborhood than one far away. We included the distance to the closest majority white census tract to test this hypothesis.

The "buffering" hypothesis suggests that whites will be less reluctant to share space with blacks if other minorities are present. This thesis has typically been put forth to explain lower segregation in multiethnic metropolises between blacks and whites (Frey and Farley 1996). Perhaps whites are more likely to move into a black neighborhood that has a sizable presence of Asians and Latinos.

Using the variation in neighborhood-level crime rates would be the best way to test whether the change in crime described above was partially responsible for the increase in white invasion. Data on neighborhood-level crime rates across the entire country, however, are not available. We therefore use the metropolitan-level crime index per 100,000 residents from the FBI Uniform Crime Reports as our measure of crime. The hypothesis being tested here is that in metropolitan areas where crime rates are lower, whites will be more willing to invade black neighborhoods.

We estimate two models, one for all tracts that were 50 percent or more black in 2000 and a second for tracts that were 90 percent or more black in 2000. We estimated the relationship between white invasion and the independent variables

TABLE 1
Descriptive Statistics

	50 Percent Black in 2000	90 Percent Black in 2000
Proportion of black neighborhoods experiencing white invasion (50% black threshold)	10.4%	5.4%
Owner occupancy rate	49.8%	47.5%
Poverty rate	27.2%	30.7%
Proportion Asian and Latino	7%	1.9%
Median household income for blacks	$28,137	$26,324
HOPE VI redevelopment	1.5%	1.7%
Distance to nearest majority white neighborhood (in Euclidean degrees)	.01	.03
Distance to nearest central business district (in Euclidean degrees)	.04	.33
Ratio of median household income to median rent	86.28	85.51
Crime rate per 100,000 persons	4,415	4,411
Dissimilarity index	.66	.70
Percent black in 2010	54.7%	73.7%
Percent white in 2010	31.7%	18.1%
Percent Hispanic in 2010	9.4%	5.2%
Percent Asian in 2010	3.4%	2.3%
n	6,179	2,039

described in the preceding paragraphs using multilevel multivariate logistic regression.[2] The average crime rate proved not to be statistically significant and was not available for approximately a quarter of the observations. Therefore, the average crime rate was dropped from the final models presented here.

Table 1 presents descriptive statistics for the independent variables along with the racial composition of the neighborhoods in 2010. As the bottom rows illustrate, despite white invasion, these neighborhoods remained predominantly black in 2010, on average.

Results

Table 2 summarizes the results of the multilevel logistic regression models. The second column reports results for neighborhoods that were at least 50 percent black in 2000 and the third column for neighborhoods that were at least 90 percent black in 2000. The relationships between the dependent and independent variables are presented as odds ratios, which range from zero to infinity. Odds ratios below one indicate that increases in the independent variable are associated with a lower probability of white invasion occurring, while odds ratios

TABLE 2
Multilevel Logistic Regression Models

	50 Percent Black in 2000	90 Percent Black in 2000
Owner occupancy rate	.19°°°	.12°°
Poverty rate	.10°°°	2.93
Proportion Asian and Latino	1.21	1276602°°°
Median household income for blacks	.99°°	1.00
HOPE VI redevelopment	1.12	.26
Distance to nearest majority white neighborhood (in Euclidean degrees)	.01°°°	.01°°°
Distance to nearest central business district (in Euclidean degrees)	.64°°°	.49°°°
Ratio of median household income to median rent	.96°°°	.99
Dissimilarity index	.02°°°	.01°°°
Intercept	30.90	409.91
Random effect	.64	1.65
n	6,141	2,036
Number of groups	319	120
Average number of observations per group	21	17

°°p < .05. °°°p < .01.

greater than one indicate that increases in the variable are associated with a higher probability of white invasion. Values farther away from one indicate a stronger relationship associated with an incremental change in the respective independent variable. The results for the two models are generally similar. We discuss the results in terms of the theoretical perspectives that motivated the inclusion of the variables in the model.

If whites avoid black spaces in part because of the social distance between blacks and whites, white invasion should be more likely to occur where social distance is less. The results provide modest support for this thesis. The dissimilarity index, our proxy for black-white relations in the larger metropolitan area, shows white invasion to be less likely in metropolitan areas that are highly segregated. However, the neighborhood-level median household income of blacks and the poverty rate had inconsistent or no relationships with the probability of white invasion. The negative relationship between the homeownership rate and white invasion contradicts the notion that whites are more comfortable with and therefore more likely to move into higher status black neighborhoods. Instead, the depressive effect of homeownership on the probability of white invasion is consistent with Ellen's (2000) argument that white renters will be more comfortable taking the risk of moving into black neighborhoods because they have less at stake than owners. The depressive effect of homeownership on the probability of white invasion is also possibly due to the lower rates of mobility among homeowners, thus leaving less opportunity for white invasion.

Explanations of white invasion as a form of gentrification suggest that inner-city neighborhoods and neighborhoods in markets where housing costs are highest would be most likely to experience white invasion. Moreover, neighborhoods that were the site of HOPE VI redevelopment schemes would have been more likely to experience gentrification and consequently white invasion. The farther away a neighborhood was from the CBD, the less likely white invasion was to occur. The greater the ratio of median household income to median rent, indicative of relatively cheaper housing, the less likely white invasion was to occur in neighborhoods that were 90 percent black in 2000. But the relationship between white invasion and housing costs is not statistically significant for neighborhoods that were 50 percent black in 2000. Whether a HOPE VI redevelopment took place in the neighborhood was not a significant predictor of white invasion. Overall, some of the evidence is consistent with gentrification, which often entails whites moving into inner-city neighborhoods in search of convenient and affordable housing.

Other controls included in the model also proved to be important predictors of white invasion. The farther a neighborhood is away from a majority white neighborhood, the lower the probability of the neighborhood experiencing white invasion. There is some support for the buffering hypothesis. For neighborhoods that were at least 90 percent black in 2000, the larger the Asian and Latino population the more likely white invasion was to occur. This pattern was not evinced, however, for neighborhoods that were at least 50 percent black in 2000.

In general, our results indicate that black neighborhoods proximate to the CBD, with low homeownership rates, not too far from white neighborhoods, and located in metropolitan areas with lower levels of segregation were most likely to experience white invasion. The neighborhood poverty rate, presence of other minorities in the neighborhood, and the affordability of housing in the metropolitan area, however, were inconsistent predictors of white invasion.

Robustness

Included in the online supplement[3] are analyses utilizing several alternative specifications of both white invasion and the independent variables. Overall, the results presented above are robust. There was a substantial increase in white invasion however defined, and black neighborhoods proximate to the CBD, not too far from white neighborhoods, and located in metropolitan areas with lower levels of segregation were most likely to experience white invasion. We also include in the online supplement a map of the number of black neighborhoods experiencing invasion.

Conclusion

As the historical narrative at the beginning of the article illustrates, white avoidance of black spaces had been the norm since the nineteenth century, and it persisted into the end of the twentieth century. Given this historical context, the surge in white invasion documented earlier represents a sharp break with the pattern of white avoidance. If the white invasion trend continues, this would force a turning point in our assumptions about the cessation of segregation. For most, the assumption was that segregation would only ease significantly when blacks were able to integrate white neighborhoods. This assumption has guided much of the policy response to desegregating America's cities. For example, HUD's initial (albeit unsuccessful) attempt to desegregate cities called for suburban jurisdictions to build housing that would be occupied by blacks (Bonastia 2000). Many of the judicial remedies put forth to settle cases that involved intentional segregation by race, such as the Gautreaux case in Chicago or the Yonkers, NY, desegregation case, also called for opening up white neighborhoods to blacks (Briggs, Darden, and Aidala 1999; Polikoff 2005). Likewise, policies to maintain and promote integration have typically focused on white neighborhoods experiencing an influx of blacks (Demarco and Galster 1993; Saltman 1990).

The findings presented in this article, however, suggest that white invasion could play a significant role in the decline of the ghetto. Although the evidence presented here does not demonstrate that policy was a major precipitating factor—HOPE VI developments were not associated with white invasion—there still is a role for policy to play. More specifically, just as the integration created by blacks moving into white neighborhoods often proved temporary, there is the risk that these black neighborhoods will soon become predominantly white. Two concerns underscore this risk.

First, news reports and qualitative accounts describe a fear of being displaced among blacks living in black neighborhoods experiencing a white influx (Courteau 2011; Freeman 2006). This suggests a need to maintain and produce affordable housing in black neighborhoods experiencing white invasion. Second, news reports and qualitative accounts have also described feelings of resentment and being "pushed out" in a figurative sense among blacks residing in neighborhoods experiencing white invasion (Freeman 2006). This residential dissatisfaction could lead to blacks leaving or avoiding these neighborhoods. Community mobilization that fosters feelings of inclusion and social cohesion in the face of rapidly changing demographics may be necessary to ensure that black neighborhoods undergoing white invasion are perceived as welcoming for people from various backgrounds.

Finally, this article lays the foundation for explanations of the upsurge in white invasion. Speculations about the role of declining crime and the easing of white racism, while perhaps intuitively appealing, need to be more rigorously tested and submitted to critical scrutiny. Likewise, the role of gentrification in driving white invasion could be assessed with more granular case studies or even qualitative accounts from whites who have moved into predominantly black neighborhoods. Only time will tell if the 2000s were a harbinger of a new era of neighborhood change or an anomalous blip unlikely to repeat in the near future.

Notes

1. In 1982 the Census of Retail Trade identified a CBD as "the downtown retail area of an SMSA central city, or other SMSA city of 50,000 persons or more. The CBD, an area of very high land valuation; high concentration of retail business, offices, hotels, and 'service' business; and high traffic flow" (U.S. Census Bureau 1982).

2. We estimated multilevel models because census tracts within the same metropolitan areas likely share unobservable characteristics that predict white invasion, thus rendering statistical tests used with conventional logistic regression inappropriate. We estimate a random slopes model because the likelihood ratio test between this model and a random slopes model failed to reject the null hypothesis that there was no difference between the two models.

3. See http://ann.sagepub.com/supplemental.

References

Bobo, Lawrence, Howard Schuman, and Charlotte Steeh. 1986. Changing racial attitudes toward residential integration. In *Housing desegregation and federal policy*, ed. John M. Goering, 152–69. Chapel Hill, NC: University of North Carolina Press.

Bobo, Lawrence, and Camille L. Zubrinsky. 1996. Attitudes on residential integration: Perceived status differences, mere in-group preference, or racial prejudice? *Social Forces* 74 (3): 883–909.

Bonastia, Chris. 2000. Why did affirmative action in housing fail during the Nixon era? *Social Problems* 47 (4): 523–42.

Briggs, Xavier de Souza, Joe T. Darden, and Angela Aidala. 1999. In the wake of desegregation: Early impacts of scattered-site public housing on neighborhoods in Yonkers, New York. *Journal of the American Planning Association* 65 (1): 27–49.

Chaskin, Robert J., and Mark L. Joseph. 2015. Contested space: Design principles and regulatory regimes in mixed-income communities replacing public housing complexes in Chicago. *The ANNALS of the American Academy of Political and Social Science* (this volume).

Chicago Commission on Race Relations. 1922. *The Negro in Chicago: A study of race relations and a race riot*. Chicago, IL: University of Chicago Press.

Courteau, Sarah L. 2011. New to the neighborhood. *Wilson Quarterly* 35 (2): 51–56.

Cutler, David M., Edward L. Glaeser, and Jacob L. Vigdor. 1999. The rise and decline of the American ghetto. *Journal of Political Economy* 107 (3): 455–507.

Demarco, Donald L., and George C. Galster. 1993. Prointegrative policy: Theory and practice. *Journal of Urban Affairs* 15 (2): 141–60.

Duncan, Otis Dudley, and Beverly Duncan. 1957. *The Negro population of Chicago*. Chicago, IL: University of Chicago Press.

Ellen, Ingrid Gould. 2000. *Sharing America's neighborhoods: The prospects for stable racial integration*. Cambridge, MA: Harvard University Press.

Farley, Reynolds, Charlotte Steeh, Tara Jackson, Maria Krysan, and Keith Reeves. 1993. Continued residential segregation in Detroit: Chocolate city, vanilla suburbs revisited. *Journal of Housing Research* 4:1–38.

Farrell, Chad R., and Barrett A. Lee. 2011. Racial diversity and change in metropolitan neighborhoods. *Social Science Research* 40 (4): 1108–23.

Freeman, Lance. 2006. *There goes the 'hood: Views of gentrification from the ground up*. Philadelphia, PA: Temple University Press.

Frey, William H., and Reynolds Farley. 1996. Latino, Asian, and black segregation in U.S. metropolitan areas: Are multiethnic metros different? *Demography* 33 (1): 35–50.

Friedson, Michael, and Patrick Sharkey. 2015. Violence and neighborhood disadvantage after the crime decline. *The ANNALS of the American Academy of Political and Social Science* (this volume).

Greenberg, Stanley. 1996. *Middle class dreams: The politics and power of the new American majority*. New Haven, CT: Yale University Press.

Gregor, Alison. 13 July 2014. Bedford-Stuyvesant: Diverse and changing. *New York Times*.

Holin, Mary Joel, Larry Burin, Gretchen Locke, and Alvaro Cortes. 2003. *Interim assessment of the HOPE VI program cross-site report*. Bethesda, MD: Abt Associates.

Immergluck, Dan. 2009. *Foreclosed: High-risk lending, deregulation, and the undermining of America's mortgage market*. Ithaca, NY: Cornell University Press.

Kneebone, Elizabeth, Carey Nadeau, and Alan Berube. 2011. *The re-emergence of concentrated poverty: Metropolitan trends in the 2000s*. Metropolitan Opportunity Series. Washington, DC: Brookings Institution.

Krysan, Maria, Mick P. Couper, Reynolds Farley, and Tyrone A. Forman. 2009. Does race matter in neighborhood preferences? Results from a video experiment. *American Journal of Sociology* 115 (2): 527–59.

Lee, Barrett A. 1985. Racially mixed neighborhoods during the 1970s: Change or stability? *Social Science Quarterly* 66:346–64.

Lee, Barrett A., Daphne Spain, and Debra J. Umberson. 1985. Neighborhood revitalization and racial change: The case of Washington, DC. *Demography* 22 (4): 581–602.

Lee, Barrett A., and Peter B. Wood. 1991. Is neighborhood racial succession place-specific? *Demography* 28 (1): 21–40.

Levenson, Frances, and Margaret Fisher. 1962. The struggle for open housing. *The Progressive* 26 (12): 25–30.

Ley, David. 1996. *The new middle class and the remaking of the central city*. New York, NY: Oxford University Press.

Logan, John R., and Richard D. Alba. 1993. Locational returns to human capital: Minority access to suburban community resources. *Demography* 30 (2): 243–68.

Logan, John R., Zengwang Xu, and Brian Stults. 2012. Interpolating U.S. decennial census tract data from as early as 1970 to 2010: A longitudinal tract database. *Professional Geographer* 66 (3): 412–20.

Logan, John R., and Charles Zhang. 2010. Global neighborhoods: New pathways to diversity and separation. *American Journal of Sociology* 115 (4): 1069–109.

Massey, Douglas S., Gretchen A. Condran, and Nancy A. Denton. 1987. The effect of residential segregation on black social and economic well-being. *Social Forces* 66 (1): 29–56.

Massey, Douglas S., and Brendan P. Mullan. 1984. Processes of black and Hispanic spatial assimilation. *American Journal of Sociology* 89 (4): 836–73.

Muhammad, Khalil Gibran. 2010. *The condemnation of blackness: Race, crime, and the making of modern urban America*. Cambridge, MA: Harvard University Press.

Musterd, Sako. 2005. Social and ethnic segregation in Europe: Levels, causes, and effects. *Journal of Urban Affairs* 27 (3): 331–48.

Polikoff, Alex. 2005. A vision for the future: Bringing Gautreaux to scale. In *Keeping the promise: Preserving and enhancing housing mobility in the Section 8 Housing Choice Voucher Program*, eds. Philip Tegeler, Mary Cunningham, and Margery Austin Turner, 137–50. Washington, DC: Poverty and Race Research Action Council.

President's Conference on Home Building and Home Ownership. 1931. *Negro housing: Report of the committee on Negro housing*. Washington, DC: President's Conference on Home Building and Home Ownership.

Riis, Jacob A. 1890/1971. *How the other half lives*. New York, NY: Dover Publications.

Roberts, Sam. 5 August 2011. Striking change in Bedford-Stuyvesant as the white population soars. *New York Times*.

Saltman, Juliet. 1990. *A fragile movement: The struggle for neighborhood stabilization*. Berkeley, CA: University of California Press.

Smith, Tom W., Michael Hout, and Peter V. Marsden. 2013. General social survey, 1972–2012 [Cumulative file]. University of Connecticut/University of Michigan Ann Arbor Roper Center for Public Opinion Research. Storrs, CT: Inter-University Consortium for Political and Social Research.

Spain, Daphne, and Shirley Laska, eds. 1980. *Back to the city: Issues in neighborhood revitalization*. Elmsford, NY: Pergamon Press.

Taeuber, Karl E., and Alma F. Taeuber. 1965. *Negroes in cities*. Chicago, IL: Aldine Publishing Company.

U.S. Census Bureau. 1982. *1982 Census of retail trade: Major retail centers in standard metropolitan statistical areas*. Washington, DC: U.S. Commerce Department.

Weaver, Robert C. 1948. *The Negro ghetto*. New York, NY: Harcourt, Brace and Company.

Wood, Peter B., and Barrett A. Lee. 1991. Is neighborhood racial succession inevitable? Forty years of evidence. *Urban Affairs Review* 26:610–23.

Zimring, Franklin E. 2007. *The great American crime decline*. New York, NY: Oxford University Press.

Gentrification in Changing Cities: Immigration, New Diversity, and Racial Inequality in Neighborhood Renewal

By
JACKELYN HWANG

This article examines how the rise of immigration and its associated racial and ethnic changes relate to gentrification. In the decades following the 1965 Hart-Celler Act, gentrification has occurred more in cities with high levels of immigration and in neighborhoods with higher levels of immigrants. These relationships, however, vary by the ways in which a city is racially segregated and by the extent to which its immigrant population has been incorporated. Using crime data, surveys, and new gentrification measures, this article compares Chicago, a highly segregated city and predominantly Hispanic immigrant destination, with Seattle, a predominantly white city with high levels of Asian immigration. The findings show that immigration and its correlates have distinct and evolving relationships with neighborhood changes that are embedded in the racial and immigrant histories of each city, and that gentrification perpetuates racial and ethnic inequality in both cities.

Keywords: gentrification; immigration; race and ethnicity; neighborhoods; gateways; inequality

Over the past several decades, many city neighborhoods that had experienced population and socioeconomic declines during the mid-twentieth century underwent reversals. Numerous studies have sought to explain this phenomenon, commonly known as "gentrification," but few have examined the role of immigration in it. Since the passage of the 1965 Hart-Celler Act, which relaxed prior immigration restrictions, Asian and Hispanic

Jackelyn Hwang is a PhD candidate in sociology and social policy and a doctoral fellow in the Multidisciplinary Program in Inequality and Social Policy at Harvard University. Her research examines the relationship between racial and ethnic inequality and neighborhood change in U.S. cities.

NOTE: This research was supported by the National Science Foundation Graduate Research Fellowship (Grant No. DGE-1144152) and the NSF-IGERT Multidisciplinary Program in Inequality and Social Policy at Harvard University (Grant No. 0333403).

DOI: 10.1177/0002716215579823

populations have increased dramatically in the United States, particularly in a handful of cities that initially served as major destinations for these groups. The influx of these new residents drastically reshaped the character and racial and ethnic compositions of the neighborhoods in which they settled.

Gentrification—middle- and upper-class residents' movement into and renewal of neighborhoods that had experienced disinvestment and decline (Smith 1998, 189)—is a process of residential selection, where households, state and corporate actors, and institutions make decisions to invest in neighborhoods. Several studies have shown that in addition to neighborhood characteristics such as school quality, housing values, the prevalence of crime, and the availability of amenities, racial and ethnic composition is an important factor in residential selection (Charles 2003). The racial and ethnic changes to neighborhoods that come with the influx of Asians and Hispanics may attract gentrification by satisfying gentrifiers' preferences for "diversity" (Zukin 1987) or by providing favored alternatives to predominantly black, low-income neighborhoods, consistent with a racial order of residential preferences persistently found in surveys (Charles 2003). The arrival of new immigrants to these declining neighborhoods can spur local economic growth through an increased demand for housing and services, entrepreneurship, and crime declines (Kirk and Laub 2010; Lee 2002; Muller 1993; Winnick 1990), stabilizing areas and potentially making them more desirable candidates for gentrification.

In this article, I present empirical evidence documenting the relationship between immigration and gentrification from 1970 to the present. Understanding how immigration and its associated racial and ethnic changes are related to gentrification sheds light on how processes of neighborhood change are associated with broader trends of neighborhood racial inequality. Hwang and Sampson (2014) demonstrated that the share of blacks and Hispanics attenuated the degree to which neighborhoods gentrified by 2009 among Chicago neighborhoods that had shown signs of gentrification in 1995 or neighborhoods that were adjacent to those showing signs of gentrification. These findings challenge the thesis that Hispanics had a positive effect on gentrification, but they only refer to recent trajectories of gentrification in a city with large and established black and Hispanic populations. I build on this prior work by extending my inquiry to 1970, when early waves of immigration and gentrification were first taking place, and across twenty-three U.S. cities. Then, I focus on two major cities with distinct racial and immigrant compositions—Chicago and Seattle. Using these two cases, I highlight the role that broader contextual features play in shaping how racial and ethnic changes associated with immigration operate differently in relation to gentrification. I conclude with a discussion of the findings and their implications for understanding urban inequality.

How Immigration Shapes Gentrification

Revitalization

Motivating my research are four perspectives on possible linkages between immigration and gentrification. According to the revitalization perspective, the drastic rise in immigration beginning in the late 1960s, and the associated growth of Asian and Hispanic populations provided a demographic renewal to declining urban neighborhoods (Winnick 1990). Having low residential and commercial rents and high vacancy rates, these neighborhoods regained a demand for housing, commercial businesses, and amenities with the arrival of new populations, consequently revitalizing these once-declining neighborhoods (Lin 1998; Muller 1993; Vigdor 2014). While not directly connected to gentrification, the revitalization perspective suggests that immigrants contributed to the economic improvement and stability of low-income neighborhoods. As a result, these neighborhoods may have become more attractive candidates for gentrification.

Neighborhood "quality"

Additional studies point to the changes that immigrants bring to characteristics associated with the quality of neighborhoods that may attract gentrification. Recent findings show a negative relationship between the in-migration of immigrants to neighborhoods and violent crime (MacDonald, Hipp, and Gill 2012). Yet immigrant concentration is also associated with lower levels of collective efficacy and greater residential instability—characteristics associated with higher levels of crime and disorder (Sampson, Morenoff, and Earls 1999). While the mechanisms responsible for this negative immigrant-crime relationship are not yet well understood, lowering crime levels, particularly violent crime, may have made neighborhoods more likely to gentrify (Kirk and Laub 2010).

Race-based residential preferences

At the same time, the influx of immigrants (particularly Asians and Hispanics) over the past 50 years has altered the racial and ethnic compositions of neighborhoods. Scholarship on residential selection finds that a racial hierarchy exists in which people generally favor white neighbors the most, black neighbors the least, and Asian and Hispanic neighbors in the middle (Charles 2003). Using vignettes and video-computer-assisted self-interviews, research shows that whites' outgroup prejudice toward blacks and Hispanics drives residential preferences beyond factors such as social class and crime (Krysan et al. 2009; Lewis, Emerson, and Klineberg 2011). While such prejudice exists, it may take the form of implicit biases or stereotyping toward minorities and minority neighborhoods rather than explicit racial prejudice, such that residents perceive more crime and disorder in neighborhoods with greater shares of blacks and Hispanics, independent of poverty and actual crime rates (Ellen 2000; Quillian and Pager 2001; Sampson and Raudenbush 2004). Therefore, among low-income neighborhoods, the growth of Asian and Hispanic neighborhoods or the influx of Asians and Hispanics to black neighborhoods may make neighborhoods more desirable candidates for gentrification relative to predominantly black neighborhoods, though less so for Hispanics. Such patterns are also consistent with qualitative accounts of gentrifiers' taste for racial and ethnic diversity (e.g., Zukin 1987), but these accounts suggest that both white and black neighborhoods with Asian or Hispanic growth are more likely to gentrify than white and black neighborhoods with no racial change. Nonetheless, evidence of rising nativism in more recent years contradicts this view (Sanchez 1997) and suggests that these compositional changes would make neighborhoods less likely to gentrify.

Context of incorporation

Last, the structure of race relations and the history of immigration in cities may also affect how immigration influences gentrification. Immigration flows have been unevenly spread among cities. Through the latter half of the twentieth century, immigration was largely concentrated in only a handful of cities and also diverged between Asians and Hispanics (Flippen and Kim 2015). In traditional immigrant destinations, native residents may be more comfortable with immigrants given that they have settled in these cities for longer periods of time and would therefore elicit less negative responses in residential selection processes (Hall 2013). But in these settings, Asians and Hispanics might also be more likely to move to coethnic neighborhoods rather than serving as pioneers in declining neighborhoods. On the other hand, disadvantaged minority groups with long-standing histories in cities may foster reputations and stigmas that would affect residential selection. Where ethnic enclaves are not yet established, however, the racial and ethnic changes associated with immigration increase the availability of heterogeneous neighborhoods, which in turn may encourage higher mobility into these neighborhoods (Crowder, Pais, and South 2012) while satisfying gentrifiers' preferences for diversity.

Framework and Approach

Taken together, the existing research proposes various pathways through which immigration might influence the trajectories of gentrification in low-income neighborhoods. While population changes may bring economic improvements and crime declines that can make neighborhoods more likely to gentrify, they may also bring declines in neighborhood cohesion and stability, which can make neighborhoods less likely to gentrify. In addition, immigration alters the racial and ethnic composition of neighborhoods that may be more or less favorable to gentrifiers depending on the larger context into which they arrive.

To assess the immigration-gentrification relationship, I first document the relationship between foreign-born, Asian, and Hispanic populations and gentrification during the late twentieth century using census data from 1970 to 2000 and field surveys of gentrification conducted from 1994 to 2001. This timeframe allows me to assess the influence of the initial rise of immigration following the 1965 legislative changes in the wake of urban decline. Second, I focus on two cities—Chicago and Seattle—and use more detailed and recent data on neighborhood characteristics from crime records, neighborhood surveys, and recent measures of observed gentrification to examine the relationships between immigration-fueled racial, ethnic, and nativity changes and neighborhood trajectories in the last two decades. By the 1990s, Hispanics and Asians had become established in particular destinations, and gentrification arguably began to take a new form, becoming more widespread and rapid (Hackworth and Smith 2001). Both Chicago and Seattle have experienced substantial immigration since 1965, but this immigration has been predominantly Hispanic in Chicago and predominantly Asian in Seattle. By focusing on two cities with distinct racial and immigrant compositions, I examine how the relationship between immigration and neighborhood trajectories varies by the broader context in which neighborhood changes take place.

Early Pathways of Gentrification in the United States

Gentrification field surveys

I use field surveys conducted by geographers Daniel J. Hammel and Elvin K. Wyly from 1994 to 2001 in twenty-three U.S. cities as measures of early waves of gentrification. Measuring gentrification across multiple neighborhoods has been notably difficult. Most large sample studies of gentrification use census- or administrative-based variables to measure gentrification, but these strategies lack direct indicators of neighborhood upgrading and are unable to distinguish gentrification from other forms of neighborhood ascent. In particular, census-based strategies neglect the distinctly visible changes to the urban landscape that are inherent to gentrification.

Hammel and Wyly (1996; see also Wyly and Hammel 1999) addressed this issue by surveying previously low-income and thus "gentrifiable" census tracts—those having median incomes below the citywide median in 1960 in the Northeast and Midwest and in 1970 in the South and West.[1] They reviewed scholarly research, city planning documents, and local press to develop a list of gentrified neighborhoods and triangulated these sources with block-by-block field surveys conducted once in each city, in which raters walked through neighborhoods documenting visible evidence of housing reinvestment and class turnover, giving particular attention to structural improvements and new construction. They categorized census tracts as gentrifying if they had a minimum of one improved structure on a majority of blocks and at least one block in the tract with at least one-third of the structures improved, and gentrifiable tracts without visible signs of reinvestment or renewal were considered ungentrified.[2] While census variables associated with higher socioeconomic status correlated well with gentrifying tracts, about 10 percent of tracts would have been incorrectly classified as gentrified when using only these census variables to predict gentrification (Hammel and Wyly 1999). The final database consists of 2,087 "gentrifiable" census tracts with nonzero populations of which 359 were gentrifying; 2,968 tracts were not gentrifiable. Although I cannot detect when gentrification began in these tracts, census data show that gentrifying tracts had relatively higher socioeconomic indicators as early as 1970. Thus, it is likely that the majority of these tracts had been gentrifying for many years prior to Hammel and Wyly's investigation.

TABLE 1
Average Probability of Gentrification across Cities by Immigration Levels

Immigration Level	Cities	Probability of Gentrification
High (% foreign-born > 1.5 × national avg., 1970–2000)	Boston	26.2
	Chicago	
	Oakland	
	San Diego	
	San Francisco	
	San Jose	
	Seattle	
Middle (% foreign-born < 1.5 × national avg. & % foreign-born > national avg., 1970–2000)	Dallas	16.0
	Denver	
	Detroit	
	Fort Worth	
	Milwaukee	
	Minneapolis–St. Paul	
	Philadelphia	
	Phoenix	
	Washington, DC	
Low (% foreign-born < national avg., 1970–2000)	Atlanta	12.7
	Baltimore	
	Cincinnati	
	Indianapolis	
	Kansas City	
	New Orleans	
	Saint Louis	

NOTE: Percent foreign-born national averages are 4.7, 6.2, 7.9, and 11.1 in 1970, 1980, 1990, and 2000, respectively.

Immigration and gentrification across cities

Table 1 lists the twenty-three cities surveyed by Hammel and Wyly (1996), categorized into three levels of immigration and the average probabilities of tract gentrification for each of these categories. Immigration levels are based on the share of foreign-born residents in the city relative to national averages from 1970 to 2000. The average probability of gentrification was much higher in cities with relatively high levels of immigrants compared with cities with lower shares of immigrants.

Immigration and gentrification across neighborhoods

Next, I assessed the relationship between immigration and gentrification at the neighborhood level. I used logistic regression models predicting the likelihood that a neighborhood was gentrifying by the time of the field survey on the foreign-born population (logged) in the first model and populations (logged) of racial and ethnic groups in the second model. I use separate models for nativity status and racial and ethnic groups because census data before 2007 do not separate nativity status by race and ethnicity, and the Hispanic and Asian populations are highly correlated with the foreign-born population since 1970. Models 3 and 4 include interaction terms with the black population to test whether the racial and ethnic changes that these residents brought to low-income, black neighborhoods are associated with gentrification. I also controlled for socioeconomic status (poverty rates, logged household income) and residential stability (ownership rates, vacancy rates, percent over 65 years old, percent living in the same residence five years ago), as well as unobserved differences between cities using fixed effects models.

Table 2 presents results predicting the likelihood of gentrification using census data linearly interpolated to 24 years prior to the gentrification field survey—the longest period of time for which census data are available for all twenty-three cities.[3] The results show that foreign-born, Asian, and Hispanic residents are positively associated with the likelihood of gentrification, even after accounting for the preexisting differences in socioeconomic status and residential stability. Moreover, these effects are greater relative to blacks and stronger in neighborhoods with higher black populations. Alternative models show that Asians and Hispanics have smaller effects relative to whites and no effects in neighborhoods with higher white populations, which suggests that a racial hierarchy of residential selection is also at work.

The Context of Immigration and Gentrification: Past and Present Pathways

While immigration and its associated racial and ethnic changes played a role in earlier waves of gentrification across the twenty-three cities, the results show that there are differences among cities. In the following analysis, I examine two cities with high immigration levels but distinct immigrant and racial compositions—Chicago and Seattle. First, I briefly describe the demographic history of these cities and present descriptive statistics for tracts that were gentrifying and those that were not by the time Hammel and Wyly surveyed them. Then, I describe additional data used to assess the relationship between immigration and neighborhood change more closely. Finally, I use these data to examine how immigration and its related racial and ethnic changes are associated with neighborhood characteristics and trajectories of gentrification in recent years.

Chicago

Throughout the twentieth century, Chicago served as a traditional destination for immigrants and continues to do so today. Since the 1965 legislative changes

TABLE 2
Regression Results with City Fixed Effects Predicting Gentrification on Nativity, Race, and Ethnicity

	Model 1		Model 2		Model 3		Model 4	
	coef.	se	coef.	se	coef.	se	coef.	se
Total population (logged)	−0.43**	(0.16)	−0.39*	(0.15)	−0.20	(0.17)	−0.32*	(0.15)
Foreign-born population (logged)	0.31**	(0.09)			0.13	(0.10)		
Asian population (logged)			0.13**	(0.04)			0.09†	(0.05)
Hispanic population (logged)			0.21**	(0.06)			0.16*	(0.07)
Black population (logged)	−0.06	(0.05)	−0.09†	(0.05)	−0.10*	(0.05)	−0.10*	(0.05)
% below poverty	1.09	(1.03)	0.73	(1.04)	1.41	(1.05)	1.09	(1.05)
Median household income (logged)	1.30**	(0.34)	1.26**	(0.34)	1.32**	(0.34)	1.31**	(0.35)
% ownership	−1.66**	(0.37)	−1.66**	(0.37)	−1.87**	(0.38)	−1.81**	(0.38)
% vacant	6.16**	(1.36)	5.76**	(1.35)	6.17**	(1.38)	5.62**	(1.36)
% over 65 years old	9.98**	(1.16)	10.66**	(1.19)	10.13**	(1.17)	10.74**	(1.20)
% same residence 5 years ago	−4.70**	(0.61)	−4.65**	(0.62)	−4.29**	(0.62)	−4.40**	(0.63)
Foreign-born × black					0.11**	(0.03)		
Asian × black							0.06**	(0.02)
Hispanic × black							0.06**	(0.02)
AIC	1428		1426		1416		1415	

NOTE: N = 2,087. Tracts with zero populations at any point from 1970 to 2000 are excluded.
$†p < .10.$ $*p < .05.$ $**p < .01.$

on immigration, Chicago has served as a major destination for both Hispanics and Asians, but Chicago's immigrants have been predominantly Hispanic. Over half of Chicago's foreign-born population is from Latin America, with only 20 percent from Asia. As the third largest destination for Puerto Ricans in the United States and a traditional destination for Hispanic immigrants, Chicago has experienced dramatic increases in its Hispanic population. In 1970, Chicago was approximately 60 percent white, 30 percent black, 7 percent Hispanic, and less than 1 percent Asian. The white and black populations have both steadily declined and each comprise less than one-third of the city's population today. The share of Hispanics, however, has increased to more than 28 percent, of which 41 percent are foreign-born and 13 percent are Puerto Rican. The share of Asians has also increased substantially but comprises less than 6 percent of the population. In contrast to Hispanics, nearly 70 percent of Chicago's Asian population is foreign-born. Moreover, Chicago's Hispanic population is more socioeconomically disadvantaged than its Asian population, as reflected by poverty rates of 24 and 18 percent and median household incomes of $41,700 and $56,700, respectively. Nearly 60 percent of Asians in Chicago have at least a bachelor's degree, but only 12 percent of Hispanics do.

Despite Chicago's racially and ethnically diverse population, it has had persistently high levels of residential segregation by race. In 2012, the tract-level share of blacks was negatively associated with the tract-level shares of whites, Hispanics, Asians, and foreign-born residents, displaying correlations of –0.71, –0.58, –0.33, and –0.71, respectively. Nonetheless, following nationwide trends, its segregation levels have been slowly declining as its neighborhoods have become increasingly diverse (Rugh and Massey 2014).

Seattle

Like Chicago, Seattle has had a substantial share of foreign-born residents since 1965. However, it has primarily served as a major destination for Asian immigrants. Over half of Seattle's foreign-born population is from Asia, while only 12 percent are from Latin America and about 10 percent are from East Africa. Seattle boasts one of the highest shares of Asians among major cities (14 percent), and nearly 65 percent of them are foreign born. Moreover, relative to Chicago, Seattle does not have a large presence of either blacks or Hispanics. Its share of Hispanics has increased nearly threefold since 1970 but was still only 6 percent by 2012. In 1970, Seattle was more than 86 percent white. Over time, this share has declined but is still high compared with other major cities (66 percent). Its share of blacks has remained between 7 and 10 percent over the last 40 years. Despite decline and a lack of growth in the shares of whites and blacks in Seattle, respectively, the population counts of these groups has increased in recent decades, as Seattle's overall population has continued to grow since 1980, whereas Chicago has been experiencing overall population decline for the last 50 years.

In contrast to Chicago, the gap between Hispanics' and Asians' socioeconomic status in Seattle is not as large, with Hispanics being relatively more advantaged and Asians relatively less advantaged than their Chicago counterparts. More than 36 percent of Seattle's Hispanic population has at least a bachelor's degree, compared with 47 percent of Asians, and the median household income for Hispanics and Asians is $48,100 and $52,000, respectively. The poverty levels are also closer, with Hispanics and Asians having poverty rates of 21 and 18 percent, respectively. While Seattle's segregation levels are far lower than Chicago's, its African American population is primarily concentrated in the southeastern section of the city. Due to the lower shares of blacks, however, these neighborhoods are generally more heterogeneous. Moreover, unlike Chicago, Seattle's foreign-born, Asian, and Hispanic populations are much less segregated from the city's blacks, exhibiting positive tract-level correlations of 0.56, 0.45, and 0.36, respectively. Instead, these three groups are highly segregated from the city's white population, with correlations of –0.89, –0.84, and –0.50, respectively.

Early gentrification in Chicago and Seattle

Table 3 presents the average tract characteristics in 1970 and 2000 for gentrifiable tracts with nonzero populations from 1970 to 2000 that remained ungentrified and for those that were gentrifying by the time of the field surveys (which were conducted in 1995 in Chicago and in 1998 in Seattle). Only 41 of Seattle's populated tracts, about one-third, but 363, or two-fifths, of Chicago's populated tracts were gentrifiable. In 1970, tracts that eventually gentrified in both cities had much greater shares of whites and college-educated residents on average. In Chicago, their Hispanic populations were similar to those in ungentrified tracts, and their shares of Asians and foreign-born residents were greater. In Seattle, however, tracts that eventually gentrified had lower shares of blacks, Hispanics, Asians, and foreign-born residents compared with tracts that remained ungentrified. These descriptive results suggest that immigrants, particularly Asians, were associated with earlier waves of gentrification in Chicago but that all minority groups were negatively associated with gentrification in Seattle. Moreover, socioeconomic disparities between the average ungentrified tract and the average gentrified tract were much greater in Chicago by 2000. Although tracts that eventually gentrified had higher income levels and more highly educated residents as early as 1970, they had similar poverty rates, higher vacancy rates, and lower residential stability. By 2000, the socioeconomic differences between gentrifying and ungentrified tracts grew wider, and the relationship reversed for vacancy rates.

Additional data sources

Crime reports. In both cities, I used reported homicides and burglaries compiled from the Chicago Police Department and the Seattle Police Department, respectively.[4] Rates are per 100,000 residents, and I used three-year averages because crime data are highly variable from year to year.

Chicago Community Adult Health Study (CCAHS). The CCAHS was a survey conducted from 2001 through 2003 with a stratified, multistage, probability sample of 3,105 adult Chicago residents. It was designed to understand the relationship between neighborhood contexts and health. I drew from the community survey component, which asked individuals various questions about their neighborhoods. Following prior analyses (e.g., Sampson and Raudenbush 2004), I constructed a tract-level measure for collective perceptions of disorder based on questions asking residents how much the following items were a problem in the neighborhood: teenagers causing a disturbance, litter, graffiti, vacant areas, drinking in public, and selling/using drugs. In addition, I constructed a measure for informal social control based on questions about the perceived likelihood of neighbors taking action if children were skipping school, defacing a building, not showing respect, and fighting; or would organize to keep a local fire station.

TABLE 3
Average Characteristics for Gentrifiable Census Tracts in Seattle and Chicago, 1970 and 2000

	Chicago				Seattle			
	1970		2000		1970		2000	
	Ungentrified	Gentrifying	Ungentrified	Gentrifying	Ungentrified	Gentrifying	Ungentrified	Gentrifying
% non-Hispanic white	31.1	62.2	16.9	57.6	58.3	88.5	48.9	77.7
% black	56.3	23.0	60.5	24.3	25.8	6.1	19.9	7.2
% Hispanic	11.8	11.7	18.6	8.8	2.5	1.5	9.0	4.7
% Asian	0.6	2.5	3.7	8.9	11.2	2.2	19.7	9.1
% foreign-born	8.8	12.1	13.1	14.8	12.7	10.3	22.9	12.7
Household income ($)	34,819	41,024	28,059	54,965	35,354	32,525	34,156	36,028
% below poverty	24.0	23.0	32.6	15.8	19.3	17.4	22.6	17.0
% bachelor's degree	4.9	18.3	18.8	61.8	10.3	16.6	31.5	55.7
% professional/managerial	11.4	26.8	26.9	59.4	17.5	25.9	37.1	51.8
% ownership	27.9	39.6	27.5	35.7	47.6	33.6	33.5	22.9
% vacant units	8.8	11.6	13.7	7.5	11.9	9.9	6.2	6.8
% over 65 years old	8.5	13.3	9.9	7.3	15.6	23.4	11.7	10.5
% same residence 5 years ago	50.6	39.5	53.0	32.9	43.7	34.4	36.6	26.9
N	298	65	298	65	19	22	19	22

NOTE: Dollar values are adjusted to 1999 constant dollars. Income and housing values are averages for 1970 values and medians for 2000 values. Tracts with zero populations at any point between 1970 and 2000 are excluded.

Seattle Neighborhoods and Crime Survey (SNCS). The SNCS was a survey of 4,904 adults across Seattle conducted from 2002 to 2003, in which 2,200 of the respondents were from a stratified random sample. The survey was intended to examine neighborhood social organization and criminal violence. Using responses from the random sample, I constructed similar tract-level measures as in the CCAHS, but the perceived disorder measure does not include questions on drinking in public or selling/using drugs because the survey did not include these questions, and the informal social control measure does not include a question about organizing to save a local fire station. In the results presented, I use stand-ardized measures for the response scales in each city.

Google Street View Gentrification Observations (GGO). The GGO project was designed to measure the degree to which neighborhoods have gentrified, using a systematic social observational approach of street blocks that draws on Google Street View. Google Street View is free, is fully accessible to the public, and provides nearly full-rotation panoramic views at the street level that are updated every one to four years, giving viewers the virtual experience of walking down the street. This approach to measuring gentrification builds on Wyly's and Hammel (1999) work, which showed that the visible features of gentrification are not always properly identified using census-based measures. Indeed, the visible aspects of gentrification express the social transformation of a neighborhood and offer a way to observe a process that is facilitated by a complex combination of actors. In addition, visible signs of neighborhood reinvestment further facilitate upgrading.

The GGO data span 2,096 and 1,000 block faces—single sides of street seg-ments—in Chicago and Seattle, respectively, using images captured in 2009–12. The project focused on trajectories of gentrification and therefore assessed a stratified random sample of blocks in census tracts that were either identified as gentrifying in Hammel and Wyly's field surveys or were adjacent to these tracts and were below the citywide median income at any point between 1960 and 1990. These adjacent tracts were included in the study to consider the spread of gentrification over time. Coders surveyed all block faces of the sampled blocks. Collectively, the GGO data cover 42 tracts in Seattle and 144 tracts in Chicago, which comprise about one-third of Seattle's tracts and one-sixth of Chicago's tracts.

The survey instrument was designed to capture various observable aspects of theoretically driven measures associated with gentrification in a neighborhood: (1) housing structures that are either older and well-maintained, newly con-structed, or rehabilitated; (2) new public investments (e.g., new bus stops) or large-scale developments (e.g., high-rise luxury condos); (3) visible beautification efforts (e.g., lawn décor); and (4) the lack of disorder and decay (e.g., no graffiti). These characteristics provide conceptually sound measures of visible neighbor-hood transformations consistent with my working definition of gentrification, and they encompass aspects of gentrification beyond demographic changes, such as commercial and aesthetic changes and public investment. Each characteristic was measured with two to three indicators. Details on the instrument, interrater reliability, and a coding guide are available online.[5] I constructed a scale for the degree of gentrification in each census tract using the average scores for each of the four multi-item measures. The final score is intended to capture a neighborhood's stage in a life cycle of neighborhood change from decline to renewal to class turnover (c.f., Hoover and Vernon 1959). For more methodological details, see Hwang and Sampson (2014).

Neighborhood "quality" and perceptions

Using crime reports, community survey data, 1990 and 2000 U.S. Census data, and 2005–9 American Community Survey five-year estimates,[6] I examined how changes in the immigrant, Asian, and Hispanic populations are associated with neighborhood characteristics described earlier that may relate to gentrification. Table 4 presents average measures for various neighborhood characteristics in Chicago's and Seattle's gentrifiable tracts and the percent of tracts that were gentrifying when Hammel and Wyly observed them. The table compares gentrifiable tracts in the lowest third and highest third of each city's share of foreign-born, Asian, and Hispanic residents in 2000, as well as tracts with no growth (loss or no change) and tracts in the highest third of each group's population change from 1990 to 2009.

In Chicago, crime and perceived disorder are higher in tracts with low shares of all of these groups, but in Seattle, crime and perceived disorder are substantially higher in tracts with high shares of all groups and high levels of immigrant and Hispanic growth. Moreover, high Asian growth is associated with larger declines in crime rates in Chicago but not in Seattle. Nonetheless, tracts with high shares of immigrants had greater declines in burglary rates over the subsequent decade in both cities. In Chicago, the lowest levels of crime and perceived disorder are actually found among tracts in the middle levels of the foreign-born and Hispanic distributions. Given that the foreign-born and Hispanic populations are generally segregated from predominantly black neighborhoods in Chicago but not in Seattle, these middle-range tracts tend to have higher shares of whites, and the lower-range tracts tend to have higher shares of blacks. Thus, neighborhood-level crime and perceived disorder are associated with predominantly black and Hispanic neighborhoods in Chicago but with all nonwhite and nonnative neighborhoods in Seattle. See the online supplement for correlates by black and white population and changes.[7]

Average levels for informal social control and residential stability, both reflecting neighborhood collective efficacy, show that Chicago tracts with high levels of immigrants are associated with higher levels of social control. In addition, tracts with high levels of immigrant and Asian growth saw increased homeownership in Chicago. By contrast, Seattle neighborhoods with high levels of all three groups had lower levels of informal social control.

Racial changes are also associated with varying levels of these populations but differ by city. In highly segregated Chicago, high immigrant growth is associated with increased racial diversity, and tracts with low levels of each group in 2000 experienced greater increased diversity. Neighborhoods with high immigration

TABLE 4
Average Characteristics by Immigrant, Asian, and Hispanic Populations and Change for Chicago's and Seattle's Gentrifiable Tracts

Category	% foreign, 2000		% Asian, 2000		% Hispanic, 2000		Δ foreign, 1990–2009		Δ Asian, 1990–2009		Δ Hispanic, 1990–2009	
	Low	High	Low	High	Low	High	No growth	High growth	No growth	High growth	No growth	High growth
Chicago												
Logged homicide rate, 1999	2.47**	1.27	2.20**	0.62	2.25**	1.56	1.65	1.58	1.15	1.31	1.50	1.90†
Logged burglary rate, 1999	8.28**	7.86	8.26**	7.52	8.19†	7.99	8.01	8.04	7.85	7.81	7.95	8.04
Δ logged homicide rate, 1999–2011	-0.66	-0.46	-0.70**	-0.17	-0.57	-0.71	-0.52	-0.41	-0.04	-0.53*	-0.50	-0.67
Δ logged burglary rate, 1999–2011	-0.30	-0.62*	-0.37	-0.62†	-0.27	-0.46	-0.31	-0.58*	-0.45	-0.56	-0.36	-0.57
Perceived disorder, 2002	0.62**	0.55	0.63**	0.43	0.59	0.58	0.57*	0.52	0.53	0.49	0.56	0.54
Informal social control, 2002	0.51	0.55†	0.52	0.53	0.52	0.54	0.52	0.53	0.53	0.54	0.53	0.54
Δ % homeownership, 1990–2009	0.09	0.11	0.10	0.18**	0.09	0.12	0.11	0.18**	0.13	0.19*	0.13	0.17
Δ racial diversity, 1990–2009	0.07**	0.01	0.08**	-0.01	0.07**	0.01	0.00	0.07**	-0.03	0.06**	0.00	0.09**
Δ logged black population, 1990–2009	-0.32	0.39**	-0.17	-0.32	-0.38	0.71**	-0.02	-0.13	-0.56*	0.07	0.04	-0.39*
Δ logged white population, 1990–2009	0.88**	0.03	0.74*	0.34	0.86**	0.12	-0.06	1.27**	-0.16	1.05**	0.22	1.04**
% gentrifying by 1995	2.53	10.13*	3.13	41.80**	5.63	4.82	8.47	27.27**	23.21	27.59	15.69	22.45
N	158	79	160	122	160	83	118	132	56	145	153	98
Seattle												
Logged homicide rate, 1999	0.46	1.61**	0.62	1.43	0.67	1.80*	0.39	1.45*	1.26	1.29	0.81	1.55*
Logged burglary rate, 1999	7.61	8.34**	7.75	8.28*	7.67	8.43**	7.59	8.12†	8.03	8.06	7.94	8.09

(continued)

TABLE 4 (CONTINUED)

Category	% foreign, 2000		% Asian, 2000		% Hispanic, 2000		Δ foreign, 1990–2009		Δ Asian, 1990–2009		Δ Hispanic, 1990–2009	
	Low	High	Low	High	Low	High	No growth	High growth	No growth	High growth	No growth	High growth
Δ logged homicide rate, 1999–2011	−0.46	−0.94	−0.46	−0.84	−0.54	−1.03	−0.23	−0.83†	−0.95	−0.71	−0.30	−1.25
Δ logged burglary rate, 1999–2011	−0.35	−0.67*	−0.46	−0.61	−0.38	−0.56	−0.70	−0.48	−0.54	−0.51	−0.63*	−0.47
Perceived disorder, 2002	0.22	0.62**	0.32	0.63**	0.31	0.59**	0.35	0.52	0.45	0.45	0.36	0.49
Informal social control, 2002	0.62**	0.30	0.55**	0.35	0.59**	0.31	0.54	0.39	0.36	0.43	0.45	0.37
Δ % homeownership, 1990–2009	0.06	0.04	0.05	0.03	0.06	0.03	0.07	0.05	0.03	0.05	0.03	0.05
Δ racial diversity, 1990–2009	0.01	−0.03	0.01	0.00	0.00	0.00	0.01	0.00	0.01	−0.01	−0.06*	0.01
Δ logged black population, 1990–2009	−0.64	−0.13	−0.64	−0.11	−2.31**	0.08	−1.66	−0.14	−0.25	−0.65	−0.91	−0.11
Δ logged white population, 1990–2009	0.16	0.14	0.23	0.06	−0.05	0.33**	0.03	0.38**	0.21	0.31	0.01	0.48**
% gentrifying by 1998	37.50	35.29	47.37	30.77	20.00	38.46	33.33	36.36	15.38	65.00**	50.00	43.75
N	16	17	19	13	10	18	6	22	13	20	10	16

NOTE.: Indicators of statistical significance are placed on larger absolute values. Percent categories are between the lowest third and highest third of census tracts in each city. High growth tracts are in the highest third of census tracts in population changes by group. Tracts with zero populations at any point from 1990 to 2009 are excluded.

†p < .10. *p < .05. **p < .01.

and Hispanic levels also had larger increases in their black populations and smaller increases in whites. High growth in all three groups was associated with the greater growth of whites in Chicago. Seattle's tracts with low levels of these groups experienced large declines in their black populations, and tracts with higher shares of Hispanics and high levels of immigrant and Hispanic growth experienced larger increases in whites. Thus, areas with rising immigrant populations or low levels of immigrants and Hispanics are associated with white influx in Chicago, but in Seattle, where Asians are the larger immigrant population, greater increases in whites occurred in neighborhoods with relatively higher shares of Hispanics and in neighborhoods with high levels of immigrant and Hispanic growth.

Finally, tracts that gentrified by the 1990s had higher levels of foreign-born and Asian residents in 2000 and higher levels of immigrant growth in Chicago and higher levels of Asian growth in Seattle. Altogether, these results suggest that immigrant groups without established local concentrations are associated with gentrification, and other neighborhood "quality" characteristics do not necessarily map onto gentrification, particularly in Seattle.

Trajectories of gentrification

I next examine how immigrants, Hispanics, and Asians are associated with neighborhoods' recent trajectories of gentrification using the gentrification measures created with Google Street View. Therefore, this analysis only examines tracts that had gentrified according to Hammel and Wyly's field observations or tracts that were adjacent to them and low-income at any point from 1960 through 1990. Figure 1 presents maps of Chicago and Seattle and the gentrification stage scores for these tracts. Tracts with bolded outlines were gentrifying when Hammel and Wyly observed them. The overall gentrification stage scores averaged to 0.68 (sd = 0.07) in Seattle and 0.65 (sd = 0.11) in Chicago. Seattle's higher levels of gentrification and less variation are likely due to the fact that Seattle's older housing stock is relatively newer, with 39 percent of its housing stock built since 1970 compared with 23 percent in Chicago. Thus, Seattle's share of older housing that is visibly well maintained is higher. Moreover, Seattle's housing market also suffered to a lesser degree during the recession: in 2008, the foreclosure rate in the Chicago metropolitan area was nearly three times the rate in the Seattle metropolitan area.

Table 5 shows how the overall gentrification stage scores correlate with racial, ethnic, and nativity characteristics of census tracts from 2000 and changes from 2000 to 2009. In 2000, the census began providing data on race and ethnicity by nativity. The bivariate correlations in both cities reveal a racial ordering such that the share of whites has the strongest positive association and the share of blacks has the strongest negative association with gentrification levels. In both cities, Hispanics also have a negative association with gentrification levels. Although tracts with large shares of Asians in Chicago were associated with gentrification across all gentrifiable tracts, Asians are not associated with gentrification levels in either city among gentrifying tracts and their adjacent tracts. Moreover, there are

FIGURE 1
Maps of Chicago and Seattle by Gentrification Stage Scores

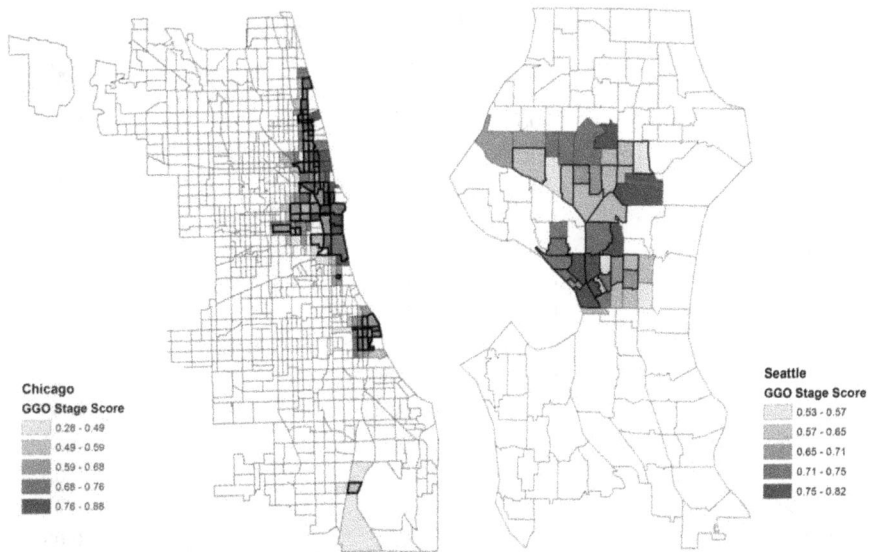

Chicago
GGO Stage Score
　0.28 - 0.49
　0.49 - 0.59
　0.59 - 0.68
　0.68 - 0.76
　0.76 - 0.88

Seattle
GGO Stage Score
　0.53 - 0.57
　0.57 - 0.65
　0.65 - 0.71
　0.71 - 0.75
　0.75 - 0.82

distinctions between native and foreign-born blacks in both cities and native and foreign-born Hispanics in Seattle.

Population changes are weakly correlated with gentrification levels for most groups. Nonetheless, changes in the native-born white and Asian populations are negatively associated with gentrification levels in Chicago, which may reflect an increasing diversity among these tracts. Changes in foreign-born blacks, who comprise a far greater share of the foreign-born population in Seattle than in Chicago, were negatively associated, and changes in foreign-born Asians were positively associated with gentrification levels in Seattle. In both cities, however, the native and foreign-born populations for each racial and ethnic group tend to concentrate in similar areas, though less so in Seattle.

As a last step, I examined these relationships after accounting for the degree to which these neighborhoods had gentrified according to Hammel and Wyly's field surveys and the simultaneous racial, ethnic, and nativity changes occurring in these neighborhoods. I used weighted least squares regression models in which the dependent variable is the gentrification stage score and weights are the square root of the number of blocks observed using Google Street View in each census tract, thus accounting for variation in the amount of coded data for each tract. To examine differential effects between cities, I also include interaction terms in all models for the main racial, ethnic, and nativity variables and a dummy variable indicating that a tract is in Seattle. Figure 2 displays the estimated effects and 95 percent confidence intervals for each city based on models estimating the association of groups by race, ethnicity, and nativity with the trajectory of gentrification. The estimated effects for the racial, ethnic, and nativity

TABLE 5
Correlations between Gentrification Stage Scores and Race, Ethnicity, and Nativity

	2000 Population Share		Δ Population (logged) 2000–2009	
	Chicago	Seattle	Chicago	Seattle
Total non-Hispanic white	0.53°°	0.21	–0.14†	0.04
Total black	–0.40°°	–0.23	0.00	–0.06
Total Hispanic	–0.19°	–0.15	0.04	0.10
Total Asian	–0.03	0.00	–0.14†	0.13
Total foreign	–0.05	0.01	0.02	–0.05
Foreign non-Hispanic white	0.43°°	0.26†	–0.05	–0.05
Foreign black	0.01	0.13	0.11	–0.27†
Foreign Hispanic	–0.21°	–0.26†	–0.05	0.09
Foreign Asian	–0.07	0.00	–0.04	0.26†
Total native	0.05	–0.01	0.03	0.20
Native non-Hispanic white	0.51°°	0.19	–0.13	0.05
Native black	–0.41°°	–0.26	–0.02	0.05
Native Hispanic	–0.16†	0.15	0.08	–0.09
Native Asian	0.10	0.00	–0.14†	–0.03

NOTE: Chicago N = 139; Seattle N = 42. Tracts with zero populations at any point between 2000 and 2009 are excluded.
†$p < .10$. °$p < .05$. °°$p < .01$.

FIGURE 2
Regression Results and Confidence Intervals Predicting Gentrification Stage Scores

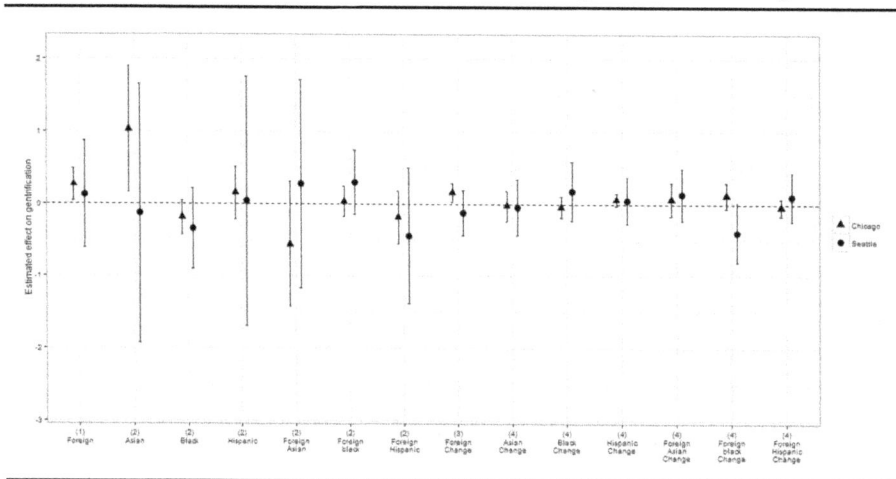

variables represent the change in the gentrification stage score given a one standard deviation change in the respective independent variable. The models are as follows: (1) 2000 nativity composition; (2) 2000 racial and ethnic by nativity composition; (3) 2000–2009 nativity composition changes; and (4) 2000–2009 changes in racial and ethnic composition by nativity. See the online supplement for distribution details for each variable and full regression tables.[8]

The results from the first model show that the 2000 foreign-born population is positively associated with the degree to which neighborhoods gentrified over the next several years in Chicago. In the second model, the Asian population in Chicago, but not in Seattle, predicts continued upward trajectories of gentrification. Results for other racial, ethnic, and nativity groups are not statistically significant.[9] The third model, which examines changes in the foreign-born population over time, shows that increases in the foreign-born population are also associated with the degree to which neighborhoods have gentrified over the last several years in Chicago. In the final model, in contrast to Chicago, changes in the foreign-born black population in Seattle are negatively associated with the trajectory of gentrification. Overall, once I account for the prior degree to which neighborhoods had gentrified during the 1990s and other population changes, the results show that foreign-born and Asian populations are positively associated with the degree of gentrification in Chicago. Moreover, the degree of gentrification is positively associated with declines of foreign-born blacks in Seattle.

Conclusion

This article demonstrates that immigration and its associated racial and ethnic changes are linked to gentrification, but this relationship is structured by the broader racial and immigration contexts of the cities in which these changes take place. Nationally, cities with higher levels of immigration had greater rates of gentrification. At the neighborhood level, the early arrival of immigrants, Hispanics, and Asians was associated with subsequent gentrification, and these effects were stronger in neighborhoods with larger shares of blacks. In Chicago, a city with high levels of residential segregation by race and high levels of Hispanic immigration, foreign-born residents and Asians were associated with early waves of gentrification and more positive neighborhood characteristics compared with black and Hispanic neighborhoods. In recent decades, however, Hispanics have become negatively associated with gentrification in Chicago: while Hispanics may have contributed to the initial stability of these neighborhoods, their continued rapid growth in the city may have resulted in ethnic enclaves to which new arrivals are more likely to move. In Seattle, a predominantly white city but with high levels of Asian immigration, nonwhite and nonnative immigrants were negatively associated with gentrification and characteristics of neighborhood quality. In recent decades, foreign-born residents, Hispanics, and Asians were not associated with the degree to which neighborhoods gentrified, but neighborhoods that gentrified to a greater degree had greater declines in their foreign-born black populations.

Although the findings presented in this article are preliminary, they suggest that racial hierarchies of inequality operate and adapt in contexts of change. The durability of neighborhood inequality and the persistence of poor, minority neighborhoods are dominant trends in most major American cities, and the patterns documented here show how gentrification unfolds along similar lines that have governed this residential stratification. Despite the rise of gentrification over the last several decades, black neighborhoods in highly segregated cities such as Chicago, particularly those not impacted by immigration, are the least likely to experience reinvestment and renewal. Even among neighborhoods that showed signs of gentrification during the 1990s or were adjacent to them, these neighborhoods are least likely to continue on upward trajectories. In predominantly white but diversifying cities such as Seattle, nonwhite/nonnative neighborhoods were least likely to gentrify; among neighborhoods that gentrified or were adjacent to them, foreign-born blacks attenuated the continued upward trajectories of these neighborhoods. Thus, the order of gentrification from one of simply whites and natives over nonwhites and nonnatives in Seattle became increasingly similar to that found in Chicago, as racial and ethnic distinctions among foreign-born residents formed.

As city leaders hope for urban revitalization through gentrification and efforts to attract immigrants, understanding the implications that these changes have for residential segregation and racial and ethnic inequality are necessary considerations to ensure that the truly disadvantaged do not continue to be left behind. In one sense, the findings provide hope for low-income, minority neighborhoods that experience influxes of immigrants and increased racial diversity, by showing that they can attract reinvestment and renewal. For the minority neighborhoods that remain in decline, policy interventions are crucial. The evidence is strong on the detrimental effects of concentrated poverty on both individual life chances and the costs on society. Disadvantaged minority neighborhoods need substantial and sustained targeted reinvestment that provides opportunities, resources, and institutions for their residents and that simultaneously protects against the loss of affordable housing as these neighborhoods improve. Based on past trends, reliance on market-based forces such as gentrification without strong policy interventions will inevitably perpetuate and perhaps worsen racial inequality.

Notes

1. The different years account for different suburbanization patterns between the regions. By focusing on low-income tracts from these years, the sample excludes tracts that became low-income after these years but prior to the field surveys.

2. Hammel and Wyly distinguished gentrification by two levels of intensity. The substantive results are similar when I incorporate the different levels into the analysis, except that blacks were not associated with high levels of gentrification.

3. Because the data do not distinguish when gentrification began in these tracts, it is possible that immigrants moved to gentrifying neighborhoods. Nonetheless, the findings hold after controlling for neighborhood socioeconomic status and residential stability.

4. Other types of crime were not consistently reported across both cities or across all years.

5. See http://scholar.harvard.edu/jackelynhwang/projects/ggo.

6. In 2010, the U.S. Census and American Community Survey changed the boundaries of census tracts. To preserve the same boundaries across the various datasets, I use 2005–9 American Community Survey estimates.

7. See http://ann.sagepub.com/supplemental.

8. Ibid.

9. When population shares are considered, the results are consistent with Hwang and Sampson's (2014) findings that blacks and Hispanics are negatively associated with the degree to which neighborhoods gentrify in Chicago (and in Seattle for blacks).

References

Charles, Camille Z. 2003. The dynamics of racial residential segregation. *Annual Review of Sociology* 29:167–207.

Crowder, Kyle, Jeremy Pais, and Scott J. South. 2012. Neighborhood diversity, metropolitan constraints, and household migration. *American Sociological Review* 77 (3): 325–53.

Ellen, Ingrid Gould. 2000. *Sharing America's neighborhoods: The prospects for stable racial integration.* Cambridge, MA: Harvard University Press.

Flippen, Chenoa, and Eunbi Kim. 2015. Immigrant context and opportunity: New destinations and socioeconomic attainment among Asians in the United States. *The ANNALS of the American Academy of Political and Social Science* (this volume).

Hackworth, Jason, and Neil Smith. 2001. The changing state of gentrification. *Tijdschrift Voor Economische En Sociale Geografie* [*The Journal of Economic and Social Geography*] 92:464–77.

Hall, Matthew. 2013. Residential integration on the new frontier: Immigrant segregation in established and new destinations. *Demography* 50 (5): 1873–96.

Hammel, Daniel J., and Elvin K. Wyly. 1996. A model for identifying gentrified areas with census data. *Urban Geography* 17 (3): 248–68.

Hoover, Edgar M., and Raymond Vernon. 1959. *Anatomy of a metropolis: The changing distribution of people and jobs within the New York metropolitan region.* Cambridge, MA: Harvard University Press.

Hwang, Jackelyn, and Robert J. Sampson. 2014. Divergent pathways of gentrification: Racial inequality and the social order of renewal in Chicago neighborhoods. *American Sociological Review* 79 (4): 726–51.

Kirk, David S., and John H. Laub. 2010. Neighborhood change and crime in the modern metropolis. *Crime & Justice* 39:441–502.

Krysan, Maria, Mick P. Couper, Reynolds Farley, and Tyrone A. Forman. 2009. Does race matter in neighborhood preferences? Results from a video experiment. *American Journal of Sociology* 115 (2): 527–59.

Lee, Jennifer. 2002. *Civility in the city: Blacks, Jews, and Koreans in urban America.* Cambridge, MA: Harvard University Press.

Lewis, Valerie A., Michael O. Emerson, and Stephen L. Klineberg. 2011. Who we'll live with: Neighborhood racial composition preferences of whites, blacks, and Latinos. *Social Forces* 89 (4): 1385–407.

Lin, Jan. 1998. Globalization and the revalorizing of ethnic places in immigration gateway cities. *Urban Affairs Review* 34 (2): 313–39.

MacDonald, John M., John R. Hipp, and Charlotte Gill. 2012. The effects of immigrant concentration on changes in neighborhood crime rates. *Journal of Quantitative Criminology* 29 (2): 191–215.

Muller, Thomas. 1993. *Immigrants and the American city.* New York, NY: New York University Press.

Quillian, Lincoln, and Devah Pager. 2001. Black neighbors, higher crime? The role of racial stereotypes in evaluations of neighborhood crime. *American Journal of Sociology* 107 (3): 717–67.

Rugh, Jacob, and Douglas Massey. 2014. Segregation in post–civil rights America: Stalled integration or end of the segregated century? *Du Bois Review* 11 (2): 205–32.

Sampson, Robert J., Jeffrey D. Morenoff, and Felton Earls. 1999. Beyond social capital: Spatial dynamics of collective efficacy for children. *American Sociological Review* 64 (5): 633–60.

Sampson, Robert J., and Stephen W. Raudenbush. 2004. Seeing disorder: Neighborhood stigma and the social construction of "broken windows." *Social Psychology Quarterly* 67 (4): 319–42.

Sanchez, George J. 1997. Face the nation: Race, immigration, and the rise of nativism in late twentieth century America. *International Migration Review* 31 (4): 1009–30.

Smith, Neil S. 1998. Gentrification. In *The encyclopedia of housing*, ed. W. V. Vliet. London: Taylor and Francis.

Vigdor, Jacob. 2014. *Immigration and New York City: The contributions of foreign-born Americans to New York's renaissance, 1975–2013*. New York, NY: Americas Society/Council of Americas.

Winnick, Louis. 1990. *New people in old neighborhoods: The role of immigrants in rejuvenating New York's communities*. New York, NY: Russell Sage Foundation.

Wyly, Elvin K., and Daniel J. Hammel. 1999. Islands of decay in seas of renewal: Housing policy and the resurgence of gentrification. *Housing Policy Debate* 10 (4): 711–71.

Zukin, Sharon. 1987. Gentrification: Culture and capital in the urban core. *Annual Review of Sociology* 13:129–47.

Violence and Neighborhood Disadvantage after the Crime Decline

Violent crime is known to be concentrated in the same urban neighborhoods as poverty and other forms of disadvantage. While U.S. violent crime has declined at an unprecedented rate over the past two decades, little is known about the spatial distribution of this decline within cities. Using longitudinal neighborhood crime data from six U.S. cities during the national crime decline, this article examines changes in (1) crime rates of neighborhoods grouped by their initial crime levels, poverty rates, and racial/ethnic makeups; (2) the neighborhood exposure to violence of urban residents classified by race/ethnicity and poverty status; and (3) the relative distribution of violent crime across urban neighborhoods. We find that crime levels declined the most in the initially most violent and disadvantaged neighborhoods and that exposure to violence fell the most among disadvantaged urban residents. Nonetheless, crime remained concentrated in cities' initially most violent and disadvantaged locales.

Keywords: crime decline; neighborhood change; concentrated disadvantage; violent crime; urban inequality

By
MICHAEL FRIEDSON
and
PATRICK SHARKEY

O ver the past 20 years, the United States has experienced the most sustained decline of violent crime in its modern history (Zimring 2008). The national rates of homicide and of all violent index crime have been cut roughly in half since the early 1990s. The drop in violence is visible in every source of data available, from official police statistics on homicides to victimization surveys conducted with

Michael Friedson is a postdoctoral fellow at New York University. His research examines crime and policing in disadvantaged urban areas, with a longitudinal focus.

Patrick Sharkey is an associate professor of sociology at New York University. His research considers the role of neighborhoods and cities in generating and maintaining inequality across multiple dimensions. He is currently examining how the decline of violent crime has affected urban life and urban inequality in America.

DOI: 10.1177/0002716215579825

ordinary Americans. No matter the data source, the decline of violent crime is staggering in its scale and duration.

Despite the evidence for how much violence has declined, little is known about where it has declined. The dearth of evidence on how the crime decline has been distributed across urban neighborhoods is a major gap in the literature. One of the most unique aspects of interpersonal violence is that it is geographically concentrated. Research conducted in multiple cities has shown that a disproportionate share of all violent crime takes place within an extremely small number of city blocks and neighborhoods (Braga, Papachristos, and Hureau 2010; Braga, Hureau, and Papachristos 2011; Weisburd et al. 2004). A converging strand of empirical research argues that the spatial concentration of violence may be a core mechanism leading to the reproduction of neighborhood inequality (Sharkey and Sampson, in press; Peterson and Krivo 2010; Sharkey 2010; Burdick-Will et al. 2011).

This research suggests that to explain and interpret the crime decline, it is necessary first to have a clear sense of the degree to which communities have been affected by it. If the decline of violent crime never reached the nation's most disadvantaged neighborhoods, then "The Great American Crime Decline" (Zimring 2008) might be seen as yet another trend that has exacerbated neighborhood inequality. Alternatively, if the decline of violent crime was concentrated in the most disadvantaged neighborhoods, then the crime drop may have weakened one of the central mechanisms by which neighborhood inequality is maintained and reproduced.

To address these questions, this study analyzes trends in neighborhood-level violent index crime rates for six cities—Chicago, Cleveland, Denver, Philadelphia, St. Petersburg (Florida), and Seattle. These municipalities are not representative of U.S. cities but were selected because all have available data on neighborhood-level violent crime over a period of at least a decade. The data allow us to describe how the crime drop was distributed across neighborhoods within each city, and to answer questions about which neighborhoods and which populations experienced the greatest declines in violent crime.

The Nexus of Concentrated Disadvantage and Violence

Multiple forms of disadvantage, from poverty to family structure to disease to homicide, tend to come bundled together in urban neighborhoods (Kasarda 1993; Sampson and Morenoff 2006; Sampson 2012). This multifaceted disadvantage tends to be durable, with a rank ordering of urban neighborhood status that is remarkably stable over time (Sampson and Morenoff 2006; Sampson 2012).

NOTE: This research was funded by a grant from the William T. Grant foundation. The authors would like to thank audience members at the Penn State Stratification Conference on Residential Inequality in American Neighborhoods and Communities as well as Mikaila Arthur, the special issue editors, and anonymous reviewers for excellent feedback on an earlier draft of the article.

The multiple dimensions of neighborhood disadvantage are mutually reinforcing, constituting what some have referred to as a "poverty trap" (Bowles, Durlauf, and Hoff 2006; Sampson and Morenoff 2006; Sampson 2012).

From the 1960s through the 1990s, the rise of violent crime and the emergence of mass imprisonment led to a new concentration of violence and a strengthened spatial link among poverty, segregation, violence, and the criminal justice system (Hagan and Peterson 1995; Krivo and Peterson 1996; Peterson and Krivo 2010; Sampson 2012). Crime and neighborhood disinvestment reinforced each other to constitute a "spiral of decay" in high-poverty areas (Skogan 1990; see also Bursik 1986; Liska and Bellair 1995). An extensive literature documents how the spatial concentration of poverty, violence, aggressive police oversight, and incarceration erode public life, compromising the capacity of neighborhood residents to achieve social cohesion and community organization (Clear 2009; Fagan and Meares 2008; Klinenberg 2003; Sampson, Raudenbush, and Earls 1997).

Considering the tight link between violent crime and urban disadvantage, the dramatic drop in violent crime that has occurred since the 1990s has potentially important implications for our understanding of urban poverty and neighborhood inequality. Pope and Pope (2012) find a highly significant, negative relationship between change in property values and change in crime rates during the 1990s in U.S. urban zip codes. Ellen and O'Regan (2008) find that the greatest rates of economic growth during the 1990s occurred in urban neighborhoods with the highest proportions of black and poor residents, while these neighborhoods experienced the greatest economic losses as crime grew in the prior two decades. These authors hypothesize that the concentration in these neighborhoods of both the crime decline and the prior crime increase may help to account for these patterns. Prior research bolsters this hypothesis, demonstrating that increasing crime rates lead to losses of relatively affluent populations at both the neighborhood and city levels (Cullen and Levitt 1999; Morenoff and Sampson 1997).

To understand whether the decline of violent crime has reversed these patterns of disinvestment, decline, and out-migration, it is essential to analyze where crime has dropped within urban areas. Much of the existing literature on the magnitude of the crime decline has focused on measuring changes in crime commission or victimization rates among groups classified by age, gender, socioeconomic status, or race/ethnicity, or in large areas such as states or regions (Cook and Laub 2002; Zimring 2008). Less attention has been devoted to the spatial dimensions of the crime decline within cities. Studying crime trends in Cleveland and Denver, Ellen and O'Regan (2009) establish that neighborhoods in these cities with the greatest proportions of minority residents experienced the greatest crime declines in the 1990s, and that neighborhood-level exposure to violent crime fell most among these cities' poor and minority residents. A recent line of research—conducted on the level of street segments and intersections (so-called microplaces)—has made groundbreaking advances in our understanding of the degree to which the crime decline and the prior incidence of violent crime were localized within cities (Braga, Papachristos, and Hureau 2010; Braga, Hureau, and Papachristos 2011; Weisburd et al. 2004). This research finds that a majority of violent incidents in Boston and Seattle were concentrated in a relative handful of microplaces. A paradoxical pattern of change and stability characterized the crime decline in Boston's most violent microplaces. Even as much of the city's declines in gun violence and robbery occurred in these locales, Boston's gun violence and robbery incidents remained concentrated in these microplaces after the declines.

This study adds to this evidence base by considering how the crime decline has been experienced across neighborhoods and across populations characterized by economic status and by race and ethnicity. We describe the degree of change in violent crime in six cities, tracking violent crime at the neighborhood level for at least a decade in all cities. As described in the following section, we consider the absolute and relative decline of violent crime, changes in different types of neighborhoods and for different segments of urban populations, and change and stability in the spatial distribution of violence. We explore implications of these patterns for future research on broader social conditions in disadvantaged neighborhoods and for policies to reverse cycles of decay and disinvestment from which these neighborhoods have previously suffered.

Data and Methods

Analytic approach

The analysis is carried out in three sections. The first section consists of a neighborhood-level analysis examining where the decline in violent crime was concentrated within cities. Each city's neighborhoods are first divided into quintiles based on their initial violent crime rates. For each quintile in each year of the data, a violent index crime rate—expressed as the number of crimes per 10,000 residents—is calculated by dividing the total number of such crimes occurring in the quintile's neighborhoods by the total number of residents of these neighborhoods, and multiplying the resulting quotient by 10,000. The amount of change in this rate, from the data's initial year to its final year, is compared for each city's most violent quintile and its remainder. Each city's neighborhoods are then divided into a group of "poor" and a group of "nonpoor" neighborhoods, based on whether at least 30 percent of their residents were in poverty during the data's initial year.[1] A violent index crime rate is calculated for each of these groups in each year of the data. The amount of change in this rate, from the data's initial to its final year, is compared for each city's poor and nonpoor neighborhoods. Finally, each city's neighborhoods are divided according to whether blacks, Hispanics, or non-Hispanic whites constituted a majority of their populations in the data's first year, or are designated as "other" if none of these groups made up a majority. Annual violent index crime rates are calculated for each set of neighborhoods, and the amount of change, from the data's initial to its final year, is compared for each set.

The second section considers change in the rate at which different groups of individuals were exposed to violent crime within their neighborhoods from the data's initial year to its final year. Changes in exposure to neighborhood violent crime are examined based on poverty status (poor and nonpoor residents), and on race/ethnicity (non-Hispanic black, Hispanic, and non-Hispanic white populations). The measure of exposure is defined as the sum of the violent crime rates of all of the city's neighborhoods, with each neighborhood rate weighted by the number of group members residing in that place. This definition can be expressed in symbolic form as $\sum_{j=1}^{n}(V_j * (P_j / P_t))$, where each V_j is the violent crime rate of one of the n neighborhoods in the city; P_j is the population in that neighborhood of the group for which the exposure rate is being calculated; and P_t is the group's population in the city as a whole.

The final section analyzes how the crime decline affected the distribution of violence across the neighborhoods of each city. The correlation is measured between the initial and final violent crime rates of the neighborhoods in each city. These rates are logged to prevent extreme or outlying values from having a disproportionate influence on results. The correlation coefficient provides one measure of the stability or change in the distribution of these neighborhoods' crime rates. Each city's neighborhoods are then divided into quintiles by their violent crime rates in the data's initial year and final year, respectively. Transition matrices are used to display the degree of change in the relative position of neighborhoods within a city in terms of violent crime. Specifically, we focus on the proportions of neighborhoods that began in the most or least violent quintile and remained there at the end of the timeframe under study.

Data

We analyze data on the violent index crime rates of the neighborhoods within six cities—Chicago, Cleveland, Denver, Philadelphia, St. Petersburg, and Seattle.[2] These cities were selected because neighborhood-level crime data covering a period of at least a decade during the national crime decline are publicly available for each. Violent index crimes, as defined by the FBI, consist of intentional homicides, robberies, rapes, and aggravated assaults. A neighborhood's violent index crime rate consists of the number of such crimes occurring within its boundaries per 10,000 neighborhood residents. Because each city's crime data are derived from a different local source, there are inconsistencies between cities in terms of the years and crimes for which data are available, as well as the definitions of "neighborhoods."

Data spanning just over a decade are given for four cities: Chicago, Philadelphia, St. Petersburg, and Seattle. The time periods covered by these municipalities' data are, respectively, 2001 to 2012, 1998 to 2009, 2000 to 2012, and 1996 to 2007. For the remaining two cities, Cleveland and Denver, data span the period from 1990 to 2010. This study's data tables and graphs display changes in Cleveland and Denver crime rates over the period from 1990 to 2010 and each of its constituent decades. This enables comparison of changes occurring in Cleveland and Denver and elsewhere over periods of similar duration. By default, results for Cleveland and Denver that are discussed in the text concern 1990 to 2010.

Neighborhoods are defined as census tracts for all cities except Denver. The source for the Denver data (the Piton Foundation) divides this city into 77 neighborhoods, whereas the 2000 U.S. Census divides Denver into 136 tracts. We believe that the benefits of adding this city to our dataset outweigh any inconsistencies introduced into our analysis by the larger size of the neighborhood units for which data are available.

Homicides and rapes are not included in the Philadelphia crime counts provided by our data source. This omission is not expected to introduce substantial inconsistencies or distortions into our analysis, given that the omitted crimes constituted just 6 percent of all violent index crimes in Philadelphia during the period under analysis,[3] and given the typically high degree of correlation between tract-level counts of the omitted crimes and of all violent index crimes. For instance, the correlation equals 0.9 (indicating an extremely strong linear relationship) between the total number of homicides and all violent index crimes occurring in Chicago tracts from 2001 to 2012, while that of the number of rapes and all violent index crimes is even stronger.

Demographic variables—namely population counts disaggregated by poverty status, race, and ethnicity—are derived from the same sources as the crime data for neighborhoods in Denver and Cleveland. For tracts in the other cities, these variables are derived from the U.S. censuses and American Community Survey five-year averages. These averages are treated as applying to the middle year in their five-year span. Annual values of demographic variables, for years between those for which they are provided by these data sources, are calculated via linear interpolation.

Longitudinal analysis of neighborhood-level data over multiple decades poses challenges due to changes in tract boundary definitions in each decennial census. These challenges do not apply to the Cleveland and Denver data, because the sources from which they are derived use consistent neighborhood boundaries throughout the periods they cover. Tract boundary changes in Seattle and St. Petersburg are minor over the periods covered by their crime data. Data for each of these cities was manually adjusted to conform to a uniform set of tract boundaries. Challenges posed by tract boundary changes in Chicago and Philadelphia are more substantial. See the online methodological appendix for a detailed discussion of adjustments made to these cities' data to facilitate the longitudinal analysis of their neighborhood crime rates.[4]

Results

Where did violent crime decline? A neighborhood-level analysis

The analysis presented in Table 1 shows the absolute and proportional changes in violent crime rates for the most violent quintile of neighborhoods (as of the baseline year) and the remainder of each city's neighborhoods. Absolute declines

TABLE 1
Change in Violent Crime Rates of Neighborhoods Grouped by Initial Violent Crime
Levels

City	Time Period	Absolute Change			Relative Change		
		Highest Quintile	Remainder	Entire City	Highest Quintile	Remainder	Entire City
Chicago	2001–2012	−109.67	−32.31	−51.36	−28.92	−32.57	−33.10
Cleveland	1990–2010	−175.83	19.27	−20.22	−43.28	18.39	−12.31
Cleveland	1990–2000	−177.00	−7.38	−41.75	−43.57	−7.05	−25.42
Cleveland[a]	2000–2010	1.17	26.65	21.53	0.51	27.36	17.57
Denver	1990–2010	−95.42	−10.77	−25.75	−47.54	−22.51	−33.14
Denver	1990–2000	−83.47	−16.19	−28.57	−41.58	−33.84	−36.76
Denver	2000–2010	−11.95	5.42	2.82	−10.19	17.12	5.73
Philadelphia	1998–2009	−62.65	−2.00	−15.40	−22.91	−2.25	−12.28
Seattle	1996–2007	−67.32	−10.47	−21.01	−28.54	−23.80	−25.55
St. Petersburg	2000–2012	−202.31	−41.31	−74.50	−42.94	−46.72	−45.88

a. Although the count of violent incidents in Cleveland was roughly stable from 2000 to 2010, the city experienced a major loss of population in these years that resulted in an increase of its overall violent crime rate.
NOTE: Neighborhoods are grouped into quintiles of roughly equal population sizes by their initial violent crime rates. The "highest quintile" consists of the neighborhoods with the highest initial rates, while the "remainder" consists of those outside the highest quintile. Absolute change is the difference between the crime rates in the last and first years of the specified time period. Relative change is this difference as a percentage of the first year's crime rate. To reduce the impact of anomalous annual crime rates on results, multiyear averages of initial crime rates are used to divide the neighborhoods into quintiles. See the online appendix (http://ann.sagepub.com/supplemental) for graphs of change in cities' neighborhood crime rates, for neighborhoods grouped by initial crime rates, poverty levels, and racial/ethnic makeup.

in violent crime in each city's most violent neighborhoods far outstripped the changes in violent crime occurring in their remainders. For example, from 2001 to 2012 the violent crime rate in Chicago's most violent neighborhoods dropped by 110 crimes per 10,000 residents, whereas the rate of violent crime dropped by 32 crimes per 10,000 residents in the remainder of the city.

Due to the large declines in the absolute levels of violent crime in each city's most violent neighborhoods, there was a convergence in violent crime rates between the most violent neighborhoods of each city and the rest of its neighborhoods. The absolute difference in violent crime rates between each city's most violent neighborhoods and all other neighborhoods shrunk by between 28 percent for Chicago and 65 percent for Cleveland during the years covered by the data.

TABLE 2

Change in Violent Crime Rates of Neighborhoods Grouped by Poverty Status

City	Time Period	Absolute Change		Relative Change	
		Poor	Nonpoor	Poor	Nonpoor
Chicago	2001–2012	−104.74	−35.38	−32.64	−30.65
Cleveland	1990–2010	−52.45	14.26	−20.32	15.89
Cleveland	1990–2000	−88.05	−2.92	−34.11	−3.26
Cleveland	2000–2010	35.60	17.18	20.93	19.79
Denver	1990–2010	−78.52	−14.50	−44.75	−25.12
Denver	1990–2000	−73.25	−20.28	−41.75	−35.13
Denver	2000–2010	−5.27	5.78	−5.16	15.43
Philadelphia	1998–2009	−32.34	−6.11	−15.71	−6.79
Seattle	1996–2007	−120.21	−15.23	−35.89	−22.60
St. Petersburg	2000–2012	−250.66	−59.79	−43.55	−46.93

NOTE: Neighborhoods are classified as "poor" if their poverty rates are at least 30 percent in the data's initial year and as "nonpoor" otherwise. See the Table 1 note for definitions of absolute and relative change.

The columns of results focusing on absolute levels of violent crime give equal weight to every incident of violent crime.[5] The second set of columns shows the declines in violent crime rates in proportional terms, or relative to initial rates. In Chicago, Seattle, and St. Petersburg, the proportional decline of violent crime was roughly equivalent in the cities' most violent neighborhoods and in the remainder of the cities' neighborhoods. In Cleveland, Denver, and Philadelphia, the proportional decline in the most violent quintile of neighborhoods was substantially larger than that in the remainder of neighborhoods. Indeed, in both Cleveland and Philadelphia, all or nearly all of the drop in violent crime was concentrated in the cities' most violent neighborhoods.

The analysis presented in Table 2 repeats that of Table 1, but for neighborhoods classified by their initial poverty status rather than their initial violent crime rates. The absolute decline in violent crime in each city's poor neighborhoods far exceeded that in its nonpoor neighborhoods. For example, the decline was more than 120 crimes per 10,000 residents in Seattle's poor tracts, compared with only about 15 crimes per 10,000 residents elsewhere. Proportional declines in violent crime were roughly similar in poor and nonpoor tracts in Chicago and St. Petersburg. In contrast, the percentage drops in violent crime in the remaining four cities were substantially greater in poor neighborhoods.

The violent crime rate was initially higher in each city's poor neighborhoods than in its remainder. The larger absolute decline in violent crime in each city's poor neighborhoods thus means that there was a convergence in crime levels among poor and nonpoor neighborhoods. The absolute difference in violent crime rates between each city's poor and nonpoor neighborhoods shrunk by between 23 percent for Philadelphia and 54 percent for Denver.

TABLE 3

Change in Violent Crime Rates of Neighborhoods Grouped by Racial/Ethnic
Composition

City	Time Period	Absolute Change				Relative Change			
		White	Black	Hispanic	Other	White	Black	Hispanic	Other
Chicago	2001–2012	–18.80	–73.55	–48.84	–37.46	–32.85	–25.66	–39.48	–40.22
Cleveland	1990–2010	20.23	–55.78	—	–96.40	20.84	–23.36	—	–40.39
Cleveland	1990–2000	2.68	–88.23	—	–104.30	2.76	–36.95	—	–43.70
Cleveland	2000–2010	17.55	32.45	—	7.90	17.60	21.55	—	5.88
Denver	1990–2010	–10.10	–60.28	–33.63	–170.71	–19.26	–54.72	–31.08	–64.98
Denver	1990–2000	–17.93	–53.67	–33.84	–131.59	–34.19	–48.72	–31.27	–50.09
Denver	2000–2010	7.83	–6.61	0.21	–39.13	22.69	–11.69	0.28	–29.84
Philadelphia[a]	1998–2009	2.53	–24.01	–76.89	–9.42	3.92	–13.91	–28.36	–6.66
Seattle	1996–2007	–15.11	—	—	–39.74	–23.43	—	—	–27.04
St. Petersburg	2000–2012	–50.83	–129.16	—	—	–51.15	–35.62	—	—

a. Although the average crime rate of a city's majority-black neighborhoods typically exceeds that of
its majority-Hispanic neighborhoods, the reverse holds in Philadelphia. The majority-Hispanic
neighborhoods in Philadelphia ($n = 16$) are exceptionally disadvantaged. Their average poverty rate
exceeded 50 percent in 2000, versus 29 percent in Philadelphia's majority-black neighborhoods ($n =
146$).
NOTE: Neighborhoods are classified according to whether non-Hispanic whites, non-Hispanic
blacks, or Hispanics make up a majority of their populations in the data's initial year, or as "other" if
none constitutes a majority. Results are excluded for categories containing two or fewer neighborhoods with available data.

The analysis in Table 3 repeats those in Table 1 and Table 2, but for neighborhoods grouped by their majority racial/ethnic compositions. Results are excluded
for a racial/ethnic category when the city contains no more than two neighborhoods (with available crime data) falling into that category. On this basis, results
are given for majority-black neighborhoods in all cities except Seattle. In these
cities, the absolute decline in violent crime in majority-black neighborhoods far
exceeded that in majority-white neighborhoods. For instance, Chicago's majority-
black neighborhoods experienced a decline of 74 violent crimes per 10,000 residents, versus 19 per 10,000 in its majority-white tracts. In two cities, Cleveland
and Philadelphia, violent crime increased in majority-white neighborhoods while
falling in majority-black tracts. Results are provided for majority-Hispanic neighborhoods in Chicago, Denver, and Philadelphia. In these cities, majority-Hispanic neighborhoods experienced greater absolute declines in violent crime than
majority-white neighborhoods.

The violent crime rate was initially higher in each city's majority-black and
majority-Hispanic neighborhoods than in its majority-white neighborhoods. The
absolute difference in violent crime rates between each city's majority-white and
majority-black neighborhoods subsequently shrunk by between 24 percent for

Chicago and 87 percent for Denver, while that between majority-white and majority-Hispanic neighborhoods shrunk by between 38 percent for Philadelphia and 45 percent for Chicago.

Although absolute declines in violent crime were greater in majority-black than majority-white neighborhoods in all cities, majority-white neighborhoods saw greater proportional declines in two cities: Chicago and St. Petersburg. In the remaining cities, proportional declines were greater in majority-black than majority-white neighborhoods. Proportional declines were greater in majority-Hispanic than in majority-white neighborhoods in all three cities that had results for majority-Hispanic neighborhoods.

The most violent, disadvantaged neighborhoods within the cities experienced declines in violent crime that were as large as, or larger than, those of the cities' remainders. This is true regardless of whether neighborhoods are divided by their initial violent crime rates, poverty levels, or racial/ethnic compositions. Still, in all cities, the most violent quintile of neighborhoods continued to experience higher crime levels, in the data's final year, than had the second most violent quintile in the initial year. The same applies to poor relative to nonpoor neighborhoods and to majority-black (in all cities but Denver) and majority-Hispanic relative to majority-white neighborhoods. In other words, even after the crime declines depicted here, the cities' worst-off neighborhoods had more violent crime than other neighborhoods had before this decline.

Who experienced the crime decline? A group-level analysis

In this section, we analyze how these changes in violent crime rates translated into shifts in individuals' exposure to violent crime in their neighborhoods.

As depicted in Figure 1, exposure to violent crime of each city's poor and nonpoor residents declined over the period covered by the data. The decline experienced by poor residents was, however, greater than that experienced by nonpoor residents in each case. For example, exposure of the poor residents of Seattle declined by 50 incidents per 10,000 residents, while that of its nonpoor residents declined by only 17 incidents. The number of violent incidents by which the exposure rate of each city's poor residents declined is at least 1.7 times greater than that by which the exposure rate of its nonpoor residents dropped.

The greater magnitude of the declines affecting poor residents means that there was a convergence in the number of violent crimes to which poor and nonpoor residents were exposed. Poor residents of each city were initially exposed to at least 54 (for Philadelphia) and at most 128 (for St. Petersburg) more violent incidents per 10,000 residents than nonpoor residents. In the data's final year, this difference exceeds 50 incidents per 10,000 residents in just one city, St. Petersburg. In four cities—Cleveland, Denver, Seattle, and St. Petersburg—this difference shrunk by at least half of its initial level. In the remaining two cities, Chicago and Philadelphia, it declined by substantial proportions, 41 percent and 26 percent, respectively. For all cities but Cleveland and Philadelphia, the poor were exposed to about as much violent crime in the data's final year as were the nonpoor in the data's initial year.

FIGURE 1

Average Neighborhood Exposure to Violent Crime of Poor and Nonpoor City Residents

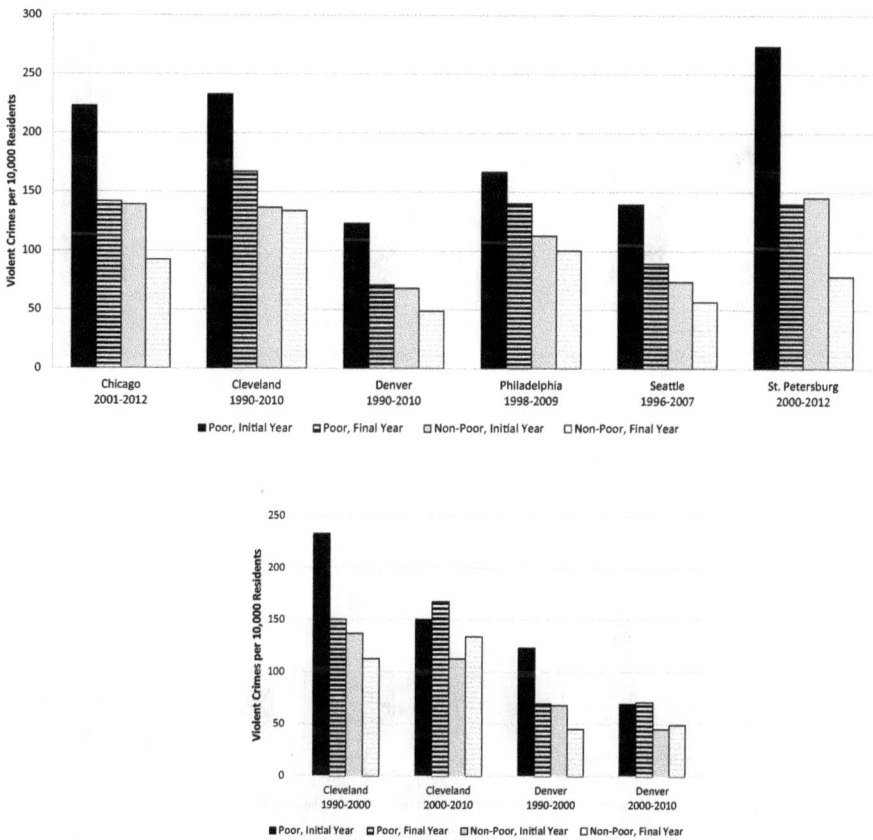

NOTE: The average neighborhood exposure rate to violent crime of a poor or nonpoor city resident is the average amount of violent crime occurring in the neighborhood of such a resident. See the main text for the formula by which this average rate is calculated. When demographic data were not available for the initial or final years of crime data, the most proximate available demographic data were used in weighting neighborhood crime rates to calculate these averages.

Each city's poor residents experienced a greater proportional drop in exposure to violent crime (measured relative to its initial rate of exposure) than its nonpoor residents. In three cities—Chicago, Philadelphia, and St. Petersburg—the proportional change for poor residents is only slightly greater than that for nonpoor residents. In the remaining cities—Cleveland, Denver, and Seattle—the proportional change is at least 1.5 times greater for poor residents than for nonpoor residents.

Initial and final exposure to violent crime of individuals classified by race/ethnicity is displayed in Figure 2. In all cities, exposure to violent crime fell for

FIGURE 2

Average Neighborhood Exposure to Violent Crime of City Residents, by Race/Ethnicity

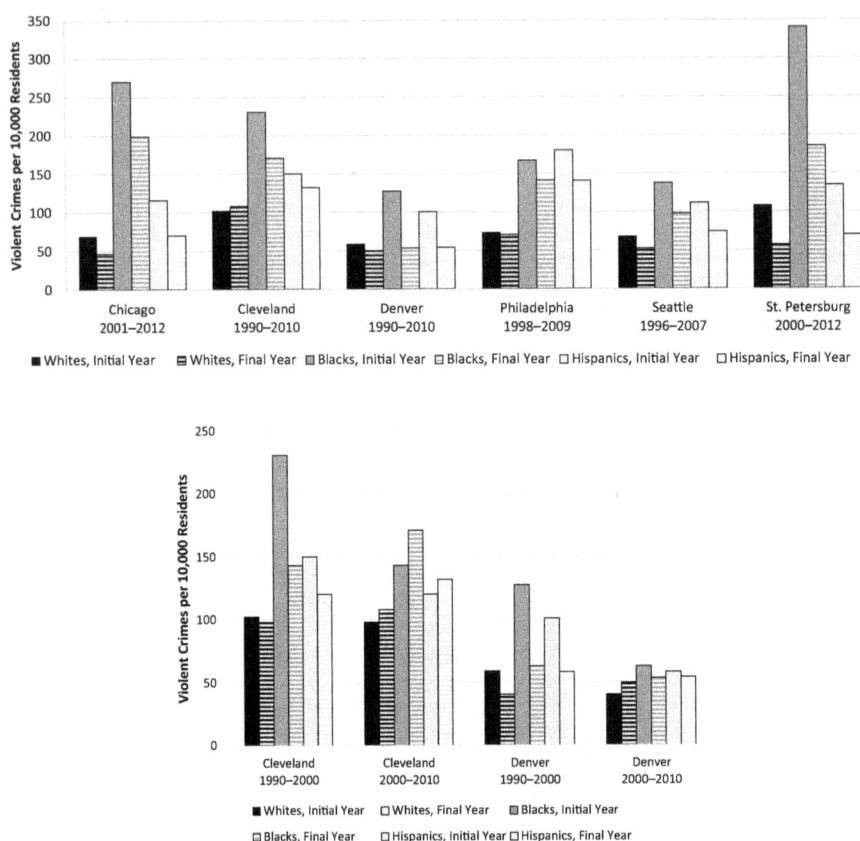

NOTE: The average neighborhood exposure rate to violent crime of a white, black, or Hispanic resident is the average amount of violent crime occurring in the neighborhood of such a resident. See the main text for the formula by which this average rate is calculated. When demographic data were not available for the initial or final years of crime data, the most proximate available demographic data were used in weighting neighborhood crime rates to calculate these averages.

both African Americans and Hispanics. Exposure of whites also fell in five of the six cities but increased in Cleveland. In the five cities, the declines experienced by African Americans were at least double the magnitude of those experienced by whites, as were the declines experienced by Hispanics in four of the five cities (the exception being St. Petersburg).

The greater magnitudes of the declines experienced by African Americans and Hispanics indicate that there was a convergence in the number of violent crimes to which African Americans and whites, as well as Hispanics and whites, were exposed. The disparity in exposure to violent crime between whites and African

Americans shrunk by between 24 percent for Chicago and 95 percent for Denver during the years covered by the data. The disparity in the exposure between whites and Hispanics shrunk by between 35 percent for Philadelphia and 90 percent for Denver. Nonetheless, in all cities except Denver, African Americans were exposed to more crime in the data's final year than were whites in their initial year. In all cities but Denver and St. Petersburg, Hispanics were likewise exposed to more crime in the data's final year than were whites in their initial year.

While absolute declines in the exposure rates of African Americans and Hispanics to violent crime exceed those for whites in all cities, the same does not hold regarding proportional declines. In St. Petersburg, for instance, the proportional decline for African Americans approximately equals that for whites. In Chicago, the proportional decline for African Americans is less than that for whites, even as the absolute decline for African Americans far exceeds that for whites. This is because Chicago's African Americans were initially exposed to violent crime at about four times the rate of whites.

The greatest beneficiaries of the crime decline, in terms of reduced exposure to violent crime, were poor and minority individuals. The neighborhood crime levels experienced by members of disadvantaged groups came to more closely resemble those experienced by everyone else. Still the exposure of each city's poor residents did not fall substantially below that which its nonpoor residents had experienced before the crime decline. The same applies to each city's black and (in most cases) Hispanic residents in relation to whites.

Change and stability in the spatial distribution of violent crime

Even as crime rates fell most in cities' most violent neighborhoods and exposure to crime fell most among disadvantaged urban residents, key relationships characterizing the spatial concentration of crime within cities, and linking crime with disadvantage, remained intact. There is a very strong correlation between the initial and final violent crime rates of the neighborhoods in each city. The coefficients (R) for these correlations range from 0.76 in Cleveland to 0.93 in Seattle and exceed 0.85 in all cities except Cleveland.[6] These results indicate that, in each city, neighborhoods' initial violent crime rates strongly predict, or explain most of the variation in, their rates in the data's final year. There was little change in the relative distribution of violent crime across the cities' neighborhoods during the periods under analysis.

As a result, the neighborhoods within each city that were most violent before that city's crime decline largely remained most violent afterward. The same holds for those that were initially least violent. Table 4 displays the proportion of each city's neighborhoods, among the quintile of its neighborhoods having the highest initial violent crime rates, that were also in the most violent quintile in the data's final period. The same information is also displayed, for comparison, with respect to each city's quintile of neighborhoods having the lowest initial rates. In four cities—Chicago, Denver, Seattle, and St. Petersburg—at least 80 percent of the neighborhoods that were initially in the most violent quintile remained in this quintile in the data's final year. In the remaining cities—Cleveland and

TABLE 4
Percentage of Neighborhoods Remaining in Cities' Highest or Lowest Quintiles of
Violent Crime Rates Across the Years of Crime Data

City	Time Period	Highest Quintile	Lowest Quintile
Chicago	2001–2012	81.18	71.11
Cleveland	1990–2010	64.79	65.63
Cleveland	1990–2000	77.46	78.13
Cleveland	2000–2010	73.68	71.88
Denver	1990–2010	80.95	68.75
Denver	1990–2000	85.71	75.00
Denver	2000–2010	85.71	68.75
Philadelphia	1998–2009	67.01	76.47
Seattle	1996–2007	85.71	73.08
St. Petersburg	2000–2012	85.71	66.67

Philadelphia—approximately two-thirds of the initially most violent neighbor-hoods remained so in the data's final year. In half of the cities—Denver, Seattle, and St. Petersburg—every neighborhood initially in the most violent quintile fell into either the most or second-most violent quintile in the data's final year. About two-thirds to three-quarters of each city's neighborhoods that were initially in the least violent quintile were also in this quintile in the data's final year.

Violent crime dropped the most in the cities' initially most violent and disad-vantaged neighborhoods, but this decline has not amounted to a substantial redistribution of violent crime within these cities. In each city, the initially most violent quintile had more violent crime in the data's final year, or after this city's crime decline, than did the second-most violent quintile in the data's initially year, prior to the decline. The neighborhoods where poor and minority popula-tions are concentrated likewise continue to be those where violent crime is con-centrated. It has been possible for violent crime to drop the most in the initially most violent and disadvantaged places, yet also remain concentrated in these places, because crime rates were initially so elevated in these places relative to the remainder of the cities. Indeed, in all cities except Philadelphia, the violent crime rate of the most violent quintile in the data's first year approached or exceeded twice that of the second-most violent quintile and was more than five times greater than the average of the three least violent quintiles' rates.

Concentrated Disadvantage and Violence after the Crime Decline

At the outset of the time periods under study, the most violent neighborhoods in each city stood apart from the city's remaining neighborhoods. The burden of violence was not spread evenly across each city, but rather was concentrated within a small segment of neighborhoods that were characterized by racial and ethnic segregation and poverty. The clear conclusion from our study is that the same places that featured the most severe concentrations of violence and poverty are the places that have changed the most. Further, the populations most affected by the problem of community violence have experienced the greatest changes over time.

Nonetheless, the analysis we present does not suggest that the spatial distribution of violence has been overturned over time. The decline of violent crime in these six cities served to ameliorate, but not to eliminate, socioeconomic and racial/ethnic disparities in community violence. The distribution of violent crime rates across each city's neighborhoods showed little change, meaning that the communities that were initially most violent largely remained so. The linkages between the spatial concentrations of violence and disadvantage remained intact even if violence decreased.

The findings from this study should be considered in relation to the study's limitations, which are primarily driven by data constraints. The study is conducted with data from a small number of cities that are not representative of American communities or cities, and the available data are not perfectly consistent across these six cities. Further, the timeframe of data availability differs across the six cities, and in some places our data do not cover the time period in which crime was declining the most. These constraints limit the types of general claims that can be made about which neighborhoods were most affected by the crime decline.

However, the consistency of patterns found across all of the cities does generate confidence in some basic claims about how the crime decline was distributed across neighborhoods. In every city we studied, the decline of violence in the most disadvantaged neighborhoods was at least as great—and usually much greater—than the decline of violence in the rest of the city. In some of these cities, like Cleveland, the drop in crime was experienced almost exclusively in the most violent neighborhoods. Across all six cities, the experience of urban disadvantage was transformed over the period under study. The connection among concentrated poverty, racial segregation, and violence was weakened, even if it was not severed.

This conclusion should provoke a shift in the study of urban poverty. For the past several decades, much of what we know about urban poverty has been learned in the context of rising crime and violence. In the aftermath of the crime decline, there are many places where daily life within areas of concentrated disadvantage is no longer dominated, to the same extent, by the threat of violence. Further research is now needed to assess whether the drop in the everyday threat of violence has altered patterns of interaction, parenting, social control, and community life within neighborhoods of concentrated disadvantage. Considering the extensive evidence base pointing to violence as a mechanism for neighborhood effects (see, e.g., Harding 2009), new research is necessary to explore whether the consequences of growing up in a poor, segregated neighborhood are different for those who have been raised in an era of declining violence.

Changes to neighborhood social life entailed by the crime decline must be assessed in the context of the massive expansion of the criminal justice system that has occurred at the same time as the crime decline. Patterns of increasing arrests and imprisonment, like those of declining crime, have been focused in urban neighborhoods of concentrated disadvantage. These punitive interventions, it has been argued, produce a range of collateral consequences entailing disruptions to family formation and cohesion, economic marginalization, and reductions in neighborhoods' capacity to exercise informal social control (Clear 2009; Defina and Hannon 2013; Fagan and Meares 2008; Wildeman and Western 2010; Wildeman, Hacker, and Weaver 2014). Research on neighborhood effects after the crime decline must consider the outcomes of living in urban spaces subjected to reduced private violence but enduring formal surveillance and control.

Without losing sight of the social costs of mass incarceration, public policy discourse should focus on harnessing the potential for broader neighborhood investment and transformation made possible by declines in interpersonal violence. The public reputations of urban spaces as violent or disorderly are known to be more enduring than the actual incidence of high levels of violence or disorder in such places (Sampson 2012; see also Sampson and Raudenbush 2004). During the era of rising violent crime, areas of concentrated disadvantage experienced a vicious cycle of violence, disorder, disinvestment, and marginalization (Skogan 1990; Bursik 1986). In this era of relative peace, there is a new potential for a virtuous cycle of declining crime and disorder, reinvestment, and greater integration of disadvantaged neighborhoods into the urban social fabric.

Notes

1. This is a common threshold for classifying neighborhoods according to whether they have high poverty levels (see, e.g., Kingsley and Pettit 2003; Mather and Dupuis 2012).

2. See the online methodological appendix for a list of the data's sources: http://ann.sagepub.com/supplemental.

3. This percentage is calculated using UCR crime counts for Philadelphia; see www.ucrdatatool.gov.

4. See http://ann.sagepub.com/supplemental.

5. It may be unsurprising that the absolute decline in violent crime was so much greater in each city's initially most violent quintile, given that the initial violent crime rate of this quintile, in each city, was several times greater than that of the city's remainder. The magnitude of the absolute decline in the latter is, of course, constrained by its relatively low initial crime level.

6. These correlations are of the logged violent crime rates, in the data's initial and final years, of all neighborhoods with more than 200 residents. To reduce the impact on the results of anomalous annual crime rates, initial and final crime rates consist of multiyear averages of annual rates. Across each of the decades making up the time period of the Cleveland data (1990–2010), the coefficient equals approximately 0.85 for the correlation of initial and final neighborhood violent crime rates.

References

Bowles, Samuel, Stephen Durlauf, and Karla Hoff, eds. 2006. *Poverty traps*. Princeton, NJ: Princeton University Press.

Braga, Anthony, David Hureau, and Andrew Papachristos. 2011. The relevance of micro places in citywide robbery trends: A longitudinal analysis of robbery incidents at street corners and block faces in Boston. *Journal of Research in Crime and Delinquency* 48 (1): 7–32.

Braga, Anthony, Andrew Papachristos, and David Hureau. 2010. The concentration and stability of gun violence at micro places in Boston, 1980–2008. *Journal of Quantitative Criminology* 26 (1): 33–53.

Burdick-Will, Julia, Jens Ludwig, Stephen Raudenbush, Robert J. Sampson, Lisa Sanbonmatsu, and Patrick Sharkey. 2011. Converging evidence for neighborhood effects on children's test scores: An experimental, quasi-experimental, and observational comparison. In *Whither opportunity? Rising inequality, schools, and children's life chances*, eds. Greg Duncan and Richard Murdane, 255–76. New York, NY: Russell Sage Foundation.

Bursik, Robert. 1986. Delinquency rates as sources of ecological change. In *The social ecology of crime*, eds. James Byrne and Robert J. Sampson, 63–76. New York, NY: Springer-Verlag.

Clear, Todd. 2009. *Imprisoning communities: How mass incarceration makes disadvantage worse*. New York, NY: Oxford University Press.

Cook, Phillip, and John Laub. 2002. After the epidemic: Recent trends in youth violence in the United States. *Crime and Justice: A Review of Research* 29:1–37.

Cullen, Julie Berry, and Steven Levitt. 1999. Crime, urban flight, and the consequences for cities. *The Review of Economics and Statistics* 81 (2): 159–69.

Defina, Robert, and Lance Hannon. 2013. The impact of mass incarceration on poverty. *Crime & Delinquency* 59 (4): 562–86.

Ellen, Ingrid Gould, and Katherine O'Regan. 2008. Reversal of fortunes: Lower-income neighborhoods in the urban US in the 1990s. *Urban Studies* 45 (4): 845–69.

Ellen, Ingrid Gould, and Katherine O'Regan. 2009. Crime and U.S. cities: Recent patterns and implications. *The ANNALS of the American Academy of Political and Social Science* 626 (1): 22–38.

Fagan, Jeffrey, and Tracey Meares. 2008. Punishment, deterrence, and social control: The paradox of punishment in minority communities. *Ohio State Journal of Criminal Law* 6:173–229.

Hagan, John, and Ruth D. Peterson. 1995. Criminal inequality in America: Patterns and consequences. In *Crime and inequality*, eds. John Hagan and Ruth D. Peterson, 14–36. Stanford, CA: Stanford University Press.

Harding, David. 2009. Collateral consequences of violence in disadvantaged neighborhoods. *Social Forces* 88 (2): 757–84.

Kasarda, John. 1993. Inner-city concentrated poverty and neighborhood distress: 1970 to 1990. *Housing Policy Debates* 4 (3): 253–302.

Kingsley, G. Thomas, and Kathryn L. S. Pettit. 2003. *Concentrated poverty: A change in course*. Washington, DC: Urban Institute.

Klinenberg, Eric. 2003. *Heat wave: A social autopsy of disaster in Chicago*. Chicago, IL: The University of Chicago Press.

Krivo, Lauren, and Ruth D. Peterson. 1996. Extremely disadvantaged neighborhoods and urban crime. *Social Forces* 75 (2): 619–48.

Liska, Allen, and Paul Bellair. 1995. Violent-crime rates and racial composition: Convergence over time. *Criminology* 101 (3): 578–610.

Mather, Mark, and Genevieve Dupuis. 2012. *Rising share of U.S. children living in high-poverty neighborhoods*. Washington, DC: Population Reference Bureau.

Morenoff, Jeffrey D., and Robert J. Sampson. 1997. Violent crime and the spatial dynamics of neighborhood transition: Chicago, 1970–1990. *Social Forces* 76 (1): 31–64.

Peterson, Ruth, and Lauren Krivo. 2010. *Divergent social worlds: Neighborhood crime and the racial-spatial divide*. New York, NY: Russell Sage Foundation.

Pope, Devin, and Jaren Pope. 2012. Crime and property values: Evidence from the 1990s crime drop. *Regional Science and Urban Economics* 42 (1): 177–88.

Sampson, Robert J. 2012. *Great American city: Chicago and the enduring neighborhood effect*. Chicago, IL: The University of Chicago Press.

Sampson, Robert J., and Jeffrey D. Morenoff. 2006. Durable inequality: Spatial dynamics, social processes, and the persistence of poverty in Chicago neighborhoods. In *Poverty traps*, eds. Samuel Bowles, Stephen N. Durlauf, and Karla Hoff, 176–203. Princeton, NJ: Princeton University Press.

Sampson, Robert J., and Stephen W. Raudenbush. 2004. Seeing disorder: Neighborhood stigma and the social construction of "broken windows." *Social Psychology Quarterly* 67 (4): 319–42.

Sampson, Robert J., Stephen W. Raudenbush, and Felton Earls. 1997. Neighborhoods and violent crime: A multilevel study of collective efficacy. *Science* 277 (5328): 918–24.

Sharkey, Patrick. 2010. The acute effects of local homicides on children's cognitive performance. *Proceedings of the National Academy of Sciences* 107 (26): 11733–38.

Sharkey, Patrick, and Robert J. Sampson. In press. Violence, cognition, and neighborhood inequality in America. In *Social neuroscience: Brain, mind, and society*, eds. Russell K. Schutt, Larry J. Seidman, and Matcheri S. Keshavan. Cambridge, MA: Harvard University Press.

Skogan, Wesley G. 1990. *Disorder and decline: Crime and the spiral of decay in American neighborhoods.* Berkeley, CA: University of California Press.

Weisburd, David, Shawn Bushway, Cynthia Lum, and Sue-Ming Yang. 2004. Trajectories of crime at places: A longitudinal study of street segments in the city of Seattle. *Criminology* 42 (2): 283–322.

Wildeman, Christopher, Jacob S. Hacker, and Vesela Weaver, eds. 2014. *Detaining democracy? Criminal justice and American civic life. The ANNALS of the American Academy of Political and Social Science* 651.

Wildeman, Christopher, and Bruce Western. 2010. Incarceration in fragile families. *The Future of Children* 20 (2): 157–77.

Zimring, Franklin E. 2008. *The great American crime decline.* New York, NY: Oxford University Press.

Conclusion

Residential Inequality: Significant Findings and Policy Implications

By
GLENN FIREBAUGH,
JOHN ICELAND,
STEPHEN A. MATTHEWS,
and
BARRETT A. LEE

A lot has changed since Thomas Pettigrew (1979, 114) identified residential segregation (the "maldistribution of blacks and whites" in metropolitan areas) as the "structural linchpin" of race relations in America. The Hispanic population has more than tripled since 1979, and Hispanic Americans have now replaced African Americans as the largest minority group in the nation. The Asian American population has also grown rapidly, increasing from 1.5 percent of the U.S. population in 1980 to 4.8 percent of the population in 2010. Non-Hispanic whites now constitute well less than two-thirds of the population, and their share continues to shrink.

When Pettigrew and others wrote about "Race and Residence" in a volume of *The ANNALS* over three decades ago (Roof 1979), they of course had no way to know of the dramatic population shifts that were to follow. Nor could they have foreseen the election of an African American U.S. president 28 years later, or the financial crisis that led to unprecedented levels of housing foreclosures. And it is only in hindsight that we can see that they were writing about residential segregation and inequality at a point in time when income inequality was beginning to grow in the United States after many decades of stability or decline.

Glenn Firebaugh is Roy Buck Professor of Sociology and Demography at The Pennsylvania State University and author of Seven Rules for Social Research *(Princeton 2008). From 1997 to 2000, he served as editor of the* American Sociological Review. *Currently, he is using U.S. census data to investigate neighborhood inequality and data on cause of death to investigate sources of change in life expectancy.*

John Iceland is head of the Department of Sociology and a professor of sociology and demography at The Pennsylvania State University. His research focuses on social demography, poverty, residential segregation, and immigration. He has written three books on these issues and published numerous articles in top journals in sociology and demography.

DOI: 10.1177/0002716215580060

They could not have foreseen the surge in violent crime rates in America that peaked in the early 1990s before falling sharply, nor could they have foreseen the economic renaissance of some central cities and the continued devolution of others.

Given these watershed events, as well as the continuation of longstanding racial and residential trends in American society, it is not surprising that a 2015 *ANNALS* volume devoted to *Residential Inequality in American Neighborhoods and Communities* would focus heavily on population change, on the income divide, on housing insecurity, on change (or lack thereof) in residential discrimination against minorities, on neighborhood differences in crime, and on neighborhood change and gentrification. In this concluding section we gather together some significant findings and key policy implications that can be drawn from this volume with respect to spatial sorting, life chances, the income divide, foreclosures and housing insecurity, residential attainment, crime in the city, gentrification, and the effects of population shifts. We begin with a sample of the significant findings from the articles in this volume.

Significant Findings

Race contributes to residential inequality independent of income

Most studies of residential segregation focus either on segregation by race (the uneven distribution of different racial groups across neighborhoods) or segregation by income (the uneven distribution of different income groups across neighborhoods). Studying racial and income segregation together reveals large racial differences in neighborhood context among households of the same annual income. Sean Reardon, Lindsay Fox, and Joseph Townsend find that black and Hispanic households must have substantially higher incomes than white or Asian households to live in similar economic contexts. They find, for example, that neighborhood economic conditions are the same for the average white household earning $11,800, the average Hispanic household earning $45,000, and the average black household earning $60,000.

A second study, by S. Michael Gaddis and Raj Ghoshal, provides suggestive evidence that racial discrimination likely accounts, at least in part, for the underrepresentation of middle- and upper-class minorities in affluent neighborhoods. Housing discrimination against African Americans and Hispanic Americans is of course well documented. What is new about the Gaddis and Ghoshal study is that

Barrett A. Lee is a professor of sociology and demography at The Pennsylvania State University. He studies community diversity, racial segregation, neighborhood change, residential mobility and displacement, and urban homelessness. A general interest in spatial manifestations of social inequality runs throughout his research.

Stephen A. Matthews is a professor of sociology, anthropology, and demography and director of the Graduate Program in Demography at The Pennsylvania State University. He has published in the leading journals in demography, sociology, public health, and geography on topics such as residential segregation, neighborhood change, access to health care, and food environments.

they investigate housing discrimination against Arab Americans. In line with prior audit studies that focus on other minority groups, Gaddis and Ghoshal find that individuals inquiring about roommate-wanted advertisements in Detroit, Houston, Los Angeles, and New York were less likely to receive positive responses when they had Arab-sounding names than when they had white-sounding names.

The adverse effects of segregation are felt by both Latinos and blacks

Prior research has shown that in metropolitan areas that are highly segregated residentially, blacks have lower high school graduation rates, higher rates of teenage motherhood, and a greater likelihood of being simultaneously out of school and out of work. Justin Steil, Jorge De la Roca, and Ingrid Gould Ellen find that these associations are as strong for Latinos as for African Americans, if not more so, indicating that the harmful effects of residential segregation are at least as large for Latinos as for African Americans. Note how much more significant this finding is today than it would have been when *The ANNALS* volume on "Race and Residence" was published in 1979 about an America where Latinos still constituted a relatively small fraction of the U.S. population.

The influx of immigrants likely contributes to the formation of mixed-income neighborhoods

The issue of mixed-income neighborhoods is attracting more attention as a counterweight to the rising economic segregation of America's neighborhoods. In their study drawing on data from Chicago as well as national data, Robert Sampson, Robert Mare, and Kristin Perkins find that, while mixed-income neighborhoods are often hard to sustain, Latino Americans are much more likely than blacks or whites to live in mixed middle-income neighborhoods. In a separate study, Jackelyn Hwang discovers that immigration is also associated with gentrification in Chicago and Seattle but cautions that this gentrification has failed to undermine the durability of neighborhood inequality and the persistence of poor, minority neighborhoods in those cities.

Rates of foreclosure varied greatly across regions, cities, and racial and ethnic groups

While there are many popular accounts of neighborhoods and communities that were particularly hard hit by the housing crisis—places where virtually whole neighborhoods were in foreclosure—we know much less about how the housing crisis varied across broad regions of the country. Based on the painstaking collation and analysis of the 9.5 million public auctions and bank repossessions from 2005 to 2012, Matthew Hall, Kyle Crowder, and Amy Spring find that rates of foreclosure varied tremendously across regions of the United States, from cumulative rates of about 2 percent for the New England and Middle Atlantic regions to a rate of greater than 16 percent for the Mountain region. They also document higher foreclosure rates in neighborhoods and metropolitan areas with large black and Latino populations as well as in racially mixed neighborhoods. Latino households were particularly hard hit. In their study of housing precarity, Rachel Dwyer and Lora Phillips Lassus also show that high levels of black-white segregation were associated with higher foreclosure rates.

The inheritance of poverty applies to residential turnover for specific housing units

Brett Theodos, Claudia Coulton, and Rob Pitingolo find that new tenants of a specific housing unit are more likely to be poor if former tenants were poor, independent of measures of the value of the house and economic characteristics of the neighborhood. They conclude that this poverty "stickiness" at the level of individual housing units likely contributes to the difficulty of upgrading neighborhood economic conditions.

Policy Implications

Policies need to be refashioned in light of how Hispanics are redefining U.S. settlement patterns

Scholars and policy-makers have been slow to react to the dramatic increase in the Hispanic population and its increasing dispersion across all regions of the country. There is, for example, a critical need for adult language training for immigrants, since English is essential to success in the labor and housing markets (Singer 2005). Yet research and policies continue to be based on a black-white, assimilation versus place stratification model that is increasingly obsolete and needs to be replaced by newer models that better reflect the rising economic, racial, linguistic, and spatial diversity of the American population. The article by Daniel Lichter, Domenico Parisi, and Michael Taquino illustrates why it is important to update research approaches in light of America's increasing diversity. Most studies focus only on metropolitan areas and generally find that Hispanic residential segregation is declining slowly, if at all. When smaller communities are also included, however, Lichter and colleagues find evidence of the spatial assimilation of Hispanics in America, both nationally and in new immigrant destinations (see also the article by Chenoa Flippen and Eunbi Kim for a description of how Asian Americans are faring in new destinations). Lichter and colleagues also report that, while Hispanics are more segregated from blacks than from whites, this too might be changing since Hispanic-black segregation is declining faster than Hispanic-white segregation. Because of such rapid changes in the demographic composition of the American population, the timely collection of population data for the United States is more critical than ever. It is essential, then, that Congress adequately fund the U.S. Census Bureau. Incomplete information often leads to inaccurate information, and inaccurate information about a fast-changing U.S. population is sure to result in bad policy.

Build public support for reducing residential inequality by focusing on the reduction of insecurity

The housing crisis that led to the Great Recession brought to the fore the "great risk shift" in American society, where responsibility for managing the ordinary risks in modern society (finding and keeping a job, saving for retirement) has increasingly shifted to individuals. In their study of the conditions that contributed to the housing crisis, Dwyer and Lassus make the important point that policies for reducing inequality are likely to be more palatable if they are targeted toward reducing precarity rather than inequality. Discussions of inequality too often are framed as zero-sum, "us vs. them," and not everyone sees inequality as a problem. However, virtually everyone is economically insecure, at least to some extent, and that insecurity can be used as a basis to build public support for government policies. Consider, for example, the rationale for extending public funding of education by two years, so the new norm would be 14 years of publicly funded education for every student. By raising the educational level of those who otherwise would not attend college, the policy presumably would compress inequality. Such a policy nonetheless is more likely to garner broad-based support if policy-makers note that educational underachievement in America is a threat to economic growth than if the policy is presented as a way to redress inequality.

Researchers and policy-makers should devote more attention to the perceptions and selection processes that individuals use in determining where to live

Two articles highlight this issue. The first, by Michael Bader and Maria Krysan, describes the results of a study asking Chicago residents to indicate, from a list of Chicago communities, the places they would "seriously consider" living and those they would "never consider." They found that, relative to whites, blacks generally had a longer list of communities they would consider and a shorter list of communities they would exclude. Members of every racial group viewed nearby communities as more attractive than more distant ones. This is one way, then, that segregation begets segregation, since existing segregation implies that nearby places are more likely to be made up of those like oneself. To overcome the centrifugal forces of residential segregation, we must know more about how individuals arrive at their perceptions of different communities and how that thinking can be expanded to consider more distant and more diverse communities. Yet, as Bader and Krysan note, most existing research fails to probe the source of individual perceptions; researchers are instead content to focus on stated preferences "unmoored" from the reality of the decision-making process.

The second study, by Lincoln Quillian, describes a method for studying mobility decisions that is more firmly anchored to the decision-making process. In the discrete-choice method that Quillian proposes, decisions about where to move are modeled as a series of dyadic choices. In line with Bader and Krysan's call for more realistic models of the decision-making process, the discrete-choice method incorporates the fact that outcomes depend in large part on the destinations that are perceived to be available. Because each area is different, local officials are in the best position to fashion policies to ensure that local inhabitants are adequately informed about their residential possibilities.

Assisted housing is not a quick fix for the economic segregation of neighborhoods

The dramatic geographic deconcentration of assisted housing units over the past 40 years has created greater opportunities for low-income assisted renters to live in neighborhoods alongside higher-income neighbors. This deconcentration might or might not reduce the economic segregation of neighborhoods because the arrival of low-income families is likely to prompt the exodus of high-income residents. To reduce that exodus, Ann Owens suggests closer cooperation between local policymakers, developers, and residents in siting project-based assisted housing. Changes in the voucher program, such as higher voucher values in more costly areas and property tax rebates for landlords who accept vouchers, might also be necessary. Owens concludes that to this point the deconcentration of assisted housing units has produced only modest economic residential integration for very low-income residents while high-income residents have become even more segregated.

In their ethnographic study of public housing residents in Chicago, Robert Chaskin and Mark Joseph conclude that even if we stemmed the exodus of high-income residents that would not be enough to produce effective neighborhood integration where there are public housing residents. (According to Sampson and colleagues, about 2 percent of Chicago's residents live in census tracts that contain public housing.) Chaskin and Joseph find that relocated public housing residents experience what they call "incorporated exclusion" in which "physical integration reproduces marginalization and leads more to withdrawal and alienation than the engagement and inclusion of relocated public housing residents." To diminish the marginalization and alienation of relocated housing residents without jeopardizing public safety, Chaskin and Joseph suggest the replacement of paternalistic regulatory policies with more participatory governance. They also suggest the provision of more public civic space for all community residents and that residents should be empowered to establish their own collective norms about the use of that space. Readers should compare Chaskin and Joseph's account with *Climbing Mount Laurel* (Massey et al. 2013), which describes how affordable housing for low-income minority families was successfully integrated into the white suburban community of Mount Laurel, New Jersey.

Reducing crime also reduces residential segregation and inequality

Lance Freeman and Tiancheng Cai also present a more hopeful picture than Chaskin and Joseph of the future of residential segregation and inequality in America. Their key finding is the discovery of a surge in the number of majority-black neighborhoods (census tracts) that have experienced a "white invasion"—an increase in the white population that exceeded 5 percent of the total tract population in the initial period. Significantly, white invasion appears to be accelerating, as the proportion of black neighborhoods experiencing a white invasion was greater from 2000 to 2010 than in the two previous decades combined.

While Freeman and Cai are careful to caution that "only time will tell if the 2000s were a harbinger of a new era of neighborhood change or an anomalous blip unlikely to repeat in the near future," their hypothesis—that white invasion is due largely to diminished crime rates in black neighborhoods—implies that white invasion is unlikely to be a blip. Although Freeman and Cai do not have the data to test the hypothesis, Michael Friedson and Patrick Sharkey provide indirect support in their study of the spatial distribution of the unprecedented decline in violent crime rates over the past two decades. They find, significantly, that violent crime declined the most in cities' initially most violent and disadvantaged locales (very often majority-black areas). Their conclusion is worth repeating here: "During the era of rising violent crime, areas of concentrated disadvantage experienced a vicious cycle of violence, disorder, disinvestment, and marginalization. ... In this era of relative peace, there is a new potential for a virtuous cycle of declining crime and disorder, and reinvestment and greater integration of disadvantaged neighborhoods into the urban social fabric."

Finally, the article by John Logan and colleagues reminds us that residential inequality is a longstanding feature of American society. Using unique historical data, they show that black ghettos existed in embryonic form in northern cities as far back as the 1880s. As this example shows, neighborhoods and communities may change, and residential inequalities may wax and wane, but some forms of inequality are stubbornly persistent. The challenge for researchers is to identify residential inequalities that are detrimental to individuals and to society, reveal their causes, describe their trends, and estimate their effects. The challenge for policy-makers is to uproot those inequalities where possible and at least to ameliorate their most harmful consequences.

References

Massey, Douglas S., Len Albright, Rebecca Casciano, Elizabeth Derickson, and David N. Kinsey. 2013. *Climbing Mount Laurel: The struggle for affordable housing and social mobility in an American suburb*. Princeton, NJ: Princeton University Press.

Pettigrew, Thomas F. 1979. Racial change and social policy. *The ANNALS of the American Academy of Political and Social Science* 441:114–41.

Roof, Wade Clark, ed. 1979. Race and residence in American cities. *The ANNALS of the American Academy of Political and Social Science* 441.

Singer, Audrey. 2005. The rise of new immigrant gateways: Historical flows, recent settlement trends. In *Redefining urban and suburban America: Evidence from census 2000, vol. 1*, eds. Alan Berube, Bruce Katz, and Robert E. Lang, 41–86. Washington, DC: Brookings Institution Press.

SAGE Premier

Online access to the most comprehensive collection of interdisciplinary journal content SAGE offers

SAGE Premier is an invaluable investment for your library. It includes leading international peer-reviewed journals and high-impact research titles published on behalf of more than 225 scholarly and professional societies. Our interdisciplinary coverage is unparalleled, spanning subject areas including business, humanities, social sciences, science, technology, medicine and many more.

Benefits for your patrons:

- Electronic access to most SAGE journals, with content dating back to 1999.

- **High-quality, interdisciplinary content** – all SAGE journals are peer-reviewed and more than half of the content in the **SAGE Premier** package is ranked in the Thomson Reuters Journal Citation Reports® (JCR).

- **Access via *SAGE Journals* (SJ)**, SAGE's award-winning online journal delivery platform, powered by HighWire Press

Benefits for librarians:

- COUNTER 3 compliant usage statistics available

- Content ownership of the subscribed content published by SAGE during the term of the agreement

- A site license designed with and for librarians

- Unlimited access via IP recognition

- Remote user login

- Athens-authenticated access

- Access rights for walk-in users

- The right to download articles and make copies for course packs and electronic reserve collections—no extra fee or permission required

- Inter-library loan

Need something more specific?

Titles included in **SAGE Premier** are also available in smaller, discipline-specific packages:

- **Humanities and Social Science (HSS) Package**
- **Science, Technology, Medicine**
- **Health Sciences**
- **Clinical Medicine**

For more information, contact
librarysales@sagepub.com

CONTENTS *continued*

www.ingramcontent.com/pod-product-compliance
Lightning Source LLC
Chambersburg PA
CBHW080244030426
42334CB00023BA/2691